BIRKHÄUSER

Applied and Numerical Harmonic Analysis

Steven G. Krantz
With the assistance of Lina Lee

Explorations in Harmonic Analysis

*With Applications to Complex Function
Theory and the Heisenberg Group*

Birkhäuser
Boston • Basel • Berlin

Steven G. Krantz
Department of Mathematics
Washington University in St. Louis
Campus Box 1146
St. Louis, MO 63130
sk@math.wustl.edu

ISBN 978-0-8176-4668-4   eISBN 978-0-8176-4669-1
DOI 10.1007/978-0-8176-4669-1

Library of Congress Control Number: 2009926530

Mathematics Subject Classification (2000): 43A80, 42B05, 42B20, 42B25, 42B30, 31B10, 31B25

Cover Design by Joseph Sherman

Printed on acid-free paper.

Birkhäuser Boston is part of Springer Science+Business Media (www.birkhauser.com)

To the memory of Alberto P. Calderón

Contents

Preface

Harmonic analysis is a venerable part of modern mathematics. Its roots began, perhaps, with late eighteenth-century discussions of the wave equation. Using the method of separation of variables, it was realized that the equation could be solved with a data function of the form $\varphi(x) = \sin jx$ for $j \in \mathbb{Z}$. It was natural to ask, using the philosophy of superposition, whether the equation could then be solved with data on the interval $[0, \pi]$ consisting of a finite linear combination of the $\sin jx$. With an affirmative answer to that question, one is led to ask about *infinite* linear combinations.

This was an interesting venue in which physical reasoning interacted with mathematical reasoning. Physical intuition certainly suggests that *any* continuous function φ can be a data function for the wave equation. So one is led to ask whether *any* continuous φ can be expressed as an (infinite) superposition of sine functions. Thus was born the fundamental question of Fourier series.

No less an *eminence gris* than Leonhard Euler argued against the proposition. He pointed out that some continuous functions, such as

$$
\varphi(x) =
\begin{cases}
\sin(x - \pi) & \text{if } 0 \le x < \pi/2, \\[2mm]
\dfrac{2(x - \pi)}{\pi} & \text{if } \pi/2 \le x \le \pi,
\end{cases}
$$

are actually not one function, but the juxtaposition of *two* functions. How, Euler asked, could the juxtaposition of two functions be written as the sum of single functions (such as $\sin jx$)? Part of the problem, as we can see, is that mathematics was nearly 150 years away from a proper and rigorous definition of function.[1] We were also more than 25 years away from a rigorous definition (to be later supplied by Cauchy and Dirichlet) of what it means for a series to converge.

---

[1] It was Goursat, in 1923, who gave a fairly modern definition of function. Not too many years before, no less a figure than H. Poincaré lamented the sorry state of the function concept. He pointed out that each new generation created bizarre "functions" only to show that the preceding generation did not know what it was talking about.

Fourier[2] in 1822 finally provided a means for producing the (formal) Fourier series for virtually any given function. His reasoning was less than airtight, but it was calculationally compelling and it seemed to work.

Fourier series live on the interval $[0, 2\pi)$, or even more naturally on the circle group $\mathbb{T}$. The Fourier analysis of the real line (i.e., the Fourier transform) was introduced at about the same time as Fourier series. But it was not until the mid-twentieth century that Fourier analysis on $\mathbb{R}^N$ came to fruition (see [BOC2], [STW]). Meanwhile, abstract harmonic analysis (i.e., the harmonic analysis of locally compact abelian groups) had developed a life of its own. And the theory of Lie group representations provided a natural crucible for noncommutative harmonic analysis.

The point here is that the subject of harmonic analysis is a point of view and a collection of tools, and harmonic analysts continually seek new venues in which to ply their wares. In the 1970s, E.M. Stein and his school introduced the idea of studying classical harmonic analysis—fractional integrals and singular integrals—on the Heisenberg group. This turned out to be a powerful device for developing sharp estimates for the integral operators (the Bergman projection, the Szegő projection, etc.) that arise naturally in the several complex variables setting. It also gave sharp subelliptic estimates for the $\overline{\partial}_b$ problem.

It is arguable that modern harmonic analysis (at least *linear harmonic analysis*) is the study of integral operators. Zygmund and Stein have pioneered this point of view, and Stein's introduction of Heisenberg group analysis validated it and illustrated it in a vital context. Certainly the integral operators of several complex variables are quite different from those that arise in the classical setting of one complex variable. And it is not just the well-worn differences between one-variable analysis and several-variable analysis. It is the nonisotropic nature of the operators of several complex variables. There is also a certain noncommutativity arising from the behavior of certain key vector fields. In appropriate contexts, the structure of the Heisenberg group very naturally models the structure of the canonical operators of several complex variables, and provides the means for obtaining sharp estimates thereof.

The purpose of the present book is to exposit this rich circle of ideas. And we intend to do so in a context for students. The harmonic analysis of several complex variables builds on copious background material. We provide the necessary background in classical Fourier series, leading up to the Hilbert transform. That will be our entree into singular integrals. Passing to several real-variables, we shall meet the Riesz fractional integrals and the Calderón–Zygmund singular integrals. The aggregate of all the integral operators encountered thus far will provide motivation (in Appendix 3) for considering pseudodifferential operators.

The material on Euclidean integral operators that has been described up to this point is a self-contained course in its own right. But for us it serves as an introduction to analysis on the Heisenberg group. In this new arena, we must first

---

[2] In his book *The Analytical Theory of Heat* [FOU].

provide suitable background material on the function theory of several complex variables. This includes analyticity, the Cauchy–Riemann equations, pseudoconvexity, and the Levi problem. All of this is a prelude to the generalized Cayley transform and an analysis of the automorphism group of the Siegel upper half-space. From this venue the Heisenberg group arises in a complex-analytically natural fashion.

Just to put the material presented here into context: We develop the ideas of integral operators up through pseudodifferential operators *not* because we are going to use pseudodifferential operators as such. Rather, they are the natural climax for this study. For us these ideas are of particular interest because they put into context, and explain, the idea of "order" of an integral operator (and of an error term). This material appears in Appendix 3. In addition, when we later make statements about asymptotic expansions for the Bergman kernel, the pseudodifferential ideas will help students to put the ideas into context. The pseudodifferential operator ideas are also lurking in the background when we discuss subelliptic estimates for the $\bar{\partial}$ problem in the last section of the book.

In addition, we present some of the ideas from the real-variable theory of Hardy spaces *not* because we are going to use them in the context of the Heisenberg group. Rather, they are the natural culmination of a study of integral operators in the context of harmonic analysis. Thus Chapters 1–5 of this book constitute a basic instroduction to harmonic analysis. Chapters 6–8 provide a bridge between harmonic analysis and complex function theory. And Chapters 9 and 10 are dessert: They introduce students to some of the cutting-edge ideas about the Siegel upper half-space and the Heisenberg group.

Analysis on the Heisenberg group still smacks of Euclidean space. But now we are working in a step-one nilpotent Lie group. So dilations, translations, convolutions, and many other artifacts of harmonic analysis take a new form. Even such a fundamental idea as fractional integration must be rethought. Certainly one of the profound new ideas is that the critical dimension for integrability is no longer the topological dimension. Now we have a new idea of *homogeneous dimension*, which is actually one greater than the topological dimension. And there are powerful analytic reasons why this must be so.

We develop the analysis of the Heisenberg group in some detail, so that we may define and calculate bounds on both fractional and singular integrals in this new setting. We provide applications to the study of the Szegő and Poisson–Szegő integrals. The book concludes with a treatment of domains of finite type—which is the next development in this chain of ideas, and is the focus of current research.

We provide considerable background here for the punch line, which is analysis on the Heisenberg group. Much of this basic material in Fourier and harmonic analysis has been covered in other venues (some by this author), but it would be a disservice to the reader were we to send him or her running off to a number of ancillary sources. We want this book to be as self-containted as possible.

We do not, however, wish the book to be boring for the experienced reader. So we put the most basic material on Fourier series in an appendix. Even there, proofs are isolated so that the reader may review this material quickly and easily. The first chapter of the book is background and history, and may be read quickly. Chapters 2 and 3 provide basic material on Fourier analysis, particularly the Fourier transform (although the ideas about singular integrals in Chapter 2 are seminal and should be absorbed carefully). In these two chapters we have also exploited the device of having proofs isolated at the end of the chapter. Many readers will have seen some of this material in a graduate real-variables course. They will want to move on expeditiously to the more exciting ideas that pertain to the task at hand. We have made every effort to aid in this task.

We introduce in this graduate text a few didactic tools to make the reading stimulating and engaging for students:

1. Each chapter begins with a *Prologue*, introducing students to the key ideas that will unfold in the text that follows.
2. Each section begins with a *Capsule*, giving a quick preview of that unit of material.
3. Each key theorem or proposition is preceded by a *Prelude*, putting the result in context and providing motivation.
4. At key junctures we include an *Exercise for the Reader* to encourage the neophyte to pick up a pencil, do some calculations, and get involved with the material.

We hope that these devices will break up the usual dry exposition of a research monograph and make this text more like an invitation to the subject.

I have taught versions of this material over the years, most recently in spring of 2006 at Washington University in St. Louis. I thank the students in the course for their attention and for helping me to locate many mistakes and misstatements. Lina Lee, in particular, took wonderful notes from the course and prepared them as TEX files. Her notes are the basis for much of the second part of this book. I thank the American Institute of Mathematics for hospitality and support during some of the writing.

In total, this is an ambitious introduction to a particular direction in modern harmonic analysis. It presents harmonic analysis *in vitro*—in a context in which it is actually applied: complex variables and partial differential equations. This will make the learning experience more meaningful for graduate students who are just beginning to forge a path of research. We expect the readers of this book to be ready to take a number of different directions in exploring the research literature and beginning his or her own investigations.

— Steven G. Krantz

1

Ontology and History of Real Analysis

Prologue: Real analysis as a subject grew out of struggles to understand, and to make rigorous, Newton and Leibniz's calculus. But its roots wander in all directions—into real analytic function theory, into the analysis of polynomials, into the solution of differential equations.

Likewise, the proper study of Fourier series began with the work of Joseph Fourier in the early nineteenth century. But threads of the subject go back to Euler, Bernoulli, and others. There is hardly any part of mathematics that has sprung fully armed from a single mathematician's head (Georg Cantor's set theory may be the exception); rather, mathematics is a flowing process that is the product of many tributaries and many currents.

It should be stressed—and the present book expends some effort to make this case—that real analysis is *not* a subject in isolation. To the contrary, it interacts profitably with complex analysis, Lie theory, differential equations, differential geometry, dynamical systems, and many other parts of mathematics. Real analysis is a basic tool for, and lies at the heart of, a good many subjects. It is part of the *lingua franca* of modern mathematics.

The present book is a paean to real analysis, but it is also a vivid illustration of how real-variable theory arises in modern, cutting-edge research (e.g., the Heisenberg group). The reader will gain an education in how basic analytic tools see new light when put in a fresh context. This is an illustration of the dynamic of modern mathematics, and of the kind of energy that causes it to continue to grow.

1.1 Deep Background in Real Analytic Functions

Capsule: There is hardly any more basic, yet more thoroughly misunderstood, part of mathematics than the theory of real analytic functions. First, it is arguably the oldest part of real analysis. Second, there is only one book on the subject (see [KRP3]). Third, since everyone learns about Taylor series

S.G. Krantz, *Explorations in Harmonic Analysis*, Applied and Numerical
Harmonic Analysis, DOI 10.1007/978-0-8176-4669-1_1,
© Birkhäuser Boston, a part of Springer Science+Business Media, LLC 2009

in calculus class, it follows that everyone thinks that an arbitrary C^∞ function has a power series expansion. Nothing could be further from the truth. The real analytic functions form a rather thin subset of the C^∞ functions, but they are still dense (in a suitable sense). Properly understood, they are a powerful and versatile tool for all analysts.

While real analysis certainly finds its roots in the calculus of Newton and Leibniz, it can be said that the true spirit of analysis is the decomposition of arbitrary functions into fundamental units. That said, analysis really began in the early nineteenth century. It was then that Cauchy, Riemann, and Weierstrass laid the foundations for the theory of real analytic functions, and also Fourier set the stage for Fourier analysis.

In the present chapter we give an overview of key ideas in the history of modern analysis. We close the chapter with a sort of history of Fourier series, emphasizing those parts that are most germane to the subject matter of this book. Chapter 2 begins the true guts of the subject.

A function is *real analytic* if it is locally representable by a convergent power series. Thus a real analytic function f of a single real-variable can be expanded about a point p in its domain as

$$f(x) = \sum_{j=0}^{\infty} a_j(x - p)^j.$$

A real analytic function F of several real-variables can be expanded about a point p in its domain as

$$F(x) = \sum_{\alpha} b_\alpha(x - p)^\alpha,$$

where α here denotes a multi-index (see [KRP3] for this common and useful notation). It is noteworthy that any harmonic function is real analytic (just because the Poisson kernel is real analytic), and certainly any holomorphic function (of either one or several complex variables) is real analytic. Holomorphic functions have the additional virtue of being *complex analytic*, which means that they can be expanded in powers of a *complex variable*.

The basic idea of a real analytic function f is that f can be broken down into elementary units—these units being the integer powers of x. The theory is at first a bit confusing, because Taylor series might lead one to think that *any* C^∞ function can be expanded in terms of powers of x. In fact, nothing could be further from the truth. It is true that, if $f \in C^k(\mathbb{R})$, then we may write

$$f(x) = \sum_{j=0}^{k} \frac{f^{(j)}(p) \cdot (x - p)^j}{j!} + R_k(x),$$

where R_k is an error term. What must be emphasized is that the error term here is of fundamental importance. In fact, the Taylor expansion converges to f at x if and only if $R_k(x) \to 0$. This statement is of course a tautology, but it is the heart of the matter.

It is a fact, which can be proved with elementary category theory arguments, that "most" C^∞ functions are *not* real analytic. Furthermore, even if the power series expansion of a given C^∞ function f *does* converge, it typically will not converge back to f. A good example to bear in mind is

$$f(x) = \begin{cases} e^{-1/x^2} & \text{if } x > 0, \\ 0 & \text{if } x \leq 0. \end{cases}$$

This f is certainly C^∞ (as may be verified using l'Hôpital's rule), but its Taylor series expansion about 0 is identically 0. So the Taylor series converges to the function

$$g(x) \equiv 0,$$

and that function does *not* agree with f on the entire right half-line.

The good news is that real analytic functions are generic in some sense. If φ is a continuous function on the unit interval $I = [0, 1]$, then the Weierstrass approximation theorem tells us that φ may be uniformly approximated on I by polynomials. Of course a polynomial is real analytic. So the theorem tells us that the real analytic functions are uniformly dense in $C([0, 1])$. A simple integration argument shows that in fact the real analytic functions are dense in $C^k([0, 1])$ equipped with its natural topology.

Real analytic functions may be characterized by the following useful and very natural condition. A C^∞ function on the interval $J = (a - \epsilon, a + \epsilon)$ is real analytic if there is an $M > 0$ such that for any integer $k \geq 0$,

$$|f^{(k)}(x)| \leq \frac{M \cdot k!}{\epsilon^k}$$

for all $x \in J$. The converse is true as well (see [KRP3] for details). By contrast, consider the following venerable result of E. Borel: If $a \in \mathbb{R}$, $\epsilon > 0$, and $c_0, c_1, \ldots$ is *any* sequence of real numbers then there is a C^∞ function f on the interval $J = (a - \epsilon, a + \epsilon)$ such that

$$f^{(j)}(p) = j! \cdot c_j$$

for every j. In other words, the power series expansion of f about p is

$$\sum_j c_j (x - p)^j.$$

1.2 The Idea of Fourier Expansions

Capsule: Fourier analysis as we know it today has its genesis in the book [FOU]. What is special about that book is that it finally gives an explicit formula—and the *right* formula— for the Fourier coefficients of an "arbitrary" function. The epistemological issue here must be clearly understood. Mathematicians had been debating for years whether an arbitrary function can be expanded in terms of sines and cosines. The discussion was hobbled by the fact that there was not yet any precise definition of "function." People were thinking in terms of expressions that could be initial data for the heat equation. Generally speaking, people thought of a function as a "curve." Certainly potential data functions were rather more general than signs and cosines. Fourier's contribution came as a revelation. His derivation of his formula(s) was suspect, but it seemed to give rise to viable Fourier series for a very broad class of functions. Of course techniques were not available to sum the resulting series and see that they converged to the initial function. Even so, Fourier's arguments were found to be compelling and definitive. The subject of Fourier analysis was duly born.

Discussions in the late eighteenth century about solutions to the wave equation led to the question whether "any" function may be expanded as a sum of sines and cosines. Thus mathematicians were led to consider a different type of building block for "arbitrary" functions. The considerations were hindered by the fact that, at the time, there was no generally accepted definition of "function," nor of convergence of a series. Many mathematicians thought of a function as a curve, or perhaps finitely many curves pieced together. Could such a function be written as the superposition of sines and cosines, each of which was a single real analytic curve and *not* pieced together?

Joseph Fourier essayed to lay the matter to rest with his classic book *The Analytical Theory of Heat* [FOU]. This book gives a not-very-rigorous derivation of the formula that is so well-known today for the Fourier series of a given function on $[0, 2\pi)$:

$$\widehat{f}(j) = \frac{1}{2\pi} \int_0^{2\pi} f(t)e^{-2\pi ij}\, dt.$$

It should be stressed that Fourier does *not* address the question whether the Fourier series

$$Sf \sim \sum_j \widehat{f}(j)e^{2\pi ijt}$$

actually converges to f. Indeed, a rigorous definition of convergence of series was yet to be formulated by Dirichlet and others. Being a practical man, however, Fourier did provide a number of concrete examples of explicit functions with their Fourier series computed in detail.

Fourier's contribution must be appreciated for the epistemological breakthrough that it was. He certainly did not prove that a "fairly arbitrary" function has a convergent Fourier series (this statement is true, and we shall prove it later in the book). But he *did* provide a paradigm for—at least in principle—expanding any given function in a sum of sines and cosines. This was a fundamentally new idea.

It took a number of years, and some struggle, for Fourier to get his ideas published. In fact, he finally published the book himself when he served as secretary of the National Academy in France. Even the scientists of the early nineteenth century—somewhat naive by today's standards—could see the logical flaws in Fourier's reasoning. But Fourier's work has had an immense impact, and it is certainly appropriate that the subject of Fourier analysis is named after him.

1.3 Differences between the Real Analytic Theory and the Fourier Theory

> **Capsule:** If f is a C^∞ function on some interval $I = (x_0 - \epsilon, x_0 + \epsilon)$, then of course there is a Taylor expansion about the point x_0. The probability that the Taylor series converges, or if it converges that it converges back to f, is 0. This assertion can be made precise in a number of ways. We omit the details, but see [KRP3]. Of course the series converges at x_0 to f, but there is no reason to suppose that it converges to f on an entire interval. Functions that *do* have this convergence property are very special. They are called *real analytic*. By contrast, most any function for which the Fourier coefficients can be computed has a convergent Fourier series—and it converges back to the initial function! Even when the Fourier series itself does not converge, some reasonable summation method may be applied to get a convergent trigonometric series. So there is a decided contrast between the two theories. Fourier series are much more flexible, and considerably more powerful, than power series.

There are fundamental and substantial differences between the theory of real analytic functions and the theory of Fourier analysis. As already noted, the real analytic functions form a rather thin set (indeed, a set of the first category) in the space of all C^∞ functions. By contrast, any continuously differentiable function has a convergent Fourier expansion. For this reason (and other reasons as well), Fourier analysis has proved to be a powerful tool in many parts of mathematical analysis and engineering.

Fourier analysis has been particularly effective in the study of signal processing and the creation of filters. Of course sines and cosines are good models for the waves that describe sound. So the analysis by Fourier expansion is both apt and accurate. If a signal has noise in it (pops and clicks for example), then one may expand the signal in a (convergent) Fourier series, extract those terms that describe the pops and clicks, and reassemble the remaining terms into a signal that is free of noise. Certainly this is the basic idea behind all the filters constructed in the 1950s and 1960s. An analogous construction in the real analytic category really would make no sense—either mathematically or physically.

Fourier analysis has proved to be a powerful tool in the study of differential equations because of the formula

$$\widehat{f'}(j) = 2\pi i j \, \widehat{f}(j).$$

Of course a similar formula holds for higher derivatives as well (and certainly there are analogous formulas in the several-variable theory). As a result, a differential equation may, by way of Fourier analysis, be converted to an algebra problem (involving a polynomial in j). This is one of the most basic techniques in the solution of linear differential equations with constant coefficients. For equations with variable coefficients, pseudodifferential operators are a natural outgrowth that addresses the relevant issues (see Appendix 3).

1.4 Modern Developments

> **Capsule:** Analysis of several real-variables is a fairly modern development. Certainly analysts in the pre-World-War-II period contented themselves with functions of one variable, and G.H. Hardy was quite aggressive in asserting that the analysis of several-variables offered nothing new (except some bookkeeping issues). It is only with the school of Calderón–Zygmund, Stein, Fefferman, and others (since 1952) that we have learned all the depth and subtlety hidden in the analysis of several real-variables. It is a tapestry rich with ideas, and continues to be developed.

The nineteenth century contented itself, by and large, with Fourier series of one variable on the interval $[0, 2\pi)$ and the Fourier transform on the real line $\mathbb{R}$. But we live in a higher-dimensional world, and there is good reason to want analytic tools that are adapted to it. In the twentieth century, especially following World War II, there has been considerable development of the theory of Fourier series of several-variables, of spherical harmonics (which is another way to generalize Fourier series to many variables), and of the multivariable Fourier transform. This is not simply a matter (as G.H. Hardy suspected) of developing multivariable notation and establishing the bookkeeping techniques needed to track the behavior of many variables. In fact, higher-dimensional Fourier analysis contains many surprises, and harbors many new phenomena. Books like [STG1] and [STE2] give a thoroughgoing survey of some of the new discoveries. The present text will give a solid introduction to the multivariable Fourier transform, singular integrals, and pseudodifferential operators. These ideas lie at the heart of many of the modern developments.

1.5 Wavelets and Beyond

> **Capsule:** The theory of wavelets is less than twenty-five years old. It frees Fourier analysis from an artificial dependence on sines and cosines, and shows that a viable Fourier analysis may be built on a much more general basis (or even a frame, which can have redundancies) that can be localized both in the space and in the time variables. Wavelet analysis has had a profound impact in engineering, particularly in signal processing, image

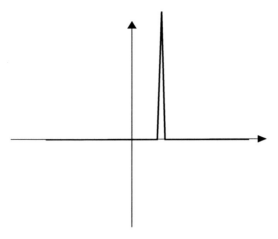

Figure 1.1. A "pop" or "click."

analysis, and other applications where localization is important. The fast Fourier transform is still important, but wavelet theory is gradually supplanting it.

While Fourier series, and classical Fourier analysis in general, are useful tools of wide applicability, they have their limitations. It is naive and unreasonable to represent a "pop" or a "click"—see Figure 1.1—as a sum of sines and cosines. For a sine or a cosine function is supported on the entire real line, and is ill suited to approximate a locally supported function that is just a spike.

Enter the modern theory of wavelets. A *wavelet* is a basis element—much like a sine or a cosine—that can be localized in its support. There can be localization both in the space variable and in the time variable. Thus, in signal processing for instance, one gets much more rapid (and more accurate) convergence and therefore much more effective filters. The reference [KRA5] gives a quick introduction, with motivation, to wavelet theory. The monographs [MEY1], [MEY2], [MEY3], and [HERG] give more thoroughgoing introductions to the subject. These new and exciting ideas are not really germane to the present book, and we shall say no more about them here.

1.6 History and Genesis of Fourier Series

Capsule: We have already alluded to the key role of the heat equation (and also the wave equation) in providing a context for the key questions of basic Fourier series. Thus Fourier analysis has had intimate connections with differential equations from its very beginnings. The big names in mathematics—all of whom were also accomplished applied mathematicians and physicists—made contributions to the early development of Fourier

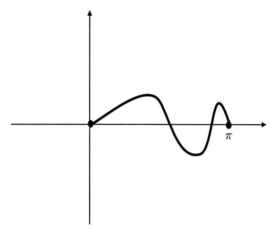

Figure 1.2. The vibrating string.

series. In this section we exhibit some of the mathematics that led to Fourier's seminal contribution.

The classical wave equation (see Figure 1.2) describes the motion of a plucked string of length π with endpoints pinned down as shown:

$$\frac{\partial^2 u}{\partial t^2} = a^2 \frac{\partial^2 u}{\partial x^2}.$$

Here a is a real parameter that depends on the tension of the string. Typically (just by a rescaling) we take $a = 1$.

In 1747, d'Alembert showed that solutions of this equation have the form

$$u(x, t) = \frac{1}{2}[f(t + x) + g(t - x)], \tag{1.6.1}$$

where f and g are "any" functions of one variable on $[0, \pi]$. [The function is extended in an obvious way, by odd reflection, to the entire real line.]

It is natural to equip the wave equation with two boundary conditions:

$$u(x, 0) = \phi(x),$$

$$\partial_t u(x, 0) = \psi(x).$$

These specify the initial position and velocity respectively.

If D'Alembert's formula is to provide a solution of this initial value problem, then f and g must satisfy

$$\frac{1}{2}[f(x) + g(-x)] = \phi(x) \tag{1.6.2}$$

and

$$\frac{1}{2}[f'(x) + g'(-x)] = \psi(x). \tag{1.6.3}$$

Integration of (1.6.3) gives a formula for $f(x) - g(-x)$. Thus, together with (1.6.2.), we may solve for f and g.

The converse statement holds as well: for any functions f and g, a function u of the form (1.6.1) satisfies the wave equation.

Daniel Bernoulli solved the wave equation by a different method (separation of variables) and was able to show that there are infinitely many solutions of the wave equation having the form

$$\phi_j(x, t) = \sin jx \cos jt.$$

This solution procedure presupposes that the initial configuration is $\phi_j(x, 0) = \sin jx$. Proceeding formally, Bernoulli hypothesized that all solutions of the wave equation satisfying $u(0, t) = u(\pi, t) = 0$ and $\partial_t u(x, 0) = 0$ will have the form

$$u = \sum_{j=1}^{\infty} a_j \sin jx \cos jt.$$

Setting $t = 0$ indicates that the initial form of the string is $f(x) \equiv \sum_{j=1}^{\infty} a_j \sin jx$. In d'Alembert's language, the initial form of the string is $\frac{1}{2}(f(x) - f(-x))$, for we know that

$$0 \equiv u(0, t) = f(t) + g(t)$$

(because the endpoints of the string are held stationary), hence $g(t) = -f(t)$. If we suppose that d'Alembert's function is odd (as is $\sin jx$, each j), then the initial position is given by $f(x)$. Thus the problem of reconciling Bernoulli's solution with d'Alembert's reduces to the question whether an "arbitrary" function f on $[0, \pi]$ may be written in the form $\sum_{j=1}^{\infty} a_j \sin jx$.

This is of course a fundamental question. It is at the heart of what we now think of as Fourier analysis. The question of representing an "arbitrary" function as a (possibly infinite) linear combination of sine functions fascinated many of the top mathematicians in the late eighteenth and early nineteenth centuries.

In the 1820s, the problem of representation of an "arbitrary" function by trigonometric series was given a satisfactory (at least satisfactory according to the standards of the day) answer as a result of two events. First there is the sequence of papers by Joseph Fourier culminating with the tract [FOU]. Fourier gave a formal method of expanding an "arbitrary" function f into a trigonometric series. He computed some partial sums for some sample f's and verified that they gave very good approximations to f. Second, Dirichlet proved the first theorem giving sufficient (and very general) conditions for the Fourier series of a function f to converge pointwise to f. *Dirichlet was one of the first, in 1828, to formalize the notions of partial sum and convergence of a series*; his ideas certainly had antecedents in work of Gauss and Cauchy.

It is an interesting historical note that Fourier had a difficult time publishing his now famous tome [FOU]. In fact, he finally published it himself after he was elected secretary of the French Academy.

For all practical purposes, these events mark the beginning of the mathematical theory of Fourier series (see [LAN]).

1.6.1 Derivation of the Heat Equation

Fourier approached the basic questions of what we now call Fourier series by way of the heat equation rather than the wave equation. We devote this subsection to a consideration of that partial differential equation.

Let there be given an insulated, homogeneous rod of length π with initial temperature at each $x \in [0, \pi]$ given by a function $f(x)$ (Figure 1.3). Assume that the endpoints are held at temperature 0, and that the temperature of each cross-section is constant. The problem is to describe the temperature $u(x, t)$ of the point x in the rod at time t.

We shall use three elementary physical principles to derive the heat equation:

(1.6.4) The density of heat energy is proportional to the temperature u; hence the amount of heat energy in any interval $[a, b]$ of the rod is proportional to $\int_a^b u(x, t)\, dx$.

(1.6.5) **[Newton's Law of Cooling]** The rate at which heat flows from a hot place to a cold one is proportional to the difference in temperature. The infinitesimal version of this statement is that the rate of heat flow across a point x (from left to right) is some negative constant times $\partial_x u(x, t)$.

(1.6.6) **[Conservation of Energy]** Heat has no sources or sinks.

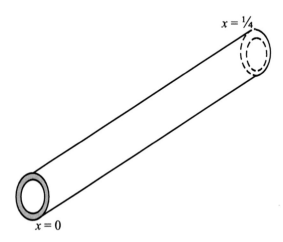

$$x = \tfrac{1}{4}$$

$$x = 0$$

Figure 1.3. An insulated rod.

Now **(1.6.6)** tells us that the only way that heat can enter or leave any interval portion $[a, b]$ of the rod is through the endpoints. And **(1.6.5)** tells us exactly how this happens. Using **(1.6.4)**, we may therefore write

$$\frac{d}{dt} \int_a^b u(x, t)\, dx = \eta^2 [\partial_x u(b, t) - \partial_x u(a, t)].$$

Here η^2 is a positive constant. We may rewrite this equation as

$$\int_a^b \partial_t u(x, t)\, dx = \eta^2 \int_a^b \partial_x^2 u(x, t)\, dx.$$

Differentiating in b, we find that

$$\partial_t u = \eta^2 \partial_x^2 u, \tag{1.6.7}$$

and that is the heat equation.

Suppose for simplicity that the constant of proportionality η^2 equals 1. Fourier guessed that the equation (1.6.7) has a solution of the form $u(x, t) = \alpha(x)\beta(t)$. Substituting this guess into the equation yields

$$\alpha(x)\beta'(t) = \alpha''(x)\beta(t),$$

or

$$\frac{\beta'(t)}{\beta(t)} = \frac{\alpha''(x)}{\alpha(x)}.$$

Since the left side is independent of x and the right side is independent of t, it follows that there is a constant K such that

$$\frac{\beta'(t)}{\beta(t)} = K = \frac{\alpha''(x)}{\alpha(x)},$$

or

$$\beta'(t) = K\beta(t),$$
$$\alpha''(x) = K\alpha(x).$$

We conclude that $\beta(t) = Ce^{Kt}$. The nature of β, and hence of α, thus depends on the sign of K. But physical considerations tell us that the temperature will dissipate as time goes on, so we conclude that $K \leq 0$. Therefore $\alpha(x) = \cos\sqrt{-K}x$ and $\alpha(x) = \sin\sqrt{-K}x$ are solutions of the differential equation for α. The initial conditions $u(0, t) = u(\pi, t) = 0$ (since the ends of the rod are held at constant temperature 0) eliminate the first of these solutions and force $K = -j^2$, $j \in \mathbb{Z}$. Thus Fourier found the solutions

$$u_j(x, t) = e^{-j^2 t} \sin jx, \quad j \in \mathbb{N},$$

of the heat equation. Observe that the exponential factor in front of the expression for u_j gives decay with increasing time.

By linearity, any finite linear combination

$$\sum_j b_j e^{-j^2 t} \sin jx$$

of these solutions is also a solution. It is physically plausible to extend this assertion to infinite linear combinations. Using the initial condition $u(x, 0) = f(x)$ again raises the question whether "any" function $f(x)$ on $[0, \pi]$ can be written as an (infinite) linear combination of the functions $\sin jx$.

Fourier used intricate but (by modern standards) logically specious means to derive the formula

$$b_j = \frac{2}{\pi} \int_0^\pi f(x) \sin jx \, dx \qquad (1.6.8)$$

for the coefficients.

Whatever the defect of Fourier's mathematical methodology, his formula gives an actual *procedure* for expanding *any* given f in a series of sine functions. This was a major breakthrough.

Of course we now realize, because of our modern understanding of Hilbert space concepts (such as orthogonality), that there is a more direct route to Fourier's formula. If we assume in advance that

$$f(x) = \sum_j b_j \sin jx$$

and that the convergence is in L^2, then we may calculate

$$\frac{2}{\pi} \int_0^\pi f(x) \sin kx \, dx = \frac{2}{\pi} \int_0^\pi \left(\sum_j b_j \sin jx \right) \sin kx \, dx$$

$$= \frac{2}{\pi} \sum_j b_j \int_0^\pi \sin jx \sin kx \, dx$$

$$= b_k.$$

[We use here the fact that $\int \sin jx \sin kx \, dx = 0$ if $j \neq k$, i.e., the fact that $\{\sin jx\}$ are orthogonal in $L^2[0, \pi]$. Also $\int_0^\pi \sin^2 kx \, dx = \pi/2$, each $k \neq 0$.]

Classical studies of Fourier series were devoted to expanding a function on either $[0, 2\pi)$ or $[0, \pi)$ in a series of the form

$$\sum_{j=0}^\infty a_j \cos jx$$

or

$$\sum_{j=1}^{\infty} b_j \sin jx$$

or as a combination of these

$$\sum_{j=0}^{\infty} a_j \cos jx + \sum_{j=1}^{\infty} b_j \sin jx.$$

The modern theory tends to use the more elegant notation of complex exponentials. Since

$$\cos jx = \frac{e^{ijx} + e^{-ijx}}{2} \qquad \text{and} \qquad \sin jx = \frac{e^{ijx} - e^{-ijx}}{2i},$$

we may seek instead to expand f in a series of the form

$$\sum_{j=-\infty}^{\infty} c_j e^{ijx}.$$

In this book we shall confine ourselves almost exclusively to the complex exponential notation. Engineers are still fond of the notation of sines and cosines.

2

The Central Idea: The Hilbert Transform

Proofs in this chapter are presented at the end of the chapter.

Prologue: The Hilbert transform is, without question, the most important operator in analysis. It arises in many different contexts, and all these contexts are intertwined in profound and influential ways. What it all comes down to is that there is only one singular integral in dimension 1, and it is the Hilbert transform. The philosophy is that all significant analytic questions reduce to a singular integral; and in the first dimension there is just one choice.

The most important fact about the Hilbert transform is that it is bounded on L^p for $1 < p < \infty$. It is also bounded on various Sobolev and Lipschitz spaces. And also on H^1_{Re} and the space of functions of bounded mean oscillation (*BMO*). We discuss many of these properties in the present chapter and later on in Chapters 4, 5, and 9. See also [KRA5] and [STE2].

Even though the Hilbert transform is well understood today, it continues to be studied intensely. Boundedness properties of the "maximum Hilbert transform" are equivalent to pointwise convergence results for Fourier series. In higher dimensions, the Hilbert transform is used to construct analytic disks. Analytic disks are important in cosmology and other parts of physics.

From our point of view in the present book, the Hilbert transform is important because it is the inspiration and the role model for higher-dimensional singular integrals. Singular integrals in $\mathbb{R}^N$ are about 55 years old. Singular integrals on the Heisenberg group and other more general settings are much newer. We shall study the former in some detail and provide some pointers to the latter. Chapters 9 and 10 develop integral operators on $\mathbb{H}^n$ in some detail—that is one of the main points of this book.

We take it for granted that the reader is familiar with the most basic ideas of Fourier series. Appendix 1 provides a review or quick reference.

S.G. Krantz, *Explorations in Harmonic Analysis*, Applied and Numerical
Harmonic Analysis, DOI 10.1007/978-0-8176-4669-1_2,
© Birkhäuser Boston, a part of Springer Science+Business Media, LLC 2009

2.1 The Notion of the Hilbert Transform

Capsule: Our first approach to the Hilbert transform will be by way of complex variable theory. The idea is to seek a means of finding the boundary function of the harmonic conjugate of a given function (which in turn is the Poisson integral of some initial boundary function). This very natural process gives rise to a linear operator that may be identified as the Hilbert transform. Later on we shall see that the Hilbert transform arises rather naturally in the context of partial summation operators for Fourier series. Most any question of convergence of Fourier series may be reduced to an assertion about mapping properties of the Hilbert transform. Thus the Hilbert transform becomes a central player in the theory of Fourier series.

Now we study the Hilbert transform H, which is one of the most important linear operators in analysis. It is essentially the *only* singular integral operator in dimension 1, and it comes up decisively in a variety of contexts. The Hilbert transform is the key player—from a certain point of view—in complex variable theory. And it is the key player in the theory of Fourier series. It also comes up in the Cauchy problem and other aspects of partial differential equations.

Put in slightly more technical terms, the Hilbert transform is important for these reasons (among others):

- It interpolates between the real and imaginary parts of a holomorphic function.
- It is the key to all convergence questions for the partial sums of Fourier series.
- It is a paradigm for all singular integral operators on Euclidean space (we shall treat these in Chapter 3).
- It is (on the real line) uniquely determined by its invariance properties with respect to the groups that act naturally on 1-dimensional Euclidean space.

One can *discover* the Hilbert transform by way of complex analysis. As we know, if f is holomorphic on D and continuous up to ∂D, we can calculate f at a point $z \in D$ from the boundary value of f by the following formula:

$$f(z) = \frac{1}{2\pi i} \int_{\partial D} \frac{f(\zeta)}{\zeta - z} \, d\zeta, \quad z \in D.$$

We call

$$\frac{1}{2\pi i} \cdot \frac{d\zeta}{\zeta - z} \tag{2.1.1}$$

the *Cauchy kernel*.

If we let $\zeta = e^{i\psi}$ and $z = re^{i\theta}$, the expression (2.1.1) can be rewritten as follows:

$$\frac{1}{2\pi i} \cdot \frac{d\zeta}{\zeta - z} = \frac{1}{2\pi} \cdot \frac{-i\bar{\zeta}d\zeta}{\bar{\zeta}(\zeta - z)}$$

$$= \frac{1}{2\pi} \cdot \frac{-ie^{-i\psi} \cdot ie^{i\psi}d\psi}{e^{-i\psi}(e^{i\psi} - re^{i\theta})}$$

$$= \frac{1}{2\pi} \cdot \frac{d\psi}{1 - re^{i(\theta - \psi)}}$$

$$= \frac{1}{2\pi} \cdot \frac{1 - re^{-i(\theta - \psi)}}{|1 - re^{i(\theta - \psi)}|^2} d\psi$$

$$= \left(\frac{1}{2\pi} \cdot \frac{1 - r\cos(\theta - \psi)}{|1 - re^{i(\theta - \psi)}|^2} d\psi \right)$$

$$+ i \left(\frac{1}{2\pi} \cdot \frac{r\sin(\theta - \psi)}{|1 - re^{i(\theta - \psi)}|^2} d\psi \right). \tag{2.1.2}$$

If we subtract $\frac{1}{4\pi} d\psi$ from the real part of the Cauchy kernel, we get

$$\mathrm{Re}\left(\frac{1}{2\pi i} \cdot \frac{d\zeta}{\zeta - z} \right) - \frac{d\psi}{4\pi} = \frac{1}{2\pi} \left(\frac{1 - r\cos(\theta - \psi)}{|1 - re^{i(\theta - \psi)}|^2} - \frac{1}{2} \right) d\psi$$

$$= \frac{1}{2\pi} \left(\frac{\frac{1}{2} - \frac{1}{2}r^2}{1 - 2r\cos(\theta - \psi) + r^2} \right) d\psi$$

$$\equiv \frac{1}{2} P_r(e^{i(\theta - \psi)}) d\psi. \tag{2.1.3}$$

Note that in the last line we have, in effect, "discovered" the classical (and well-known) Poisson kernel.

This is an important lesson, and one to be remembered as the book develops: The real part of the Cauchy kernel is (up to a small correction) the Poisson kernel. That is, the kernel that reproduces harmonic functions is the real part of the kernel that reproduces holomorphic functions.

In the next section we shall examine the imaginary part of the Cauchy kernel and find the Hilbert transform revealed.

2.2 The Guts of the Hilbert Transform

Now let us take the reasoning that we used above (to discover the Poisson kernel) and turn it around. Suppose that we are given a real-valued function $f \in L^2(\partial D)$. Then we can use the Poisson integral formula to produce a function u on D such that $u = f$ (almost everywhere) on ∂D. We may find a harmonic conjugate of u, say $u^\dagger$, such that $u^\dagger(0) = 0$ and $u + iu^\dagger$ is holomorphic on D. What we hope to do is to produce a boundary function $f^\dagger$ for $u^\dagger$. This will create some symmetry in the picture. For we began with a function f from which we created u; now we are extracting $f^\dagger$ from $u^\dagger$. Our ultimate goal is to study the linear operator $f \mapsto f^\dagger$.

The following diagram illustrates the idea:

$$L^2(\partial D) \ni f \longrightarrow u$$

$$\downarrow$$

$$f^\dagger \longleftarrow u^\dagger$$

If we define a function h on D as

$$h(z) \equiv \frac{1}{2\pi i} \int_{\partial D} \frac{f(\zeta)}{\zeta - z} \, d\zeta, \quad z \in D,$$

then obviously h is holomorphic in D. We know from calculations in the last section that the real part of h is (up to adjustment by an additive real constant) the Poisson integral u of f. Therefore Re h is harmonic in D and Im h is a harmonic conjugate of Re h. Thus, if h is continuous up to the boundary, then we will be able to say that $u^\dagger = \operatorname{Im} h$ and $f^\dagger(e^{i\theta}) = \lim_{r \to 1^-} u^\dagger(re^{i\theta})$.

So let us look at the imaginary part of the Cauchy kernel in (2.1.2):

$$\frac{r \sin(\theta - \psi)}{2\pi \, |1 - re^{i(\theta - \psi)}|^2}.$$

If we let $r \to 1^-$, then we obtain

$$\frac{\sin(\theta - \psi)}{2\pi \, |1 - e^{i(\theta - \psi)}|^2} = \frac{\sin(\theta - \psi)}{2\pi \, (1 - 2\cos(\theta - \psi) + 1)}$$

$$= \frac{\sin(\theta - \psi)}{4\pi \, (1 - \cos(\theta - \psi))}$$

$$= \frac{2 \sin(\frac{\theta - \psi}{2}) \cos(\frac{\theta - \psi}{2})}{4\pi \cdot 2 \cos^2(\frac{\theta - \psi}{2})}$$

$$= \frac{1}{4\pi} \cot\left(\frac{\theta - \psi}{2}\right).$$

Hence we obtain the Hilbert transform[1] $H : f \to f^\dagger$ as follows:

$$Hf(e^{i\theta}) = \int_0^{2\pi} f(e^{it}) \cot\left(\frac{\theta - t}{2}\right) dt. \tag{2.1.4}$$

---

[1] There are subtle convergence issues—both pointwise and operator-theoretic—which we momentarily suppress. Details may be found, for instance, in [KAT].

[We suppress the multiplicative constant here because it is of no interest.] Note that we can express the kernel as

$$\cot\frac{\theta}{2} = \frac{\cos\frac{\theta}{2}}{\sin\frac{\theta}{2}} = \frac{1 - \frac{(\theta/2)^2}{2!} \pm \cdots}{\frac{\theta}{2}\left(1 - \frac{(\theta/2)^2}{3!} \pm \cdots\right)} = \frac{2}{\theta}\left(1 + \mathcal{O}(|\theta|^2)\right) = \frac{2}{\theta} + E(\theta),$$

where $E(\theta) = \mathcal{O}(|\theta|)$ is a bounded continuous function. Therefore, we can rewrite (2.1.4) as

$$Hf(e^{i\theta}) \equiv \int_0^{2\pi} f(e^{it})\cot\left(\frac{\theta - t}{2}\right) dt$$

$$= \int_0^{2\pi} f(e^{it})\frac{2}{\theta - t}dt + \int_0^{2\pi} f(e^{it})E(\theta - t)\,dt.$$

Note that the first integral is singular at θ and the second one is bounded and trivial to estimate—just by applying Schur's lemma (see [SAD] and our Lemma A1.5.5).[2]

In practice, we usually write

$$\cot\left(\frac{\theta - t}{2}\right) \approx \frac{2}{\theta - t},$$

simply ignoring the trivial error term. Both sides of this "equation" are called the kernel of the Hilbert transform. When we study the Hilbert transform, we generally use the kernel on the right; and we omit the 2 in the numerator.

2.3 The Laplace Equation and Singular Integrals on $\mathbb{R}^N$

Let us look at Laplace equation in $\mathbb{R}^N$ for $N > 2$:

$$\Delta u(x) = \left(\sum_{j=1}^N \frac{\partial^2}{\partial x_j^2}\right) u(x) = 0.$$

The fundamental solution[3] (see [KRA4]) for the Laplacian is

$$\Gamma(x) = c_N \cdot \frac{1}{|x|^{N-2}}, \quad N > 2,$$

where c_N is a constant that depends on N.

---

[2] Schur's lemma, in a very basic form, says that convolution with an L^1 kernel is a bounded operator on L^p. This assertion may be verified with elementary estimates from measure theory—exercise.

[3] It must be noted that this formula is not valid in dimension 2. One might guess this, because when $N = 2$ the formula in fact becomes trivial. The correct form for the fundamental solution in dimension 2 is

$$\Gamma(x) = \frac{1}{2\pi}\log|x|.$$

Details may be found in [KRA4].

Exercise for the Reader: Prove the defining property for the fundamental solution, namely, that $\Delta\Gamma(x) = \delta_0$, where δ_0 is the Dirac mass at 0. (**Hint:** Use Green's theorem, or see [KRA4].)

We may obtain one solution u of $\Delta u = f$ by convolving f with Γ:

$$u = f * \Gamma.$$

For notice that $\Delta u = f * \Delta\Gamma = f * \delta_0 = f$.

In the ensuing discussion we shall consider the integrability of expressions like $|x|^\beta$ near the origin (our subsequent discussion of fractional integrals in Chapter 5 will put this matter into context). We shall ultimately think of this kernel as a fractional power (positive or negative) of the fundamental solution for the Laplacian.

The correct way to assess such a situation is to use polar coordinates:

$$\int_{|x|<1} |x|^\beta \, dx = \int_\Sigma \int_0^1 r^\beta \cdot r^{N-1} \, dr \, d\sigma(\xi).$$

A few comments are in order: The symbol Σ denotes the unit sphere in $\mathbb{R}^N$, and $d\sigma$ is rotationally invariant area measure (see Chapter 9 for a consideration of Hausdorff measure on a general surface) on Σ. The factor r^{N-1} is the Jacobian of the change of variables from rectangular coordinates to spherical coordinates. Of course the integral in the rotational variable ξ is trivially a constant. The integral in r converges precisely when $\beta > -N$. Thus we think of $-N$ as the "critical index" for integrability at the origin.

Now let us consider the following transformation:

$$T : f \longmapsto f * \Gamma.$$

The kernel Γ is singular at the origin to order $-(N-2)$. Studying L^p mapping properties of this transformation is easy because Γ is locally integrable. We can perform estimates with easy techniques such as the generalized Minkowski inequality and Schur's lemma (see [SAD] and our Lemmas A1.5.5, A1.5.8). In fact, the operator T is a special instance of a "fractional integral operator." We shall have more to say about this family of operators as the book develops.

The first derivative of Γ is singular at the origin to order $-(N-1)$ and is therefore also locally integrable:

$$\frac{\partial\Gamma}{\partial x_j} = C \cdot \frac{x_j}{|x|^N}.$$

Again, we may study this "fractional integral" using elementary techniques that measure only the *size* of the kernel.

But if we look at the second derivative of Γ, we find that

$$\frac{\partial^2\Gamma}{\partial x_j \partial x_k} = C_{jk} \frac{x_j x_k}{|x|^{N+2}} \equiv K(x)$$

is singular at the origin of order $-N$ and the integral has a critical singularity at 0. Hence, to analyze the transformation

$$\widetilde{T} : f \longmapsto \int f(t)K(x - t)dt,$$

we use the *Cauchy principal value*, denoted by P.V. and defined as follows:

$$\text{P.V.} \int f(t)K(x - t)dt = \lim_{\epsilon \to 0^+} \int_{|x-t|>\epsilon} f(t)K(x - t)dt.$$

We shall be able to see, in what follows, that $\widetilde{T}$ (defined using the Cauchy principal value) induces a distribution. It will also be bounded on $L^p(\mathbb{R}^N)$, $1 < p < \infty$. The operator $\widetilde{T}$ is unbounded on L^1 and unbounded on L^∞. When specialized down to dimension 1, the kernel for the operator $\widetilde{T}$ takes the form

$$K(t) = \frac{1}{t}.$$

This is of course the kernel of the Hilbert transform. In other words, the Hilbert transform is a special case of these fundamental considerations regarding the solution operator for the Laplacian.

In the next section we return to our consideration of the Hilbert transform as a linear operator on function spaces.

2.4 Boundedness of the Hilbert Transform

The Hilbert transform induces a distribution

$$\phi \longmapsto \int \frac{1}{x - t}\phi(t)dt, \qquad \text{for all } \phi \in C_c^\infty.$$

But why is this true? On the face of it, this mapping makes no sense. The integral is not convergent in the sense of the Lebesgue integral (because the kernel $1/(x - t)$ is not integrable). Some further analysis is required in order to understand the claim that this displayed line defines a distribution.

We understand this distribution by way of the Cauchy principal value:

$$\text{P.V.} \int \frac{1}{x - t}\phi(t)dt = \text{P.V.} \int \frac{1}{t}\phi(x - t)dt$$

$$= \lim_{\epsilon \to 0^+} \int_{|t|>\epsilon} \frac{1}{t}\phi(x - t)dt$$

$$= \lim_{\epsilon \to 0^+} \left[\int_{1>|t|>\epsilon} \frac{1}{t}\phi(x - t)dt + \int_{|t|>1} \frac{1}{t}\phi(x - t)dt \right]$$

$$= \lim_{\epsilon \to 0^+} \int_{1>|t|>\epsilon} \frac{1}{t}[\phi(x - t) - \phi(x)]dt + \int_{|t|>1} \frac{1}{t}\phi(x - t)dt.$$

In the first integral we have used the key fact that the kernel is odd, so has mean value
0. Hence we may subtract off a constant (and it integrates to 0). Of course the second
integral does *not* depend on ϵ, and it converges by Schwarz's inequality.

Since $\phi(x - t) - \phi(x) = \mathcal{O}(|t|)$, the limit in the first integral exists. That is to
say, the integrand is bounded so the integral makes good sense. We may perform the
following calculation to make this reasoning more rigorous:

For $\epsilon > 0$ define

$$I_\epsilon = \int_{\epsilon < |t| < 1} \frac{1}{t} O(|t|) dt.$$

Now if $0 < \epsilon_1 < \epsilon_2 < \infty$ we have

$$I_{\epsilon_2} - I_{\epsilon_1} = \int_{\epsilon_2 < |t| < 1} \frac{1}{t} O(|t|) dt - \int_{\epsilon_1 < |t| < 1} \frac{1}{t} O(|t|) dt = \int_{\epsilon_1 < |t| < \epsilon_2} \mathcal{O}(1) dt \to 0$$

as $\epsilon_1, \epsilon_2 \to 0$. This shows that our principal value integral converges.

Let $\mathcal{S}$ denote the standard Schwartz space from distribution theory (for which
see [STG1], [KRA5]). If $f \in \mathcal{S}$, we have

$$Hf(x) = \text{P.V.} \int \frac{1}{x - t} f(t) dt$$

and

$$\widehat{Hf} = \left(\frac{1}{t}\right)\widehat{} \cdot \hat{f}.$$

Since $\frac{1}{t}$ is homogeneous of degree -1, we find that $\left(\frac{1}{t}\right)\widehat{}$ is homogeneous of degree
0 (see Chapter 3 on the Fourier transform and Chapter 4 on multipliers). Therefore it
is bounded.

Now

$$\|Hf\|_{L^2} = \|\widehat{Hf}\|_{L^2} = \left\|\left(\frac{1}{t}\right)\widehat{} \cdot \hat{f}\right\|_{L^2} \leq C\|\hat{f}\|_{L^2} = c\|f\|_{L^2}.$$

By dint of a tricky argument that we shall detail below, Marcel Riesz (and, in its
present form, Salomon Bochner) proved that $H : L^p \to L^p$ when p is a positive even integer. By what is now known as the Riesz–Thorin interpolation theorem
(stated below), he then showed that H is bounded on $p > 2$. Then a simple duality
argument guarantees that H is also bounded on L^p for $1 < p < 2$.

Prelude: Interpolation theory is now an entire subject unto itself. For many years
it was a collection of isolated results known only to a few experts. The seminal
paper [CAL] cemented the *complex method* of interpolation (the one used to prove
Riesz–Thorin) as an independent entity. In the same year, Lions and Peetre [LIP]
inaugurated the real method of interpolation. The book [BERL] gives an overview of
the subject of interpolation.

In general the setup is this: One has Banach spaces X_0, X_1 and Y_0, Y_1 and an operator

$$T : X_0 \cap X_1 \to Y_0 \cup Y_1.$$

One hypothesizes that

$$\|Tf\|_{Y_j} \le C_j \|f\|_{X_j}$$

for $j = 0, 1$. The job then is to identify certain "intermediate spaces" and conclude that T is bounded in norm on those intermediate spaces.

Theorem 2.4.1 (Riesz–Thorin interpolation theorem) *Let $1 \le p_0 < p_1 \le \infty$. Let T be a linear operator that is bounded on L^{p_0} and L^{p_1}, i.e.,*

$$\|Tf\|_{L^{p_0}} \le C_0 \|f\|_{L^{p_0}},$$

$$\|Tf\|_{L^{p_1}} \le C_1 \|f\|_{L^{p_1}}.$$

Then T is a bounded operator on L^p, $\forall p_0 \le p \le p_1$, and

$$\|Tf\|_{L^p} \le C_0^{\frac{p_1-p}{p_1-p_0}} \cdot C_1^{\frac{p-p_0}{p_1-p_0}} \cdot \|f\|_{L^p}.$$

Now let us relate the Hilbert transform to Fourier series. We begin by returning to the idea of the Hilbert transform as a multiplier operator. Indeed, let $\mathbf{h} = \{h_j\}$, with $h_j = -i \operatorname{sgn} j$; here the convention is that

$$\operatorname{sgn} x = \begin{cases} -1 & \text{if } x < 0, \\ 0 & \text{if } x = 0, \\ 1 & \text{if } x > 0. \end{cases}$$

Then the Hilbert transform H is given by the multiplier h. This means that for $f \in L^1(\mathbb{T})$,

$$Hf = \sum_j h_j \widehat{f}(j) e^{ijt}.$$

[How might we check this assertion? You may calculate both the left-hand side and the right-hand side of this last equation when $f(t) = \cos jt$. The answer will be $\sin jt$ for every j, just as it should be—because $\sin jt$ is the boundary function for the harmonic conjugate of the Poisson integral of $\cos jt$. Likewise when $f(t) = \sin jt$ (then Hf as written here is $\cos jt$). That is enough information—by the Stone–Weierstrass theorem—to yield the result.] In the sequel we shall indicate this relationship by $H = \mathcal{M}_h$.

So defined, the Hilbert transform has the following connection with the partial sum operators:

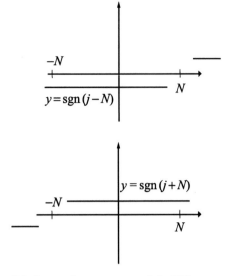

Figure 2.1. Summation operators and the Hilbert transform.

$$\chi_{[-N,N]}(j) = \frac{1}{2}[1 + \mathrm{sgn}(j+N)] - \frac{1}{2}[1 + \mathrm{sgn}(j-N)]$$

$$+ \frac{1}{2}[\chi_{\{-N\}}(j) + \chi_{\{N\}}(j)]$$

$$= \frac{1}{2}[\mathrm{sgn}(j+N) - \mathrm{sgn}(j-N)] + \frac{1}{2}[\chi_{\{-N\}}(j) + \chi_{\{N\}}(j)].$$

See Figure 2.1. Therefore, letting $e_k g(t) \equiv e^{ikt} g(t)$ and letting P_j be orthogonal projection onto the space spanned by e^{ijt}, we have

$$S_N f(e^{it}) = \mathcal{M}_{\chi_{[-N,N]}} f(e^{it})$$

$$= i e_{-N} H[e_N f] - i e_N H[e_{-N} f] + \frac{1}{2}[P_{-N} f + P_N f]. \qquad (2.1.5)$$

To understand this last equality, let us examine a piece of it. We look at the linear operator corresponding to the multiplier

$$m(j) \equiv \mathrm{sgn}(j+N).$$

Let $f(t) \sim \sum_{j=-\infty}^{\infty} \widehat{f}(j)e^{itj}$. Then

$$\mathcal{M}_m f(t) = \sum_j \mathrm{sgn}(j+N)\widehat{f}(j)e^{ijt}$$

$$= \sum_j \mathrm{sgn}(j)e^{-iNt}\widehat{f}(j-N)e^{ijt}$$

$$= i e^{-iNt} \sum_{j} (-i)\text{sgn}(j)\widehat{f}(j-N)e^{ijt}$$

$$= i e^{-iNt} \sum_{j} (-i)\text{sgn}(j)(e_N f)\widehat{}(j)e^{ijt}$$

$$= i e^{-iNt} H[e_N f](t).$$

This is of course precisely what is asserted in the first half of the right-hand side of (2.1.5).

We know that the Hilbert transform is bounded on L^2 because it is a multiplier operator coming from a bounded sequence. It also turns out to be bounded on L^p for $1 < p < \infty$. [We shall discuss this fact about H below, and eventually prove it.] Similar remarks apply to the projection operators P_j. Taking these boundedness assertions for granted, we now reexamine equation (2.1.5). Multiplication by a complex exponential does not change the size of an L^p function (in technical language, it is an *isometry* of L^p). So (2.1.5) tells us that S_N is a difference of compositions of operators, all of which are bounded on L^p. And the norm is plainly bounded independent of N. In conclusion, if we assume that H is bounded on L^p, $1 < p < \infty$, then Functional Analysis Principle I (see Appendix 1) tells us (since trigonometric polynomials are dense in L^p for $1 \leq p < \infty$) that norm convergence holds in L^p for $1 < p < \infty$. We now state this as a theorem:

Prelude: What is remarkable about this next theorem is that it reduces a question of convergence of a sequence of operators to the question of the boundedness of a *single* operator. This illustrates the power of functional analysis—a power that was virtually *discovered* in the context of Fourier analysis. From our modern perspective, the uniform boundedness principle makes this all quite natural.

Theorem 2.4.2 *Fix* $1 < p < \infty$ *and assume (to be proved below) that the Hilbert transform H is bounded on $L^p(\mathbb{T})$. Let $f \in L^p(\mathbb{T})$. Then* $\|S_N f - f\|_{L^p} \to 0$ *as $N \to \infty$. Explicitly,*

$$\lim_{N \to \infty} \left[\int_{\mathbb{T}} |S_N f(x) - f(x)|^p \, dx \right]^{1/p} = 0.$$

The converse of this theorem is true as well, and can be proved by even easier arguments. We leave the details to the reader—or see [KAT].

It is useful in the study of the Hilbert transform to be able to express it explicitly as an integral operator. The next lemma is of great utility in this regard.

Prelude: The next lemma is one of the key ideas in Laurent Schwartz's [SCH] distribution theory. It is an intuitively appealing idea that any translation-invariant operator is given by convolution with a kernel, but if one restricts attention to just *functions*, then one will not always be able to find this kernel. Distributions make possible a new, powerful statement.

Lemma 2.4.3 *If the Fourier multiplier* $\Lambda = \{\lambda_j\}_{j=-\infty}^{\infty}$ *induces a bounded operator* $\mathcal{M}_\Lambda$ *on* L^p, *then the operator is given by a convolution kernel* $K = K_\Lambda$. *In other words,*

$$\mathcal{M}_\Lambda f(x) = f * K(x) = \frac{1}{2\pi} \int_0^{2\pi} f(t) K(x - t) dt.$$

This convolution kernel is specified by the formula

$$K(e^{it}) = \sum_{j=-\infty}^{\infty} \lambda_j e^{it}.$$

[In actuality, the sum that defines this kernel may have to be interpreted using a summability technique, or using distribution theory, or both. In practice we shall always be able to calculate the kernel with our bare hands when we need to do so. So this lemma will play a tacit role in our work.]

If we apply Lemma 2.4.3 directly to the multiplier for the Hilbert transform, we obtain the formal series

$$k(e^{it}) \equiv \sum_{j=-\infty}^{\infty} -i \cdot \operatorname{sgn} j \cdot e^{ijt}.$$

Of course the terms of this series do not tend to zero, so this series does not converge in any conventional sense. Instead we use Abel summation (i.e., summation with factors of $r^{|j|}$, $0 \le r < 1$) to interpret the series: For $0 \le r < 1$ let

$$k_r(e^{it}) = \sum_{j=-\infty}^{\infty} -i r^{|j|} \cdot \operatorname{sgn} j \cdot e^{ijt}.$$

The sum over the positive indices is

$$-i \sum_{j=1}^{\infty} r^j \cdot e^{ijt} = -i \sum_{j=1}^{\infty} [re^{it}]^j$$

$$= -i \left[\frac{1}{1 - re^{it}} - 1 \right]$$

$$= \frac{-ire^{it}}{1 - re^{it}}.$$

Similarly, the sum over negative indices can be calculated to be equal to

$$\frac{ire^{-it}}{1 - re^{-it}}.$$

Adding these two pieces yields that

$$k_r(e^{it}) = \frac{-ire^{it}}{1 - re^{it}} + \frac{ire^{-it}}{1 - re^{-it}}$$

$$= \frac{-ir[e^{it} - e^{-it}]}{|1 - re^{it}|^2}$$

$$= \frac{2r \sin t}{|1 - re^{it}|^2}$$

$$= \frac{2r \sin t}{1 + r^2 - 2r \cos t}$$

$$= \frac{2r \cdot 2 \cdot \sin \frac{t}{2} \cos \frac{t}{2}}{(1 + r^2 - 2r) + 2r(1 - \cos^2 \frac{t}{2} + \sin^2 \frac{t}{2})}$$

$$= \frac{4r \sin \frac{t}{2} \cos \frac{t}{2}}{(1 + r^2 - 2r) + 2r(2 \sin^2 \frac{t}{2})}.$$

We formally let $r \to 1^-$ to obtain the kernel

$$k(e^{it}) = \frac{\sin \frac{t}{2} \cos \frac{t}{2}}{\sin^2 \frac{t}{2}} = \cot \frac{t}{2}. \tag{2.1.6}$$

This is the standard formula for the kernel of the Hilbert transform—just as we derived it by different means in the context of complex analysis. Now we have given a second derivation using Fourier analysis ideas. It should be noted that we have suppressed various subtleties about the validity of Abel summation in this context, as well as issues concerning the fact that the kernel k is not integrable (near the origin, $\cot \frac{t}{2} \approx 2/t$). For the full story, consult [KAT].

Just to repeat, we resolve the nonintegrability problem for the integral kernel k in (2.1.6) using the so-called *Cauchy principal value*, and it will now be defined again. Thus we usually write

$$\text{P.V.} \frac{1}{2\pi} \int_{-\pi}^{\pi} f(x - t) \cot \left(\frac{t}{2}\right) dt,$$

and we interpret this to mean

$$\lim_{\epsilon \to 0^+} \frac{1}{2\pi} \int_{\epsilon < |t| \leq \pi} f(x - t) \cot \left(\frac{t}{2}\right) dt. \tag{2.1.7}$$

Observe in (2.1.7) that for $\epsilon > 0$ fixed, $\cot(t/2)$ is actually *bounded* on the domain of integration. Therefore the integral in (2.1.7) makes sense, by Hölder's inequality, as long as $f \in L^p$ for some $1 \leq p \leq \infty$. The deeper question is whether the limit exists, and whether that limit defines an L^p function.

We will prove the L^p-boundedness of the Hilbert transform, using a method of S. Bochner, below.

The reduction of norm convergence of Fourier series to the study of the Hilbert transform is fundamental to the study of Fourier series. But it also holds great philosophical significance in the modern history of analysis. For it shows that we may reduce the study of the (infinitely many) partial sums of the Fourier series of a function to the study of a *single* integral operator. The device for making this reduction is—rather than study one function at a time—to study an entire space of functions at once. This is what functional analysis is all about.

Many of the basic ideas in functional analysis—including the uniform boundedness principle, the open mapping theorem, and the Hahn–Banach theorem—grew out of questions of Fourier analysis. Even today, Fourier analysis has led to many new ideas in Hilbert and Banach space theory—see [STE2], especially the Cotlar–Knapp–Stein lemma (see Section 9.10).

In the next section we shall examine the Hilbert transform from another point of view.

In the present section, we have taken the validity of Theorem 2.1.2 for granted. The details of this result, and its proof, will be treated as the book develops. Our intention in the next section is to discuss these theorems, and to look at some examples. In the next section we prove the L^p-boundedness of the Hilbert transform.

2.5 L^p Boundedness of the Hilbert Transform

Now we shall prove (at the end of the chapter) that the Hilbert transform is bounded on $L^p(\mathbb{T})$, $1 < p < \infty$. We will present an argument due to S. Bochner. This will allow us to make good use of the Riesz–Thorin interpolation theorem that we discussed in Section 2.4.

Prelude: Next we present the famous result of Marcel Riesz from 1926. People had been struggling for years to prove that the Hilbert transform was bounded on the L^p spaces other than $p = 2$, so Riesz's result must be considered a true breakthrough. The actual argument that we now present is due to Salomon Bochner. But Riesz had slightly different tricks that also yielded a boundedness result just for the even, integer values of p. It requires an extra idea, namely interpolation of linear operators, to get the result for all p, $1 < p < \infty$ (as in the ensuing theorem).

Proposition 2.5.1 *The Hilbert transform is bounded on $L^p(\mathbb{T})$ when $p = 2k$ is a positive, even integer.*

Theorem 2.5.2 *The Hilbert transform is bounded on L^p, $1 < p < \infty$.*

Remark: The argument at the end of the proof of the last theorem (see Appendix 1) is commonly called a "duality argument." Later in the book, when this idea is needed, it will be invoked without further comment or detail.

We complete our consideration of the Hilbert transform by treating what happens on the spaces L^1 and L^∞.

Prelude: The failure of singular integrals on the extreme spaces L^1 and L^∞ is a fundamental part of the theory. The former fact gave rise, in part, to the relatively new idea of real-variable H^1_{Re} (the real-variable Hardy space—see Section 8.8). Singular integrals *are* bounded on H^1_{Re}. The latter fact gave rise to the space *BMO* of functions of bounded mean oscillation (also see Chapter 8). Singular integrals are also bounded on *BMO*. The book [KRA5] gives a sketch of some of these ideas. Stein's early work [STE5] on the space $L \log L$ (the space of functions f such that $\int |f| \log^+ |f| dx$ is finite) was another attempt to deal with the failure of singular integrals on L^1.

Proposition 2.5.3 *Norm summability for Fourier series fails in both L^1 and L^∞.*

The proof of this last fact is just another instance of the concept of duality, as noted earlier.

We conclude this discussion by noting that the Hilbert transform of the characteristic function of the interval $[0, 1]$ is a logarithm function—do the easy calculation yourself. Thus the Hilbert transform does *not* map L^∞ to L^∞. By duality, it does not map L^1 to L^1. That completes our treatment of the nonboundedness of the Hilbert transform on these endpoint spaces.

2.6 The Modified Hilbert Transform

Capsule: The Hilbert transform, in its raw form, is a convolution operator with kernel $\cot \frac{t}{2}$. This is an awkward kernel to handle, just because it is a transcendental function. We show in this section that the kernel may be replaced by $1/t$. Most any question about the operator given by convolution with $\cot \frac{t}{2}$ may be studied by instead considering the operator given by convolution with $1/t$. Thus the latter operator has come (also) to be known as the Hilbert transform.

We repeat here a basic lesson from this chapter. We note that in practice, people do not actually look at the operator consisting of convolution with $\cot \frac{t}{2}$. This kernel is a transcendental function, and is tedious to handle. Thus what we do instead is to look at the operator

$$\widetilde{H} : f \longmapsto \mathrm{P.V.} \frac{1}{2\pi} \int_{-\pi}^{\pi} f(x-t) \cdot \frac{2}{t} \, dt. \qquad (2.6.1)$$

Clearly the kernel $2/t$ is much easier to think about than $\cot \frac{t}{2}$. It is also homogeneous of degree -1, a fact that will prove significant when we adopt a broader point of view later.

Prelude: In the literature, people discuss variants of the Hilbert transform and still call these the "Hilbert transform." Once one understands the basic idea, it is a trivial matter to pass back and forth among all the different realizations of this fundamental singular integral.

Lemma 2.6.2 *If the modified Hilbert transform $\mathcal{H}$ is bounded on L^1, then it is bounded on L^∞.*

We end this section by recording what is perhaps the deepest result of basic Fourier analysis. Formerly known as the Lusin conjecture, and now as Carleson's theorem, this result addresses the pointwise convergence question for L^2. We stress that the approach to proving something like this is to study the *maximal Hilbert transform*—see Functional Analysis Principle I in Appendix 1.

Prelude: The next theorem is the culmination of more than fifty years of effort by the best mathematical analysts. This was *the* central question of Fourier analysis. Carleson's proof of the theorem was a triumph. Subsequently Fefferman [FEF4] produced another, quite different proof that was inspired by Stein's celebrated limits of sequences of operators theorem [STE6]. And there is now a third approach by Lacey and Thiele [LAT]. It must be noted that this last approach derives from ideas in Fefferman's proof.

Theorem 2.6.3 (Carleson [CAR]) *Let $f \in L^2(\mathbb{T})$. Then the Fourier series of f converges almost everywhere to f.*

The next result is based on Carleson's theorem, but requires significant new ideas.

Prelude: It definitely required a new idea for Richard Hunt to extend Carleson's result from L^2 to L^p for $1 < p < 2$ (of course the case L^p for $2 < p < \infty$ comes for free since then $L^p \subseteq L^2$). P. Sjölin [SJO1] has refined Hunt's theorem even further to obtain spaces of functions that are smaller than L^1, yet larger than L^p for every $p > 1$, on which pointwise convergence of Fourier series holds. The sharpest result along these lines is due to Hunt and Taibleson [HUT].

Theorem 2.6.4 (Hunt [HUN]) *Let $f \in L^p(\mathbb{T})$, $1 < p \leq \infty$. Then the Fourier series of f converges almost everywhere to f.*

A classical example of A. Kolmogorov (see [KAT], [ZYG]) provides an L^1 function whose Fourier series converges[4] *at no point* of $\mathbb{T}$. This phenomenon provides significant information: If instead the example were of a function with Fourier series diverging a.e., then we might conclude that we were simply using the wrong measure to detect convergence. But since there is an L^1 function with *everywhere diverging Fourier series*, we conclude that there is no hope for pointwise convergence in L^1.

Proofs of the Results in Chapter 2

Proof of Lemma 2.1.3: A rigorous proof of this lemma would involve a digression into distribution theory and the Schwartz kernel theorem. We refer the interested reader to either [STG1] or [SCH].   □

---

[4] It may be noted that Kolmogorov's original construction was very difficult. Nowadays, using functional analysis, this result may be had with little difficulty—see [KAT].

Proof of Proposition 2.1.4:   Let f be a continuous real function on $[0, 2\pi)$. We normalize f (by subtracting off a constant) so that $\int f \, dx = 0$. Let u be its Poisson integral, so u is harmonic on the disk D and vanishes at 0. Let v be that harmonic conjugate of u on D such that $v(0) = 0$. Then $h = u + iv$ is holomorphic and $h(0) = 0$.

Fix $0 < r < 1$. Now we write

$$0 = 2\pi h^p(0)$$

$$= \int_0^{2\pi} h^{2k}(re^{i\theta}) \, d\theta$$

$$= \int_0^{2\pi} [u(re^{i\theta}) + iv(re^{i\theta})]^{2k} \, d\theta$$

$$= \int_0^{2\pi} u^{2k} \, d\theta + i \binom{2k}{1} \int u^{2k-1} v \, d\theta - \binom{2k}{2} \int u^{2k-2} v^2 \, d\theta + \cdots$$

$$+ i^{2k-1} \binom{2k}{2k-1} \int uv^{2k-1} \, d\theta + i^{2k} \int v^{2k} \, d\theta.$$

We rearrange the last equality as

$$\int_0^{2\pi} v^{2k} \, d\theta \leq \binom{2k}{2k-1} \int_0^{2\pi} |uv^{2k-1}| \, d\theta$$

$$+ \binom{2k}{2k-2} \int_0^{2\pi} |u^2 v^{2k-2}| \, d\theta + \cdots$$

$$+ \binom{2k}{2} \int_0^{2\pi} |u^{2k-2} v^2| \, d\theta + \binom{2k}{1} \int_0^{2\pi} |u^{2k-1} v| \, d\theta$$

$$+ \int_0^{2\pi} |u^{2k}| \, d\theta.$$

We apply Hölder's inequality to each composite term on the right—using the exponents $2k/j$ and $2k/[2k - j]$ on the jth term, for $j = 1, 2, \ldots, 2k - 1$. It is convenient to let $S = [\int u^{2k} \, d\theta]^{1/2k}$ and $T = [\int v^{2k} \, d\theta]^{1/2k}$, and we do so. The result is

$$T^{2k} \leq \binom{2k}{2k-1} ST^{2k-1} + \binom{2k}{2k-2} S^2 T^{2k-2} + \cdots$$

$$+ \binom{2k}{2} S^{2k-2} T^2 + \binom{2k}{1} S^{2k-1} T + S^{2k}.$$

Now define $U = T/S$ and rewrite the inequality as

$$U^{2k} \leq \binom{2k}{2k-1} U^{2k-1} + \binom{2k}{2k-2} U^{2k-2} + \cdots + \binom{2k}{2} U^2 + \binom{2k}{1} U + 1.$$

Divide through by U^{2k-1} to obtain

$$U \le \binom{2k}{2k-1} + \binom{2k}{2k-2}U^{-1} + \cdots + \binom{2k}{2}U^{-2k+3} + \binom{2k}{1}U^{-2k+2} + U^{-2k+1}.$$

If $U \ge 1$, then it follows that

$$U \le \binom{2k}{2k-1} + \binom{2k}{2k-2} + \cdots + \binom{2k}{2} + \binom{2k}{1} + 1 \le 2^{2k}.$$

We conclude, therefore, that

$$\|v\|_{L^{2k}} \le 2^k \|u\|_{L^{2k}}.$$

But of course the function $v(re^{i\theta})$ is the Hilbert transform of $u(re^{i\theta})$. The proof is therefore complete. $\qquad\square$

Proof of Theorem 2.1.5: We know that the Hilbert transform is bounded on L^2, L^4, L^6, ... We may immediately apply the Riesz–Thorin theorem (Section 2.1.3) to conclude that the Hilbert transform is bounded on L^p for $2 \le p \le 4$, $4 \le p \le 6$, $6 \le p \le 8$, etc. In other words, the Hilbert transform is bounded on L^p for $2 \le p < \infty$.

Now let $f \in L^p$ for $1 < p < 2$. Let φ be any element of $L^{p/[p-1]}$ with norm 1. Notice that $2 < p/[p-1] < \infty$. Then

$$\int Hf \cdot \varphi \, d\theta = \int \left[\int f(\psi) \cot \frac{\theta - \psi}{2} \, d\psi \right] \varphi(\theta) \, d\theta$$

$$= \iint \varphi(\theta) \cot \frac{\theta - \psi}{2} \, d\theta f(\psi) \, d\psi$$

$$= -\int \left[\int \varphi(\theta) \cot \frac{\psi - \theta}{2} \, d\theta \right] f(\psi) \, d\psi$$

$$= -\int H\varphi(\psi) f(\psi) \, d\psi.$$

Using Hölder's inequality together with the fact that we know that the Hilbert transform is bounded on $L^{p/[p-1]}$, we may bound the right-hand side by the expression $C\|\varphi\|_{L^{p/[p-1]}}\|f\|_{L^p} \le \|f\|_{L^p}$. Since this estimate holds for any such choice of φ, the result follows. $\qquad\square$

Proof of Proposition 2.1.6: It suffices for us to show that the modified Hilbert transform (as defined in Section 10.2) fails to be bounded on L^1 and fails to be bounded on L^∞. In fact, the following lemma will cut the job by half:

Proof of Lemma 2.2.1: Let f be an L^∞ function. Then

$$\|\mathcal{H}f\|_{L^\infty} = \sup_{\substack{\phi \in L^1 \\ \|\phi\|_{L^1}=1}} \left| \int \mathcal{H}f(x) \cdot \phi(x) \, dx \right| = \sup_{\substack{\phi \in L^1 \\ \|\phi\|_{L^1}=1}} \left| \int f(x)(\mathcal{H}^*\phi)(x) \, dx \right|.$$

But an easy formal argument (as in the proof of Theorem 10.5) shows that

$$\mathcal{H}^* \phi = -\mathcal{H}\phi.$$

Here $\mathcal{H}^*$ is the adjoint of $\mathcal{H}$. [In fact, a similar formula holds for *any* convolution operator—exercise.] Thus the last line gives

$$\|\mathcal{H}f\|_{L^\infty} = \sup_{\substack{\phi \in L^1 \\ \|\phi\|_{L^1}=1}} \left| \int f(x)\mathcal{H}\phi(x)\,dx \right|$$

$$\leq \sup_{\substack{\phi \in L^1 \\ \|\phi\|_{L^1}=1}} \|f\|_{L^\infty} \cdot \|\mathcal{H}\phi\|_{L^1}$$

$$\leq \sup_{\substack{\phi \in L^1 \\ \|\phi\|_{L^1}=1}} \|f\|_{L^\infty} \cdot C\|\phi\|_{L^1}$$

$$= C \cdot \|f\|_{L^\infty}.$$

Here C is the norm of the modified Hilbert transform acting on L^1. We have shown that if $\mathcal{H}$ is bounded on L^1, then it is bounded on L^∞. That completes the proof. $\square$

3

Essentials of the Fourier Transform

Elementary ideas about the Fourier transform appear in Appen- dix 2. Proofs for the first two sections of this Chapter are presented at the end of those two sections.

> **Prologue:** The nature of Fourier analysis on the circle $\mathbb{T}$ is determined by the collection of *characters* on the circle: these are the continuous, multiplicative homomorphisms of $\mathbb{T}$ into $\mathbb{T}$ itself—namely the functions $t \mapsto e^{int}$ for $n \in \mathbb{Z}$. The Peter–Weyl theorem tells us that a viable Fourier analysis may be built on these characters.
>
> The nature of Fourier analysis on the real line $\mathbb{R}$ is determined by the collection of characters on the line: these are the functions $t \mapsto e^{i\xi t}$ for $\xi \in \mathbb{R}$. Again, the Peter–Weyl theorem tells us that a natural Fourier analysis (i.e., the Fourier transform) may be based on these functions. For $\mathbb{R}^N$ the characters are $t \mapsto e^{i\xi \cdot t}$ for $\xi \in \mathbb{R}^N$ and the Fourier transform is defined analogously.
>
> It is interesting to note that Fourier introduced a version of the Fourier transform in his studies of the heat equation in 1811. The paper was not published until 1824. Cauchy and Poisson also studied the matter in 1815 and 1816 respectively. Camille Deflers gave a proof of Fourier inversion in 1819. Deflers also proved a version of the Riemann–Lebesgue lemma.
>
> Using the Poisson summation formula, many questions about Fourier series (even multiple Fourier series!) may be reduced to questions about the Fourier transform (see [KRA5] for the details). As a result, today, the Fourier transform is the primary object of study for classical Fourier analysts.

3.1 Quadratic Integrals and Plancherel

> **Capsule:** Certainly one of the beautiful features of Fourier analysis is the Plancherel formula. As a consequence, the Fourier transform is an isometry

S.G. Krantz, *Explorations in Harmonic Analysis*, Applied and Numerical
Harmonic Analysis, DOI 10.1007/978-0-8176-4669-1_3,
© Birkhäuser Boston, a part of Springer Science+Business Media, LLC 2009

on L^2. The theory of Sobolev spaces is facilitated by the action of the Fourier transform on L^2. Pseudodifferential operators and Fourier integral operators are made possible in part by the nice way that the Fourier transform acts on L^2. The spectral theory of the action of the Fourier transform on L^2 is very elegant.

We earlier made some initial remarks about the quadratic Fourier theory. Now we give a more detailed treatment in the context of the Fourier transform.

Prelude: This next result is the celebrated theorem of Plancherel. As presented here, it seems to be a natural and straightforward consequence of the ideas we have been developing. But it is a truly profound result, and has serious consequences. The entire L^2 theory of the Fourier transform depends on this proposition.

Proposition 3.1.1 (Plancherel) *If $f \in C_c^\infty(\mathbb{R}^N)$, then*

$$(2\pi)^{-N} \int |\widehat{f}(\xi)|^2 \, d\xi = \int |f(x)|^2 \, dx.$$

Definition 3.1.2 For any $f \in L^2(\mathbb{R}^N)$, the Fourier transform of f can be defined in the following fashion: Let $f_j \in C_c^\infty$ satisfy $f_j \to f$ in the L^2 topology. It follows from the proposition that $\{\widehat{f_j}\}$ is Cauchy in L^2. Let g be the L^2 limit of this latter sequence. We set $\widehat{f} = g$.

It is easy to check that this definition of $\widehat{f}$ is independent of the choice of sequence $f_j \in C_c^\infty$ and that

$$(2\pi)^{-N} \int |\widehat{f}(\xi)|^2 \, d\xi = \int |f(x)|^2 \, dx. \tag{3.1.3}$$

We next record the "polarized form" of Plancherel's formula:

Prelude: Of course the next result is a "polarized" version of Plancherel's theorem. It follows just from algebra. But it is an extremely useful identity, and makes possible many of the advanced ideas in the subject. Certainly one should compare this results with Proposition A2.2.7. In view of Fourier inversion, the two formulas are logically equivalent.

Proposition 3.1.4 *If $f, g \in L^2(\mathbb{R}^N)$, then*

$$\int f(t) \cdot \overline{g(t)} \, dt = (2\pi)^{-N} \int \widehat{f}(\xi) \overline{\widehat{g}(\xi)} \, d\xi.$$

Exercises for the Reader: Restrict attention to dimension 1. Let $\mathcal{F}$ be the Fourier transform. Consider $\mathcal{G} \equiv (2\pi)^{1/2} \cdot \mathcal{F}$ as a bounded linear operator (indeed an isometry) on the Hilbert space $L^2(\mathbb{R})$. Prove that the four roots of unity (suitably scaled) are eigenvalues of $\mathcal{G}$. [*Hint:* What happens when you compose $\mathcal{G}$ with itself four times?]

Which functions in L^2 are invariant (up to a scalar factor) under the Fourier transform? We know that $(ixf)^\vee(\xi) = (\widehat{f})'(\xi)$ and $(f')\widehat{\ }(\xi) = -i\xi\widehat{f}(\xi)$. As a result, the differential operator $d^2/dx^2 - x^2 I$ is invariant under the Fourier transform. It seems plausible that any solution of the differential equation

$$\frac{d^2}{dx^2}\phi - x^2\phi = \lambda\phi, \tag{3.1.5}$$

for λ a suitable constant, will also be mapped by the Fourier transform to itself. Since the function $e^{-x^2/2}$ is (up to a constant) mapped to itself by the Fourier transform, it is natural to guess that equation (3.1.5) would have solutions involving this function. Thus perform the change of notation $\phi(x) = e^{-x^2/2} \cdot \Phi(x)$. Guess that Φ is a polynomial, and derive recursions for the coefficients of that polynomial. These polynomials are called the *Hermite polynomials*. A full treatment of these matters appears in [WIE, pp. 51–55] and in [FOL3, p. 248].

You may also verify that the polynomials Φ that you find form (after suitable normalization) an orthonormal basis for L^2 when calculated with respect to the measure $d\mu = \sqrt{2}e^{-x^2/2}\,dx$. For details, see [WIE].

We now know that the Fourier transform $\mathcal{F}$ has the following mapping properties:

$$\mathcal{F}: L^1 \to L^\infty,$$

$$\mathcal{F}: L^2 \to L^2.$$

These are both bounded operations. It follows that $\mathcal{F}$ is defined on L^p for $1 < p < 2$ (since for such p, $L^p \subseteq L^1 + L^2$).

The Riesz–Thorin interpolation theorem (see [STG1]) allows us to conclude that

$$\mathcal{F}: L^p \to L^{p'}, \quad 1 \le p \le 2,$$

where $p' = p/(p-1)$. If $p > 2$, then $\mathcal{F}$ does not map L^p into any nice function space. The theory of distributions is required in order to obtain any satisfactory theory. We shall not treat the matter here. The precise norm of $\mathcal{F}$ on L^p, $1 \le p \le 2$, has been computed by Beckner [BEC].

3.2 Sobolev Space Basics

Capsule: Sobolev spaces were invented as an extension of the L^2 theory of the Fourier transform. It is awkward to study the action of the Fourier transform on C^k (even though these spaces are very intuitively appealing). But it is very natural to study the Fourier transform acting on the Sobolev space H^s. The Rellich lemma and the restriction and extension theorems for Sobolev spaces make these objects all the more compelling to study—see [KRA3] for details.

Definition 3.2.1 If $\phi \in D$ (the space of testing functions, or C^∞ functions with compact support), then we define the norm

$$\|\phi\|_{H^s} = \|\phi\|_s \equiv \left(\int |\widehat{\phi}(\xi)|^2 (1 + |\xi|^2)^s \, d\xi \right)^{1/2}.$$

We let the space $H^s(\mathbb{R}^N)$ be the closure of D with respect to $\|\quad\|_s$.

In the case that s is a nonnegative integer, then

$$(1 + |\xi|^{2s}) \approx (1 + |\xi|^2)^s \approx \sum_{|\alpha| \le 2s} |\xi|^\alpha \approx \left(\sum_{|\alpha| \le s} |\xi|^\alpha \right)^2.$$

Therefore

$$\phi \in H^s \quad \text{if and only if} \quad \widehat{\phi} \cdot \left(\sum_{|\alpha| \le s} |\xi|^\alpha \right) \in L^2.$$

This last condition means that $\widehat{\phi} \xi^\alpha \in L^2$ for all multi-indices α with $|\alpha| \le s$. That is,

$$\left(\frac{\partial}{\partial x} \right)^\alpha \phi \in L^2 \quad \forall \alpha \text{ such that } |\alpha| \le s.$$

Thus we have the following proposition:

Prelude: Although it is intuitively appealing to think of an element of the Sobolev space H^s as an L^2 function with derivatives up to order s also lying in L^2, this is not in practice the most useful characterization of the space. In actual applications of the Sobolev spaces, the norm in the preceding definition is what gives the most information.

Proposition 3.2.2 *If s is a nonnegative integer then*

$$H^s = \left\{ f \in L^2 : \frac{\partial^\alpha}{\partial x^\alpha} f \in L^2 \text{ for all } \alpha \text{ with } |\alpha| \le s \right\}.$$

Here derivatives are interpreted in the sense of distributions.

Notice in passing that if $s > r$ then $H^s \subseteq H^r$ because

$$|\widehat{\phi}|^2 (1 + |\xi|^2)^r \le C \cdot |\widehat{\phi}|^2 (1 + |\xi|^2)^s.$$

The Sobolev spaces turn out to be easy to work with because they are modeled on L^2—indeed each H^s is canonically isomorphic as a Hilbert space to L^2 (exercise). But they are important because they can be related to more classical spaces of smooth functions. That is the content of the Sobolev embedding theorem:

Prelude: It is this next result that is the heuristic justification for the Sobolev spaces. The intuition is that C^k spaces are much more intuitively appealing, but they do not behave well under the integral operators that we wish to study. Sobolev spaces are less intuitive but, thanks to Plancherel, they behave very naturally under the integral operators. So this theorem provides a bridge between the two sets of ideas.

Theorem 3.2.3 (Sobolev) *Let $s > N/2$. If $f \in H^s(\mathbb{R}^N)$ then f can be corrected on a set of measure zero to be continuous.*

More generally, if $k \in \{0, 1, 2, \ldots\}$ and if $f \in H^s, s > N/2 + k$, then f can be corrected on a set of measure zero to be C^k.

Remarks:
1. If $s = N/2$ then the first part of the theorem is false (exercise).
2. The theorem may be interpreted as saying that $H^s \subseteq C^k_{\text{loc}}$ for $s > k + N/2$. In other words, the identity provides a continuous embedding of H^s into C^k_{loc}. A converse is also true. Namely, if $H^s \subseteq C^k_{\text{loc}}$ for some nonnegative integer k then $s > k + N/2$.

To see this, notice that the hypotheses $u_j \to u$ in H^s and $u_j \to v$ in C^k imply that $u = v$. Therefore the inclusion of H^s into C^k is a closed map. It is therefore continuous by the closed graph theorem. Thus there is a constant C such that

$$\|f\|_{C^k} \leq C\|f\|_{H^s}.$$

For $x \in \mathbb{R}^N$ fixed and α a multi-index with $|\alpha| \leq k$, the tempered distribution e_x^α defined by

$$e_x^\alpha(\phi) = \left(\frac{\partial^\alpha}{\partial x^\alpha}\right)\phi(x)$$

is bounded in $(C^k)^*$ with bound independent of x and α (but depending on k). Hence $\{e_x^\alpha\}$ form a bounded set in $(H^s)^* \equiv H^{-s}$. As a result, for $|\alpha| \leq k$ we have that

$$\|e_x^\alpha\|_{H^{-s}} = \left(\int \left|(\widehat{e_x^\alpha})(\xi)\right|^2 (1 + |\xi|^2)^{-s}\, d\xi\right)^{1/2}$$

$$= \left(\int \left|(-i\xi)^\alpha e^{ix\cdot\xi}\right|^2 (1 + |\xi|^2)^{-s}\, d\xi\right)^{1/2}$$

$$\leq C \left(\int (1 + |\xi|^2)^{-s+|\alpha|}\, d\xi\right)^{1/2}$$

is finite, independent of x and α. But this can happen only if $2(k - s) < -N$, that is, if $s > k + N/2$.

Exercise for the Reader: Imitate the proof of the Sobolev theorem to prove Rellich's lemma: If $s > r$ then the inclusion map $i : H^s \to H^r$ is a compact operator.

Proofs of the Results in Sections 3.1, 3.2

Proof of Proposition 3.1.1: Define $g = f * \widetilde{\overline{f}} \in C_c^\infty(\mathbb{R}^N)$. Then

$$\widehat{g} = \widehat{f} \cdot \widehat{\widetilde{\overline{f}}} = \widehat{f} \cdot \overline{\widehat{\widetilde{f}}} = \widehat{f} \cdot \overline{\widetilde{\widehat{f}}} = \widehat{f} \cdot \overline{\widehat{f}} = |\widehat{f}|^2. \tag{3.1.1.1}$$

Now

$$g(0) = f * \tilde{\overline{f}}\,(0) = \int f(-t)\overline{f}(-t)dt = \int f(t)\overline{f}(t)dt = \int |f(t)|^2\,dt.$$

By Fourier inversion and formula (3.1.1.1) we may now conclude that

$$\int |f(t)|^2\,dt = g(0) = (2\pi)^{-N}\int \widehat{g}(\xi)\,d\xi = (2\pi)^{-N}\int |\widehat{f}(\xi)|^2\,d\xi.$$

This is the desired formula.   □

Proof of Proposition 3.1.4:   The proof consists of standard tricks from algebra: First assume that f, g are real-valued. Apply the Plancherel formula that we proved to the function $(f + g)$ and then again to the function $(f - g)$ and then subtract.

The case of complex-valued f and g is treated by applying the Plancherel formula to $f + g$, $f - g$, $f + ig$, $f - ig$ and combining.   □

Proof of Theorem 3.2.3:   For the first part of the theorem, let $f \in H^s$. By definition, there exist $\phi_j \in D$ such that $\|\phi_j - f\|_{H^s} \to 0$. Then

$$\|\phi_j - f\|_{L^2} = \|\phi_j - f\|_0 \le \|\phi_j - f\|_s^2 \to 0. \qquad (3.2.3.1)$$

Our plan is to show that $\{\phi_j\}$ is an equibounded, equicontinuous family of functions. Then the Ascoli–Arzelà theorem [RUD1] will imply that there is a subsequence converging uniformly on compact sets to a (continuous) function g. But (3.2.3.1) guarantees that a subsequence of this subsequence converges pointwise to the function f. So $f = g$ almost everywhere and the required (first) assertion follows.

To see that $\{\phi_j\}$ is equibounded, we calculate that

$$|\phi_j(x)| = c \cdot \left| \int e^{-ix\cdot\xi}\widehat{\phi}_j(\xi)\,d\xi \right|$$

$$\le c \cdot \int |\widehat{\phi}_j(\xi)|(1 + |\xi|^2)^{s/2}(1 + |\xi|^2)^{-s/2}\,d\xi$$

$$\le c \cdot \left(\int |\widehat{\phi}_j(\xi)|^2(1 + |\xi|^2)^s\,d\xi \right)^{1/2} \cdot \left(\int (1 + |\xi|^2)^{-s}\,d\xi \right)^{1/2}.$$

Using polar coordinates, we may see easily that for $s > N/2$,

$$\int (1 + |\xi|^2)^{-s}\,d\xi < \infty.$$

Therefore

$$|\phi_j(x)| \le C\|\phi_j\|_{H^s} \le C'$$

and $\{\phi_j\}$ is equibounded.

To see that $\{\phi_j\}$ is equicontinuous, we write

$$|\phi_j(x) - \phi_j(y)| = c \left| \int \widehat{\phi}_j(\xi) \left(e^{-ix\cdot\xi} - e^{-iy\cdot\xi} \right) d\xi \right|.$$

Observe that $|e^{-ix\cdot\xi} - e^{-iy\cdot\xi}| \leq 2$ and, by the mean value theorem,

$$|e^{-ix\cdot\xi} - e^{-iy\cdot\xi}| \leq |x - y| \, |\xi|.$$

Then, for any $0 < \epsilon < 1$,

$$|e^{-ix\cdot\xi} - e^{-iy\cdot\xi}| = |e^{-ix\cdot\xi} - e^{-iy\cdot\xi}|^{1-\epsilon} |e^{-ix\cdot\xi} - e^{-iy\cdot\xi}|^{\epsilon} \leq 2^{1-\epsilon} |x - y|^{\epsilon} |\xi|^{\epsilon}.$$

Therefore

$$|\phi_j(x) - \phi_j(y)| \leq C \int |\widehat{\phi}_j(\xi)| |x - y|^{\epsilon} |\xi|^{\epsilon} \, d\xi$$

$$\leq C|x - y|^{\epsilon} \int |\widehat{\phi}_j(\xi)| \left(1 + |\xi|^2 \right)^{\epsilon/2} d\xi$$

$$\leq C|x - y|^{\epsilon} \|\phi_j\|_{H^s} \left(\int (1 + |\xi|^2)^{-s+\epsilon} \, d\xi \right)^{1/2}.$$

If we select $0 < \epsilon < 1$ such that $-s+\epsilon < -N/2$ then we find that $\int (1+|\xi|^2)^{-s+\epsilon} \, d\xi$ is finite. It follows that the sequence $\{\phi_j\}$ is equicontinuous and we are done.

The second assertion may be derived from the first by a simple inductive argument. We leave the details as an exercise.   □

3.3 Key Concepts of Fractional Integrals

Capsule: Fractional integrals and singular integrals are the two most basic types of integral operators on Euclidean space. A fractional integral operator is **(a)** inverse to a fractional differential operator (i.e., a power of the Laplacian) and **(b)** a smoothing operator (in the sense that it takes functions in a given function class to functions in a "better" function class). Fractional integrals act on L^p, on *BMO*, on (real-variable) Hardy spaces, on Sobolev spaces, and on many other natural function spaces. They are part of the bedrock of our theory of integral operators. They are used routinely in the study of regularity of partial differential equations, in harmonic analysis, and in many other disciplines.

For $\phi \in C_c^1(\mathbb{R}^N)$ we know (see Proposition 3.1.2) that

$$\widehat{\frac{\partial \phi}{\partial x_j}}(\xi) = -i\xi_j \cdot \widehat{\phi}(\xi). \tag{3.3.1}$$

In other words, the Fourier transform converts differentiation in the x-variable to multiplication by a monomial in the Fourier transform variable. Of course higher-order derivatives correspond to multiplication by higher-order monomials.

It is natural to wonder whether the Fourier transform can provide us with a way to think about differentiation to a fractional order. In pursuit of this goal, we begin by thinking about the Laplacian

$$\Delta\phi \equiv \sum_{j=1}^{N} \frac{\partial^2}{\partial x_j^2}\phi.$$

Of course formula (3.3.1) shows that

$$\widehat{\Delta\phi}(\xi) = -|\xi|^2\widehat{\phi}(\xi). \tag{3.3.2}$$

In the remainder of this section, let us use the notation

$$\mathcal{D}^2\phi(\xi) = -\Delta\phi(\xi). \tag{3.3.3}$$

Then we set $\mathcal{D}^4\phi \equiv \mathcal{D}^2 \circ \mathcal{D}^2\phi$, and so forth. What will be important for us is that the negative of the Laplacian is a positive operator, as (3.3.2) shows. Thus it will be natural to take roots of this operator.

Now let us examine the Fourier transform of $\mathcal{D}^2\phi$ from a slightly more abstract point of view. Observe that the operator $\mathcal{D}^2$ is translation-invariant. Therefore, by the Schwartz kernel theorem (Section 2.1 and [SCH]), it is given by a convolution kernel k_2. Thus

$$\mathcal{D}^2\phi(x) = \phi * k_2(x).$$

Therefore

$$\widehat{\mathcal{D}^2\phi}(\xi) = \widehat{\phi}(\xi) \cdot \widehat{k_2}(\xi). \tag{3.3.4}$$

If we wish to understand differentiation from the point of view of the Fourier transform, then we should calculate $\widehat{k_2}$ and then k_2. But comparison of (3.3.2), (3.3.3), and (3.3.4) tells us instantly that

$$\widehat{k_2}(\xi) = |\xi|^2.$$

[Since the expression on the right neither vanishes at infinity—think of the Riemann–Lebesgue lemma—nor is in L^2, the reader will have to treat the present calculations as purely formal. Or else interpret everything in the language of distributions.] In other words,

$$\widehat{\mathcal{D}^2\phi}(\xi) = |\xi|^2 \cdot \widehat{\phi}(\xi).$$

More generally,

$$\widehat{\mathcal{D}^{2j}\phi}(\xi) = |\xi|^{2j} \cdot \widehat{\phi}(\xi).$$

The calculations presented thus far should be considered to have been a finger exercise. Making them rigorous would require a considerable dose of the theory of Schwartz distributions, and this we wish to avoid. Now we enter the more rigorous phase of our discussion.

It turns out to be more efficient to study fractional integration than fractional differentiation. This is only a technical distinction, but the kernels that arise in the theory of fractional integration are a bit easier to study. Thus, in analogy with the operators $\mathcal{D}^2$, we define $\mathcal{I}^2\phi$ according to the identity

$$\widehat{\mathcal{I}^2\phi}(\xi) = |\xi|^{-2} \cdot \hat{\phi}(\xi).$$

Observing that this Fourier multiplier is rotationally invariant and homogeneous of degree -2, we conclude that the *convolution kernel* corresponding to the fractional integral operator $\mathcal{I}^2$ is $k_2(x) = c \cdot |x|^{-N+2}$ for some constant c. In what follows we shall suppress this constant. Thus

$$\mathcal{I}^2\phi(x) = \int_{\mathbb{R}^N} |t|^{-N+2}\phi(x - t)dt,$$

at least when $N > 2$.[1]

More generally, if $0 < \beta < N$, we *define*

$$\mathcal{I}^\beta\phi(x) = \int |t|^{-N+\beta}\phi(x - t)dt$$

for any testing function $\phi \in C_c^1(\mathbb{R}^N)$. Observe that this integral is absolutely convergent—near the origin because ϕ is bounded and $|x|^{-N+\beta}$ is integrable, and near ∞ because ϕ is compactly supported. The operators $\mathcal{I}^\beta$ are called *fractional integral operators*.

3.4 The Sense of Singular Integrals

Capsule: Of course singular integrals are the main point of this book. They are at the heart of many, if not most, basic questions of modern real analysis. Based both philosophically and technically on the Hilbert transform, singular integrals are the natural higher-dimensional generalization of a fundamental concept from complex function theory and Fourier series. The hypothesized homogeneity properties of a classical singular integral kernel are a bit primitive by modern standards, but they set the stage for the many theories of singular integrals that we have today.

---

[1] Many of the formulas of potential theory have a special form in dimension 2. For instance, in dimension 2 the Newton potential is given by a logarithm. Riesz's classical theory of potentials—for which see [HOR2]—provides some explanation for this phenomenon.

In the last section we considered Fourier multipliers of positive or negative homogeneity. Now let us look at Fourier multipliers of homogeneity 0. Let $m(\xi)$ be such a multiplier. It follows from our elementary ideas about homogeneity and the Fourier transform that the corresponding kernel must be homogeneous of degree $-N$, where N is the dimension of the ambient space. Call the kernel k.

Now we think of k as the kernel of an integral operator. We can, in the fashion of Calderón and Zygmund, write

$$k(x) = \frac{[k(x) \cdot |x|^N]}{|x|^N} \equiv \frac{\Omega(x)}{|x|^N}.$$

Here it must be that Ω is homogeneous of degree 0.

Let c be the mean value of Ω on the unit sphere. Thus

$$c = \frac{1}{\sigma(S)} \int_S \Omega(\xi) \, d\sigma(\xi),$$

where S is the unit sphere and $d\sigma$ is rotationally invariant area measure on S.

We write

$$k(x) = \frac{\Omega(x) - c}{|x|^N} + \frac{c}{|x|^N} \equiv k_1(x) + k_2(x).$$

Now if φ is a $C_c^\infty(\mathbb{R}^N)$ testing function then we may calculate

$$\int_{\mathbb{R}^N} k_1(x) \varphi(x) \, dx = \int_0^\infty \int_{\Sigma_{N-1}} [\Omega(s) - c] \varphi(\rho s) \, d\sigma(s) \rho^{-N} \cdot r^{N-1} \, d\rho$$

$$= \int_0^\infty \int_{\Sigma_{N-1}} [\Omega(s) - c][\varphi(\rho s) - \varphi(0)] \, d\sigma(s) \rho^{-N} \cdot \rho^{N-1} \, d\rho$$

$$= \int_0^\infty \int_{\Sigma_{N-1}} [\Omega(s) - c] \cdot \mathcal{O}(\rho) \, d\sigma(s) \rho^{-N} \cdot \rho^{N-1} \, d\rho,$$

and the integral converges. Thus k_1 induces a distribution in a natural way.

But integration against k_2 *fails* to induce a distribution. For if the testing function φ is identically 1 near the origin then the integrand in

$$\int_{\mathbb{R}^N} k_2(x) \varphi(x) \, dx$$

near 0 is like $c/|x|^N$, and that is not integrable (just use polar coordinates as above).

We conclude that if we want our kernel k to induce a distribution—which seems like a reasonable and minimal requirement for any integral operator that we should want to study—then we need $c = 0$. In other words, Ω must have mean value 0. It follows that the original Fourier multiplier m must have mean value 0. In fact, if you compose Ω with a rotation and then average over all rotations (using Haar measure on the special orthogonal group) then you get zero. Since the map $\Omega \mapsto m$

commutes with rotations, the same assertion holds for m. Hence m must have mean value 0.

Important examples in dimension N of degree-zero-homogeneity, mean-value-zero Fourier multipliers are the *Riesz multipliers*

$$m_j(x) = \frac{x_j}{|x|}, \quad j = 1, \ldots, N.$$

These correspond (see the calculation in [STE1]) to kernels

$$k_j(x) = c \cdot \frac{x_j}{|x|^{N+1}}, \quad j = 1, \ldots, N.$$

Notice that each of the Riesz kernels k_j is homogeneous of degree $-N$. And the corresponding $\Omega_j(x) = x_j/|x|$ has (by odd parity) mean value zero on the unit sphere.

The integral operators

$$R_j f(x) \equiv \text{P.V.} \int_{\mathbb{R}^N} k_j(x - t) f(t) dt$$

are called the *Riesz transforms* and are fundamental to harmonic analysis on Euclidean space. We shall see them put to good use in a proof of the Sobolev embedding theorem in Section 5.2. Later, in our treatment of real-variable Hardy spaces in Chapter 8, we shall hear more about these important operators. In particular, Section 8.8 treats the generalized Cauchy–Riemann equations, which are an analytic aspect of the Riesz transform theory.

Of course there are uncountably many distinct smooth functions Ω on the unit sphere that have mean value zero. Thus there are uncountably many distinct Calderón–Zygmund singular integral kernels. Singular integrals come up naturally in the regularity theory for partial differential equations, in harmonic analysis, in image analysis and compression, and in many other fields of pure and applied mathematics. They are a fundamental part of modern analysis.

3.5 Ideas Leading to Pseudodifferential Operators

Capsule: Ever since the 1930s there has been a desire for an algebra of operators that contains all parametrices for elliptic operators, and that has operations (i.e., composition, inverse, adjoint) that are naturally consistent with considerations of partial differential equations. It was not until the late 1960s that Kohn/Nirenberg, and later Hörmander, were able to create such a calculus. This has really transformed the theory of differential equations. Many constructs that were formerly tedious and technical are now smooth and natural. Today there are many different calculi of pseudodifferential operators, and also Hörmander's powerful calculus of Fourier integral operators. This continues to be an intense area of study.

We shall treat pseudodifferential operators in some detail in Appendix 3. In the present section we merely indicate what they are, and try to fit them into the context of Fourier multipliers.

Pseudodifferential operators were invented by Kohn/Nirenberg [KON1] and Hörmander [HOR6] (building on work by Mihlin, Calderón/Zygmund, and many others) in order to provide a calculus of operators for constructing parametrices for elliptic partial differential equations.

Here, by a "calculus" we mean a class of operators that are easy to compose, calculate adjoints of, and calculate inverses of—in such a way that the calculated operator is seen to fit into the given class and to have the same form. Also the "order" of the calculated operator should be easy to determine. Again, Appendix 3 on pseudodifferential operators will make it clear what all this terminology means. The terminology "parametrix" means an approximate inverse to a given elliptic operator. The manner in which an inverse is "approximate" is a key part of the theory, and is measured in terms of mapping properties of the operators.

It took a long time to realize that the most effective way to define pseudo-differential operators is in terms of their symbols (i.e., on the Fourier transform side). A preliminary definition of pseudodifferential operator, rooted in ideas of Mihlin, is this:

A function $p(x, \xi)$ is said to be a *symbol* of order m if p is C^∞, has compact support in the x variable, and is homogeneous of degree m in ξ when ξ is large. That is, we assume that there is an $M > 0$ such that if $|\xi| > M$ and $\lambda > 1$ then

$$p(x, \lambda \xi) = \lambda^m p(x, \xi). \qquad (3.5.1)$$

Thus the key defining property of the symbol of a pseudodifferential operator is its rate of decay in the ξ variable. A slightly more general, and more useful, definition of pseudodifferential operator (due to Kohn and Nirenberg) is:

Let $m \in \mathbb{R}$. We say that a smooth function $\sigma(x, \xi)$ on $\mathbb{R}^N \times \mathbb{R}^N$ is a *symbol* of order m if there is a compact set $K \subseteq \mathbb{R}^N$ such that $\operatorname{supp} \sigma \subseteq K \times \mathbb{R}^N$ and, for any pair of multi-indices α, β, there is a constant $C_{\alpha,\beta}$ such that

$$\left| D_\xi^\alpha D_x^\beta \sigma(x, \xi) \right| \le C_{\alpha,\beta} \left(1 + |\xi| \right)^{m - |\alpha|}. \qquad (3.5.2)$$

We write $\sigma \in S^m$.

We define the corresponding pseudodifferential operator to be

$$T_\sigma f \equiv \int \widehat{f}(\xi) \sigma(x, \xi) e^{-ix \cdot \xi} \, d\xi.$$

Thus we see the pseudodifferential operator being treated just like an ordinary (constant-coefficient) Fourier multiplier. The difference now is that the symbol will, in general, have "variable coefficients." This is so that we can ultimately treat elliptic partial differential equations with variable coefficients.

In the long term, what we want to see—and this will be proved rigorously in Appendix 3—is that

$$T_p \circ T_q \approx T_{pq},$$

$$[T_p]^* \approx T_{\bar{p}},$$

and

$$[T_p]^{-1} \approx T_{1/p}.$$

Of course it must be made clear what the nature of the approximation $\approx$ is, and that (highly nontrivial) matter is a central part of the theory of pseudodifferential operators to be developed below (in Appendix 3).

4

Fractional and Singular Integrals

Prologue: In some vague sense, the collection of all fractional and singular integrals forms a poor man's version of a classical calculus of pseudo-differential operators. Certainly a fractional integral is very much like the parametrix for a strongly elliptic operator.

But fractional and singular integrals do not form a calculus in any natural sense. They are certainly *not* closed under composition, and there is no easy way to calculate compositions and inverses. Calculations also reveal that the collection is not *complete* in any natural sense. Finally, fractional and singular integrals (of the most classical sort) are all convolution operators. Many of the most interesting partial differential equations of elliptic type are *not* translation-invariant. So we clearly need a *larger* calculus of operators.

In spite of their limitations, fractional and singular integrals form the bedrock of our studies of linear harmonic analysis. Many of the most basic questions in the subject—ranging from the Sobolev embedding theorem to boundary limits of harmonic conjugate functions to convergence of Fourier series—can be reduced to estimates for these fundamental operators. And any more advanced studies will certainly be based on techniques developed to study these more primitive operators.

The big names in the study of fractional integrals are Riemann, Liouville, and M. Riesz. The big names in the study of singular integrals are Mikhlin, Calderón, Zygmund, Stein, and Fefferman. More modern developments are due to Christ, Rubio de Francia, David, Journé, and many others.

The subject of fractional and singular integrals is one that continues to evolve. As new contexts for analysis develop, new tools must be created along with them. Thus the subject of harmonic analysis continues to grow.

4.1 Fractional and Singular Integrals Together

Capsule: In this section we introduce fractional and singular integrals. We see how they differ from the point of view of homogeneity, and from

S G Krantz, *Explorations in Harmonic Analysis*, Applied and Numerical
Harmonic Analysis, DOI 10.1007/978-0-8176-4669-1_4,
© Birkhäuser Boston, a part of Springer Science+Business Media, LLC 2009

the point of view of the Fourier transform. We relate the idea to fractional powers of the Laplacian, and the idea of the fractional derivative.

We have seen the Hilbert transform as the central artifact of any study of convergence issues for Fourier series. The Hilbert transform turns out to be a special instance, indeed the most fundamental instance, of what is known as a *singular integral*. Calderón and Zygmund, in 1952 (see [CALZ]), identified a very natural class of integral operators that generalized the Hilbert transform to higher dimensions and subsumed everything that we knew up until that time about the Hilbert transform.

The work of Calderón and Zygmund of course built on earlier studies of integral operators. One of the most fundamental types of integral operators—far more basic and simpler to study than singular integral operators—is the *fractional integral operators*. These operators turn out to model integrations to a fractional order. They are very natural from the point of view of partial differential equations: For instance, the Newton potential is a fractional integral operator.

In the present chapter, as a prelude to our study of singular integrals, we provide a treatment of fractional integrals. Then we segue into singular integrals and lay the basis for that subject.

Consider the Laplace equation

$$\Delta u = f,$$

where $f \in C_c(\mathbb{R}^N)$. Especially since Δ is an elliptic operator, we think of the Laplacian as a "second derivative." Thus it is natural to wonder whether u will have two more derivatives than f. We examine the matter by constructing a particular u using the fundamental solution for the Laplacian. For convenience we take the dimension $N \geq 3$.

Recall that the fundamental solution for Δ is $\Gamma(x) = c \cdot |x|^{-N+2}$. It is known that $\Delta \Gamma = \delta$, the Dirac delta mass. Thus $u = \Gamma * f$ is a solution to the Laplace equation. Note that any other solution will differ from this one by a harmonic function that is certainly C^∞. So it suffices for our studies to consider this one solution.

Begin by noticing that the kernel Γ is homogeneous of degree $-N + 2$, so it is certainly locally integrable. We may therefore conclude that u is locally bounded.

As we have noted before,

$$\frac{\partial u}{\partial x_j} = \frac{\partial \Gamma}{\partial x_j} * f$$

and

$$\frac{\partial^2 u}{\partial x_j \partial x_k} = \frac{\partial^2 \Gamma}{\partial x_j \partial x_k} * u.$$

The first kernel is homogeneous of degree $-N + 1$; the second is homogeneous of degree $-N$ and is *not* locally integrable. In fact, the convolution makes no sense as a Lebesgue integral.

One must in fact interpret the integral with the Cauchy principal value (discussed elsewhere in the present book), and apply the Calderón–Zygmund theory of singular integrals in order to make sense of this expression. By inspection, the kernel for $\partial^2 u/\partial x_j \partial x_k$ is homogeneous of degree $-N$ and has mean value 0 on the unit sphere. Since a classical singular integral is bounded on L^p for $1 < p < \infty$, we may conclude that $\partial^2 u/\partial x_j \partial x_k$ is locally in L^p, hence locally integrable.

In conclusion, the solution u of $\Delta u = f$ has roughly two more derivatives than the data function f. That is what was expected. Our ensuing discussion of fractional integrals (and singular integrals) will flesh out these ideas and make them more rigorous.

4.2 Fractional Integrals and Other Elementary Operators

Capsule: Fractional integrals are certainly more basic, and easier to study, than singular integrals. The reason is that the estimation of fractional integrals depends only on *size* (more precisely on the distribution of values of the absolute value of the kernel) and not on any cancellation properties. Thus we begin our study of integral operators with these elementary objects. The mapping properties of fractional integrals on L^p spaces are interesting. These results have been extended to the real-variable Hardy spaces and other modern contexts.

Now the basic fact about fractional integration is that it acts naturally on the L^p spaces, for p in a particular range. Indeed we may anticipate exactly what the correct theorem is by using a little dimensional analysis.

Fix $0 < \beta < N$ and suppose that an inequality of the form

$$\|\mathcal{I}^\beta \phi\|_{L^q} \leq C \cdot \|\phi\|_{L^p}$$

were true for all testing functions ϕ. It could happen that $q = p$ or $q > p$ (a theorem of Hörmander [HOR3] tells us that it cannot be that $q < p$). Let us replace ϕ in both sides of this inequality by the expression $\alpha_\delta \phi(x) \equiv \phi(\delta x)$ for $\delta > 0$. Writing out the integrals, we have

$$\left(\int_{\mathbb{R}^N} \left| \int_{\mathbb{R}^N} |t|^{\beta - N} \alpha_\delta \phi(x - t) dt \right|^q dx \right)^{1/q} \leq C \cdot \left(\int_{\mathbb{R}^N} |\alpha_\delta \phi(x)|^p dx \right)^{1/p}.$$

On the left side we replace t by t/δ and x by x/δ; on the right we replace x by x/δ. The result, after a little calculation (i.e., elementary changes of variable), is

$$\left(\delta^{-\beta q - N} \right)^{1/q} \left(\int_{\mathbb{R}^N} \left| \int_{\mathbb{R}^N} |t|^{\beta - N} \phi(x - t) dt \right|^q dx \right)^{1/q}$$

$$\leq \delta^{-N/p} C \cdot \left(\int_{\mathbb{R}^N} |\phi(x)|^p dx \right)^{1/p}.$$

In other words,

$$\delta^{N/p} \cdot \delta^{-\beta-N/q} \le C'.$$

Notice that the value of C' comes from the values of the two integrals—neither of which involves δ. In particular, C' depends on ϕ. But ϕ is fixed once and for all.

Since this last inequality must hold for any fixed ϕ and for all $\delta > 0$, there would be a contradiction either as $\delta \to 0$ or as $\delta \to +\infty$ if the two expressions in δ did not cancel out.

We conclude that the two exponents must cancel each other, or

$$\beta + \frac{N}{q} = \frac{N}{p}.$$

This may be rewritten in the more classical form

$$\frac{1}{q} = \frac{1}{p} - \frac{\beta}{N}.$$

Thus our dimensional-analysis calculation points to the correct theorem:

Prelude: Certainly one of the marvels of the Fourier transform is that it gives us a natural and intuitively appealing way to think about fractional differentiation and fractional integration. The fractional integration theorem of Riesz is the tactile device for understanding fractional integration.

Theorem 4.2.1 *Let* $0 < \beta < N$. *The integral operator*

$$\mathcal{I}^\beta \phi(x) \equiv \int_{\mathbb{R}^N} |t|^{-N+\beta} \phi(x-t)dt,$$

initially defined for $\phi \in C_c^1(\mathbb{R}^N)$, *satisfies*

$$\|\mathcal{I}^\beta \phi\|_{L^q(\mathbb{R}^N)} \le C \cdot \|\phi\|_{L^p(\mathbb{R}^N)},$$

whenever $1 < p < N/\beta$ *and* q *satisfies*

$$\frac{1}{q} = \frac{1}{p} - \frac{\beta}{N}. \tag{4.2.1.1}$$

Of course, this result requires a bona fide proof. We shall provide a proof of (a very general version of) this result in Theorem 9.8.7. It may be noted that there are versions of Theorem 4.1.1 for $p \le 1$ (see [KRA1]) and also for $p \ge N/\beta$ (see [STE2], [HOR2]).

Example 4.2.2 Let $N \ge 1$ and consider on $\mathbb{R}^N$ the kernel

$$K(x) = \frac{x_1/|x|}{|x|^{N-\beta}}$$

for $0 < \beta < N$ as usual. The integral operator

$$f \mapsto f * K$$

may be studied by instead considering

$$f \mapsto f * |K|,$$

and there is no loss of generality in doing so. Of course

$$|K| \leq \frac{1}{|x|^{N-\beta}},$$

and the latter is a classical fractional integration kernel.

The point here is that cancellation has no role when the homogeneity of the kernel is less than the critical index N. One may as well replace the kernel by its absolute value, and majorize the resulting positive kernel by the obvious fractional integral kernel.

It is also enlightening to consider the point of view of the classical work [HLP]. Its authors advocated proving inequalities by replacing each function by its non-increasing rearrangement. For simplicity let us work on the real line. If f is a nonnegative function in L^p then its nonincreasing rearrangement will be a function (roughly) of the form $\widetilde{f}(x) = x^{-1/p}$. The fractional integral kernel will have the form $k(x) = |x|^{-1+\beta}$. Thus the convolution of the two will have size about $\widetilde{f} * k = x^{-1/p+(-1+\beta)+1} = x^{-1/p+\beta}$. But this latter function is just about in L^q, where

$$\frac{1}{q} = \frac{1}{p} - \beta = \frac{1}{p} - \frac{\beta}{1}.$$

This result is consistent with (4.2.1.1).

Fractional integrals are one of the two building blocks of the theory of integral operators that has developed in the last half century. In the next section we introduce the other building block.

4.3 Lead-In to Singular Integral Theory

Capsule: We know that singular integrals, as they are understood today, are a natural outgrowth of the Hilbert transform. The Hilbert transform has a long and venerable tradition in complex and harmonic analysis. The generalization to N dimensions, due to Calderón and Zygmund in 1952, revolutionized the subject. This made a whole new body of techniques available for higher-dimensional analysis. Cauchy problems, commutators of operators, subtle boundary value problems, and many other natural contexts for analysis were now amenable to a powerful method of attack. This section provides some background on singular integrals.

We begin with a table that illustrates some of the differences between fractional integrals and singular integrals.

Type of Integral	Fractional	Singular
Linear	yes	yes
Translation-Invariant	yes	yes
Rotationally Invariant Kernel	yes	never
Is a Pseudodifferential Operator	yes	yes
Compact	yes on Sobolev spaces	never
L^p Bounded	increases index p	bounded on L^p, $1 < p < \infty$
Smoothing	always smoothing	never smoothing

The Hilbert transform (Section 1.7) is the quintessential example of a singular integral. [We shall treat singular integrals in detail in Section 9.8 through 9.11] In fact, in dimension 1 it is, up to multiplication by a constant, the only classical singular integral. This statement means that the function $1/t$ is the only integral kernel that **(i)** is smooth away from 0, **(ii)** is homogeneous of degree -1, and **(iii)** has "mean value 0" on the unit sphere[1] in $\mathbb{R}^1$.

In $\mathbb{R}^N$, $N > 1$, there are a great many singular integral operators. Let us give a formal definition (refer to [CALZ]):

Definition 4.3.1 A function $K : \mathbb{R}^N \setminus \{0\} \to \mathbb{C}$ is called a *Calderón–Zygmund singular integral kernel* if it possesses the following three properties:

(4.3.1.1) The function K is smooth on $\mathbb{R}^N \setminus \{0\}$.

(4.3.1.2) The function K is homogeneous of degree $-N$.

(4.3.1.3) $\int_{\Sigma_{N-1}} K(x)\, d\sigma(x) = 0$, where Σ_{N-1} is the $(N-1)$-dimensional unit sphere in $\mathbb{R}^N$, and $d\sigma$ is rotationally invariant surface measure on that sphere.[2]

It is worthwhile to put this definition in context. Let β be a fixed complex number and consider the functional

---

[1] We have yet to explain what the critical "mean value 0" property is. This will come in the present and ensuing sections.

[2] In some contexts this surface measure is called *Hausdorff measure*. See [FOL3] or [FED] or our Chapter 9. Calderón and Zygmund themselves were rather uncomfortable with the concept of surface measure. So they formulated condition **(4.3.1.3)** as $\int_{a<|z|<b} K(x)\, dx = 0$ for any $0 < a < b < \infty$. We leave it as an exercise for you to verify that the two different formulations are equivalent. It is property **(4.3.1.3)** that is called the mean value zero property.

$$\phi \longmapsto \int \phi(x)|x|^{\beta}\,dx,$$

which is defined on functions ϕ that are C^{∞} with compact support. When $\mathrm{Re}\,\beta > -N$, this definition makes good sense, because we may estimate the integral near the origin (taking $\mathrm{supp}(\phi) \subseteq B(0, R)$, $C = \sup|\phi|$, and $C' = C \cdot \sigma(\Sigma_{N-1})$) by

$$C \cdot \left| \int_{\{|x| \le R\}} |x|^{\beta}\,dx \right| \le C \cdot \int_{\{|x| \le R\}} |x|^{\mathrm{Re}\,\beta}\,dx = C' \cdot \int_0^R r^{\mathrm{Re}\,\beta+N-1}\,dr < \infty.$$

Now we change our point of view; we think of the testing function ϕ as being fixed and we think of $\beta \in \mathbb{C}$ as the variable. In fact, Morera's theorem shows that

$$\mathcal{G}(\beta) \equiv \int \phi(x)|x|^{\beta}\,dx$$

is well defined and is a holomorphic function of β on $\{\beta \in \mathbb{C} : \mathrm{Re}\,\beta > -N\}$. We may ask whether this holomorphic function can be analytically continued to the rest of the complex plane.

In order to carry out the needed calculation, it is convenient to assume that the testing function ϕ is a radial function: $\phi(x) = \phi(x')$ whenever $|x| = |x'|$. In this case we may write $\phi(x) = f(r)$, where $r = |x|$. Then we may write, using polar coordinates,

$$\mathcal{G}(\beta) \equiv \int \phi(x)|x|^{\beta}\,dx = c \cdot \int_0^{\infty} f(r)r^{\beta} \cdot r^{N-1}\,dr.$$

Integrating by parts in r gives

$$\mathcal{G}(\beta) = -\frac{c}{\beta + N} \int_0^{\infty} f'(r)r^{\beta+N}\,dr.$$

Notice that the boundary term at infinity vanishes since ϕ (and hence f) is compactly supported; the boundary term at the origin vanishes because of the presence of $r^{\beta+N}$.

We may continue, in this fashion, integrating by parts to obtain the formulas

$$\mathcal{G}(\beta) = \frac{(-1)^{j+1}}{(\beta + N)(\beta + N + 1) \cdots (\beta + N + j)} \int_0^{\infty} f^{(j+1)}(r)r^{\beta+N+j}\,dr.$$
$$(4.3.2_j)$$

The key fact is that any two of these formulas for $\mathcal{G}(\beta)$ are equal for $\mathrm{Re}\,\beta > -N$. Yet the integral in formula $(4.3.2_j)$ makes sense for $\mathrm{Re}\,\beta > -N - j - 1$. Hence formula $(4.3.2_j)$ can be used to *define* an analytic continuation of $\mathcal{G}$ to the domain $\mathrm{Re}\,\beta > -N - j - 1$ (except for certain poles). As a result, we have a method of analytically continuing $\mathcal{G}$ to the entire complex plane, less the poles at $\{-N, -N - 1, -N - 2, \ldots, -N - j\}$. Observe that these poles are exhibited explicitly in the denominator of the fraction preceding the integral in the formulas $(4.3.2_j)$ that define $\mathcal{G}$.

The upshot of our calculations is that it is possible to make sense of the operator consisting of integration against $|x|^\beta$ as a classical fractional integral operator provided that $\beta \neq -N, -N - 1, \ldots$. More generally, an operator with integral kernel homogeneous of degree β, where $\beta \neq -N, -N - 1, \ldots$, is amenable to a relatively simple analysis (as we shall learn below).

If instead we consider an operator with kernel homogeneous of degree β where β takes on one of these critical values $-N, -N - 1, \ldots$, then some additional condition must be imposed on the kernel. These observations give rise to the mean-value-zero condition (4.3.1.3) in the definition of the Calderón–Zygmund singular integrals that are homogeneous of degree $-N$. [The study of singular integrals of degree $-N - k$, $k > 0$, is an advanced topic (known as the theory of strongly singular integrals), treated for instance in [FEF1]. We shall not discuss it here.]

Now let K be a Calderón–Zygmund kernel. The associated integral operator

$$T_K(\phi)(x) \equiv \int K(t)\phi(x - t)dt$$

makes no formal sense, even when ϕ is a testing function. This is so because K is not absolutely integrable at the origin. Instead we use the notion of the *Cauchy principal value* to evaluate this integral. To wit, let $\phi \in C_c^1(\mathbb{R}^N)$. Set

$$T_K(\phi)(x) = \text{P.V.} \int K(t)\phi(x - t)dt \equiv \lim_{\epsilon \to 0^+} \int_{|t|>\epsilon} K(t)\phi(x - t)dt.$$

We have already shown in Section 2.1.3 that this last limit exists. We take this now for granted.

Example 4.3.3 Let $N \geq 1$ and consider on $\mathbb{R}^N$ the kernel

$$K(x) = \frac{x_1/|x|}{|x|^N}.$$

We may rewrite this kernel as

$$K(x) = \frac{\Omega(x)}{|x|^N},$$

with $\Omega(x) = x_1/|x|$. Of course this Ω is homogeneous of degree 0 and (by parity) has mean-value-zero on the unit sphere.

Thus K is a classical Calderón–Zygmund kernel, and the operator

$$f \mapsto f * K$$

is bounded on $L^p(\mathbb{R}^N)$ for $1 < p < \infty$.

Remark 4.3.4 There is an alternative to the standard mean value zero condition for a kernel $k(x) = \Omega(x)/|x|^N$ that we have been discussing. Known as *Hörmander's! condition*, it is that

$$\left| \int_{|x|>2|y|} k(x-y) - k(x)\,dx \right| \le C. \tag{4.3.4.1}$$

This new condition is rather flexible, and rather well suited to the study of the mapping properties of the integral operator defined by integration against k. It also makes good sense on manifolds. Also the condition (4.3.4.1) *is implied by* the classical mean-value-zero condition. So it is a formally weaker condition. Let us now prove this last implication.

Thus let $k(x) = \Omega(x)/|x|^N$, with Ω a C^1 function on $\mathbb{R}^N \setminus \{0\}$ and satisfying the usual mean-value-zero property on the sphere. Then

$$\left| \int_{|x|>2|y|} k(x-y) - k(x)\,dx \right| = \left| \int_{|x|>2|y|} \frac{\Omega(x-y)}{|x-y|^N} - \frac{\Omega(x)}{|x|^N}\,dx \right|$$

$$= \left| \int_{|x|>2|y|} \left[\frac{\Omega(x-y)}{|x-y|^N} - \frac{\Omega(x-y)}{|x|^N} \right] \right.$$

$$\left. + \left[\frac{\Omega(x-y)}{|x|^N} - \frac{\Omega(x)}{|x|^N} \right] dx \right|$$

$$\le \int_{|x|>2|y|} \left| \frac{\Omega(x-y)}{|x-y|^N} - \frac{\Omega(x-y)}{|x|^N} \right|$$

$$+ \left| \frac{\Omega(x-y)}{|x|^N} - \frac{\Omega(x)}{|x|^N} \right| dx$$

$$\equiv I + II.$$

We now perform an analysis of I. The estimation of II is similar, and we leave the details to the reader.

Now

$$I \le C \cdot \int_{|x|>2|y|} \left| \frac{|x|^N - |x-y|^N}{|x-y|^N \cdot |x|^N} \right| dx$$

$$\le C \cdot \int_{|x|>2|y|} \frac{|y| \cdot |x|^{N-1}}{|x-y|^N \cdot |x|^N}\,dx$$

$$\le C \cdot \int_{|x|>2|y|} \left| \frac{|y| \cdot |x|^{N-1}}{|x|^{2N}} \right| dx$$

$$\le C|y| \cdot \int_{|x|>2|y|} \frac{1}{|x|^{N+1}}\,dx$$

$$\le C.$$

In the last step of course we have used polar coordinates as usual.

When you provide the details of the estimation of II, you should keep in mind that the derivative of a function homogeneous of degree λ will in fact be homogeneous of degree $\lambda - 1$.

Building on ideas that we developed in the context of the Hilbert transform, the most fundamental question that we might now ask is whether

$$\|T_K(\phi)\|_{L^p} \le C\|\phi\|_{L^p}$$

for all $\phi \in C_c^1(\mathbb{R}^N)$. If, for some fixed p, this inequality holds, then a simple density argument will extend the operator T_K and the inequality to all $\phi \in L^p(\mathbb{R}^N)$. In fact, this inequality holds for $1 < p < \infty$ and fails for $p = 1, \infty$. The first of these two statements is called the *Calderón–Zygmund theorem*, and we prove it in Section 9.8. The second follows just as it did in Chapter 2 for the Hilbert transform; we leave the details as an exercise for the interested reader.

Here is a summary of what our discussion has revealed thus far:

Prelude: This next theorem is one of the great results in analysis of the twentieth century. For it captures the essence of why the Hilbert transform works, and generalizes it to the N-variable setting. Both the statement and its proof are profound, and the technique of the Calderón–Zygmund theorem has had a tremendous influence. Today there are many generalizations of the Calderón–Zygmund theory. We might mention particularly the $T(1)$ theorem of David–Journé [DAVJ], which gives a "variable-coefficient" version of the theory.

Theorem 4.3.5 (Calderón–Zygmund) *Let K be a Calderón–Zygmund kernel. Then the operator*

$$T_K(\phi)(x) \equiv \text{P.V.} \int K(t)\phi(x - t)dt,$$

for $\phi \in C_c^1(\mathbb{R}^N)$, is well defined. It is bounded in the L^p norm for $1 < p < \infty$. Thus it extends as a bounded operator from L^p to L^p. It is not bounded on L^1 nor is it bounded on L^∞.

It is natural to wonder whether there are spaces that are related to L^1 and L^∞, and that might serve as their substitutes for the purposes of singular integral theory. As we shall see, the correct approach is to consider the subspace of L^1 that behaves naturally under certain canonical singular integral operators. This approach yields a subspace of L^1 that is known as H^1 (or, more precisely, H^1_{Re})—this is the real-variable Hardy space of Fefferman, Stein, and Weiss. The dual of this new space is a superspace of L^∞. It is called *BMO* (the functions of bounded mean oscillation). We shall explore these two new spaces, and their connections with the questions under discussion, as the book develops.

Notice that the discussion at the end of Section 4.2 on how to construct functions of a given homogeneity also tells us how to construct Calderón–Zygmund kernels. Namely, let ϕ be any smooth function on the unit sphere of $\mathbb{R}^N$ that integrates to zero with respect to area measure.[3] Extend it to a function Ω on all of space (except the origin) so that Ω is homogeneous of degree zero, i.e., let $\Omega(x) = \phi(x/|x|)$. Then

$$K(x) = \frac{\Omega(x)}{|x|^N}$$

is a Calderón–Zygmund kernel.

We shall provide a proof of Theorem 4.3.5 in Chapter 9.

We close this section with an application of the Riesz transforms and singular integrals to an embedding theorem for function spaces. This is a version of the so-called *Sobolev embedding theorem*.

Prelude: This version of the Sobolev embedding theorem is far from sharp, but it illustrates the naturalness and the utility of fractional integrals and singular integrals. They are seen here as the obvious tools in a calculation of fundamental importance.

Theorem 4.3.5 *Fix a dimension $N > 1$ and let $1 < p < N$. Let $f \in L^p(\mathbb{R}^N)$ with the property that $(\partial/\partial x_j)f$ exists and is in L^p, $j = 1, \ldots, N$. Then $f \in L^q(\mathbb{R}^N)$, where $1/q = 1/p - 1/N$.*

Proof: As usual, we content ourselves with a proof of an a priori inequality for $f \in C_c^\infty(\mathbb{R}^N)$. We write

$$\widehat{f}(\xi) = \sum_{j=1}^{N} \frac{1}{|\xi|} \cdot \frac{\xi_j}{|\xi|} \cdot [\xi_j \, \widehat{f}(\xi)].$$

Observe that $\xi_j \, \widehat{f}(\xi)$ is (essentially, up to a trivial constant multiple) the Fourier transform of $(\partial/\partial x_j)f$. Also $\xi_j/|\xi|$ is the multiplier for the jth Riesz transform R_j. And $1/|\xi|$ is the Fourier multiplier for a fractional integral $\mathcal{I}^1$. [All of these statements are true up to constant multiples, which we omit.] Using operator notation, we may therefore rewrite the last displayed equation as

$$f(x) = \sum_{j=1}^{N} \mathcal{I}^1 \circ R_j \left(\frac{\partial}{\partial x_j} f(x) \right).$$

We know by hypothesis that $(\partial/\partial x_j)f \in L^p$. Now R_j maps L^p to L^p and I_1 maps L^p to L^q, where $1/q = 1/p - 1/N$ (see Theorem 4.2.1). That completes the proof. $\qquad\square$

Remark: In an ideal world, we would study C^k spaces because they are intuitive and natural. But the integral operators that come up in practice—the fractional integral and singular integrals, as well as the more general pseudodifferential operators— do not behave well on the C^k spaces. A C^k space is basically (the integral of) a space with uniform norm, and we have seen that singular integrals do not respect

---

[3] In fact, even if ϕ does *not* integrate to zero, we may replace ϕ by $\widetilde{\phi} \equiv \phi - c$, where c is the mean value of ϕ on the sphere.

the uniform norm. Neither do fractional integrals. The space *BMO* can provide substitute results, but they are technical and somewhat awkward. The Sobolev spaces, especially those modeled on L^2, are considerably more useful and easy to manipulate. And of course the reason for this is that the Fourier transform acts so nicely on L^2.

A Crash Course in Several Complex Variables

Prologue: The function theory of several complex variables (SCV) is—obviously—a generalization of the subject of one complex variable. Certainly some of the results in the former subject are inspired by ideas from the latter subject. But SCV really has an entirely new character.

One difference (to be explained below) is that in one complex variable *every* domain is a domain of holomorphy; but in several complex variables some domains are and some are not (this is a famous theorem of F. Hartogs). Another difference is that the Cauchy–Riemann equations in several-variables form an overdetermined system; in one variable they do not. There are also subtleties involving the $\bar{\partial}$-Neumann boundary value problem, such as subellipticity; we cannot treat the details here, but see [KRA3].

Most of the familiar techniques of one complex variable—integral formulas, Blaschke products, Weierstrass products, the argument principle, conformality—either do not exist or at least take a quite different form in the several-variable setting. New tools, such as sheaf theory and the $\bar{\partial}$-Neumann problem, have been developed to handle problems in this new setting.

Several complex variables is exciting for its interaction with differential geometry, with partial differential equations, with Lie theory, with harmonic analysis, and with many other parts of mathematics. In this text we shall see several complex variables lay the groundwork for a new type of harmonic analysis.

Next we turn our attention to the function theory of several complex variables. One of the main purposes of this book is to provide a foundational introduction to the harmonic analysis of several complex variables. So this chapter constitutes a transition. It will provide the reader with the basic language and key ideas of several complex variables so that the later material makes good sense.

As a working definition let us say that a function $f(z_1, z_2, \ldots, z_n)$ of several complex variables is holomorphic if it is holomorphic in each variable separately.

S. G. Krantz, *Explorations in Harmonic Analysis*, Applied and Numerical Harmonic Analysis, DOI 10.1007/978-0-8176-4669-1_5,
© Birkhäuser Boston, a part of Springer Science+Business Media, LLC 2009

We shall say more about different definitions of holomorphicity, and their equivalence, as the exposition progresses.

5.1 What Is a Holomorphic Function?

Capsule: There are many ways to define the notion of holomorphic function of several complex variables. A function is holomorphic if it is holomorphic (in the classical sense) in each variable separately. It is holomorphic if it satisfies the Cauchy–Riemann equations. It is holomorphic if it has a local power series expansion about each point. There are a number of other possible formulations. We explore these, and some of the elementary properties of holomorphic functions, in this section.

In the discussion that follows, a *domain* is a connected, open set. Let us restrict attention to functions $f : \Omega \to \mathbb{C}$, Ω a domain in $\mathbb{C}^n$, that are locally integrable (denoted $f \in L^1_{loc}$). That is, we assume that $\int_K |f(z)| \, dV(z) < \infty$ for each compact subset K of Ω. In particular, in order to avoid aberrations, we shall only discuss functions that are distributions (to see what happens to function theory when such a standing hypothesis is not enforced, see the work of E.R. Hedrick [HED]). Distribution theory will not, however, play an explicit role in what follows.

For $p \in \mathbb{C}^n$ and $r \geq 0$, we let

$$D^n(p, r) = \{z = (z_1, \ldots, z_n) \in \mathbb{C}^n : |p_j - z_j| < r \text{ for all } j\}$$

and

$$\overline{D}^n(p, r) = \{z = (z_1, \ldots, z_n) \in \mathbb{C}^n : |p_j - z_j| \leq r \text{ for all } j\}.$$

These are the open and closed *polydisks* of radius r.

We also define balls in complex space by

$$B(z, r) = \left\{z = (z_1, \ldots, z_n) \in \mathbb{C}^n : \sum_j |p_j - z_j|^2 < r^2\right\}$$

and

$$\overline{B}(z, r) = \left\{z = (z_1, \ldots, z_n) \in \mathbb{C}^n : \sum_j |p_j - z_j|^2 \leq r^2\right\}.$$

We now offer three plausible definitions of holomorphic function on a domain $\Omega \subseteq \mathbb{C}^n$:

DEFINITION A: A function $f : \Omega \to \mathbb{C}$ is holomorphic if for each $j = 1, \ldots, n$ and each fixed $z_1, \ldots, z_{j-1}, z_{j+1}, \ldots, z_n$ the function

$$\zeta \mapsto f(z_1, \ldots, z_{j-1}, \zeta, z_{j+1}, \ldots, z_n)$$

is holomorphic, in the classical one-variable sense, on the set

$$\Omega(z_1, \ldots, z_{j-1}, z_{j+1}, \ldots, z_n) \equiv \{\zeta \in \mathbb{C} : (z_1, \ldots, z_{j-1}, \zeta, z_{j+1}, \ldots, z_n) \in \Omega\}.$$

In other words, we require that f be holomorphic in each variable separately.

DEFINITION B: A function $f : \Omega \to \mathbb{C}$ that is continuously differentiable in each complex variable separately on Ω is said to be holomorphic if f satisfies the Cauchy–Riemann equations in each variable separately.

This is just another way of requiring that f be holomorphic in each variable separately.

DEFINITION C: A function $f : \Omega \to \mathbb{C}$ is holomorphic (viz. complex analytic) if for each $z^0 \in \Omega$ there is an $r = r(z^0) > 0$ such that $\overline{D}^n(z^0, r) \subseteq \Omega$ and f can be written as an absolutely and uniformly convergent power series

$$f(z) = \sum_\alpha a_\alpha (z - z^0)^\alpha$$

for all $z \in D^n(z^0, r)$. [Here we use multi-index notation as previously discussed in Section 3.2.]

Fortunately, Definitions A–C, as well as several other plausible definitions, are equivalent. Some of these equivalences, such as A $\Leftrightarrow$ C, are not entirely trivial to prove. We shall explain the equivalences in the discussion that follows. All details may be found in [KRA4]. Later on we shall fix one formal definition of holomorphic function of several-variables and proceed logically from that definition.

We conclude this section by reminding the reader of some of the complex calculus notation in this subject. If f is a C^1 function on a domain in $\mathbb{C}$ then we define

$$\frac{\partial f}{\partial z} = \frac{1}{2}\left(\frac{\partial f}{\partial x} - i\frac{\partial f}{\partial y}\right)$$

and

$$\frac{\partial f}{\partial \overline{z}} = \frac{1}{2}\left(\frac{\partial f}{\partial x} + i\frac{\partial f}{\partial y}\right).$$

Note that

$$\frac{\partial z}{\partial z} = 1, \qquad \frac{\partial z}{\partial \overline{z}} = 0,$$

$$\frac{\partial \overline{z}}{\partial z} = 0, \qquad \frac{\partial \overline{z}}{\partial \overline{z}} = 1.$$

Observe that for a C^1 function f on a domain $\Omega \subseteq \mathbb{C}$, $\partial f/\partial \overline{z} \equiv 0$ if and only if f satisfies the Cauchy–Riemann equations; this is true in turn if and only if f is holomorphic on Ω. The operators $\partial/\partial z_j$ and $\partial/\partial \overline{z}_j$ are defined analogously.

Now let f be a C^1 function on a domain $\Omega \subseteq \mathbb{C}^n$. We write

$$\partial f = \sum_{j=1}^{n} \frac{\partial f}{\partial z_j} dz_j$$

and

$$\overline{\partial} f = \sum_{j=1}^{n} \frac{\partial f}{\partial \overline{z}_j} d\overline{z}_j.$$

Notice that $\overline{\partial} f \equiv 0$ on Ω if and only if $\partial f/\partial \overline{z}_j \equiv 0$ for each j, and that this in turn is true if and only if f is holomorphic in each variable separately. According to the discussion above, this is the same as f being holomorphic in the several-variable sense.

It is a straightforward exercise to see that $\overline{\partial}\,\overline{\partial} f \equiv 0$ and $\partial\,\partial f \equiv 0$.

5.2 Plurisubharmonic Functions

Capsule: In some sense, just as subharmonic functions are really the guts of one-variable complex analysis, so plurisubharmonic functions are the guts of several-variable complex analysis. Plurisubharmonic functions capture a good deal of analytic information, but they are much more flexible than holomorphic functions. For example, they are closed under the operation of taking the maximum. Plurisubharmonic functions may be studied using function-theoretic techniques. Thanks to the school of Lelong, they may also be studied using partial differential equations. The singular set of a plurisubharmonic function—also called a *pluripolar set*—is also very important in pluripotential theory.

The function theory of several complex variables is remarkable in the range of techniques that may be profitably used to explore it: algebraic geometry, one complex variable, differential geometry, partial differential equations, harmonic analysis, and function algebras are only some of the areas that interact with the subject. Most of the important results in several complex variables bear clearly the imprint of one or more of these disciplines. But if there is one set of ideas that belongs exclusively to several complex variables, it is those centering on pluriharmonic and plurisubharmonic functions. These play a recurring role in any treatment of the subject; we record here a number of their basic properties.

The setting for the present section is $\mathbb{C}^n$. Let $a, b \in \mathbb{C}^n$. The set

$$\{a + b\zeta : \zeta \in \mathbb{C}\}$$

is called a *complex line* in $\mathbb{C}^n$.

Remark: Note that *not every real two-dimensional affine space in $\mathbb{C}^n$ is a complex line*. For instance, the set $\ell = \{(x + i0, 0 + iy) : x, y \in \mathbb{R}\}$ is not a complex line in

$\mathbb{C}^2$ according to our definition. This is rightly so, for the complex structures on ℓ and $\mathbb{C}^2$ are incompatible. This means the following: if $f : \mathbb{C}^2 \to \mathbb{C}$ is holomorphic then it *does not* follow that $z = x + iy \mapsto f(x + i0, 0 + iy)$ is holomorphic. The point is that a complex line is supposed to be a (holomorphic) *complex affine embedding* of $\mathbb{C}$ into $\mathbb{C}^n$.

Definition 5.2.1 A C^2 function $f : \Omega \to \mathbb{C}$ is said to be *pluriharmonic* if for every complex line $\ell = \{a + b\zeta\}$ the function $\zeta \mapsto f(a + b\zeta)$ is harmonic on the set $\Omega_\ell \equiv \{\zeta \in \mathbb{C} : a + b\zeta \in \Omega\}$.

Remark 5.2.2 Certainly a function f is holomorphic on Ω if and only if $\zeta \mapsto f(a + b\zeta)$ is holomorphic on Ω_ℓ for every complex line $\ell = \{a + b\zeta\}$. This assertion follows immediately from our definition of holomorphic function and the chain rule.

A C^2 function f on Ω is pluriharmonic iff $(\partial^2/\partial z_j \partial \bar{z}_k)f \equiv 0$ on Ω for all $j, k = 1, \ldots, n$. This in turn is true iff $\partial \bar{\partial} f \equiv 0$ on Ω.

In the theory of one complex variable, harmonic functions play an important role because they are (locally) the real parts of holomorphic functions. The analogous role in several complex variables is played by pluriharmonic functions. To make this statement more precise, we first need a "Poincaré lemma":

Lemma 5.2.3 Let $\alpha = \sum_j a_j dx_j$ be a differential form with C^1 coefficients and satisfying $d\alpha = 0$ on a neighborhood of a closed box $S \subseteq \mathbb{R}^N$ with sides parallel to the axes. Then there is a function a on S satisfying $da = \alpha$.

Proof: See [LOS]. □

The *proof* of the Poincaré lemma shows that if α has real coefficients, then a can be taken to be real. Or one can derive this fact from linear algebraic formalism. We leave the details for the interested reader.

Prelude: In several complex variables, pluriharmonic functions are for many purposes the Ersatz for harmonic functions. They are much more rigid objects than harmonic functions, and satisfy a much more complicated partial differential equation than the Laplacian. Certainly a pluriharmonic function is harmonic, but the converse is false.

Proposition 5.2.4 Let $D^n(P, r) \subseteq \mathbb{C}^n$ be a polydisk and assume that $f : D^n(P, r) \to \mathbb{R}$ is C^2. Then f is pluriharmonic on $D^n(P, r)$ if and only if f is the real part of a holomorphic function on $D^n(P, r)$.

Proof: The "if" part is trivial.

For "only if," notice that $\alpha \equiv i(\bar{\partial} f - \partial f)$ is real and satisfies $d\alpha = 0$. But then, by Poincaré, there exists a real function g such that $dg = \alpha$. In other words, $d(ig) = (\partial f - \bar{\partial} f)$. Counting degrees, we see that $\bar{\partial}(ig) = -\bar{\partial} f$. It follows that $\partial(f + ig) = \bar{\partial} f - \bar{\partial} f = 0$; hence g is the real function we seek—it is the function that completes f to a holomorphic function. □

Observe that it is automatic that the g we constructed in the proposition is pluriharmonic, just because it is the imaginary part of a holomorphic function.

Remark: If f is a function defined on a polydisk and harmonic in each variable separately, it is natural to wonder whether f is pluriharmonic. In fact, the answer is "no," as the function $f(z_1, z_2) = z_1 \bar{z}_2$ demonstrates. This is a bit surprising, since the answer is affirmative when "harmonic" and "pluriharmonic" are replaced by "holomorphic" and "holomorphic."

Questions of "separate (P)," where (P) is some property, implying joint (P) are considered in detail in Hervé [HER]. See also the recent work of Wiegerinck [WIE].

Exercises for the Reader:

1. Pluriharmonic functions are harmonic, but the converse is false.
2. If f and f^2 are pluriharmonic then f is either holomorphic or conjugate holomorphic.

Remark: It is well known how to solve the Dirichlet problem for harmonic functions on smoothly bounded domains in $\mathbb{R}^N$, in particular on the ball (see [KRA4] and [GRK12]). Pluriharmonic functions are much more rigid objects than harmonic functions; in fact, the Dirichlet problem for these functions cannot always be solved. Let ϕ be a smooth function on the boundary of the ball B in $\mathbb{C}^2$ with the property that $\phi \equiv 1$ in a relative neighborhood of $(1, 0) \in \partial B$ and $\phi \equiv -1$ in a relative neighborhood of $(-1, 0) \in \partial B$. Then any pluriharmonic function assuming ϕ as its boundary function would have to be identically equal to 1 in a neighborhood of $(1, 0)$ and would have to be identically equal to -1 in a neighborhood of $(-1, 0)$. Since a pluriharmonic function is real analytic, these conditions are incompatible.

The situation for the Dirichlet problem for pluriharmonic functions is rather subtle. In fact, there is a partial differential operator $\mathcal{L}$ on ∂B such that a smooth f on ∂B is the boundary function of a pluriharmonic function if and only if $\mathcal{L}f = 0$ (see [BED], [BEF]). The operator $\mathcal{L}$ may be computed using just the theory of differential forms. It is remarkable that $\mathcal{L}$ is of third order.

Recall that a function f taking values in $\mathbb{R} \cup \{-\infty\}$ is said to be *upper semicontinuous* (u.s.c.) if for any $\alpha \in \mathbb{R}$, the set

$$U^\alpha = \{x : f(x) > \alpha\}$$

is open. Likewise, the function is *lower semicontinuous* (l.s.c.) if for any $\beta \in \mathbb{R}$, the set

$$U_\beta = \{x : f(x) < \beta\}$$

is open.

Definition 5.2.5 Let $\Omega \subseteq \mathbb{C}^n$ and let $f : \Omega \to \mathbb{R} \cup \{-\infty\}$ be u.s.c. We say that f is *plurisubharmonic* if for each complex line $\ell = \{a + b\zeta\} \subseteq \mathbb{C}^n$, the function

$$\zeta \mapsto f(a + b\zeta)$$

is subharmonic on $\Omega_\ell \equiv \{\zeta \in \mathbb{C} : a + b\zeta \in \Omega\}$.

Remark: Because it is cumbersome to write out the word "plurisubharmonic," a number of abbreviations for the word have come into use. Among the most common are *psh*, *plsh*, and *plush*. We shall sometimes use the first of these.

Exercise for the Reader: If $\Omega \subseteq \mathbb{C}^n$ and $f : \Omega \to \mathbb{C}$ is holomorphic then $\log |f|$ is psh; so is $|f|^p$, $p > 0$. The property of plurisubharmonicity is local (exercise—what does this mean?). A real-valued function $f \in C^2(\Omega)$ is psh iff

$$\sum_{j,k=1}^{n} \frac{\partial^2 f}{\partial z_j \partial \bar{z}_k}(z) w_j \bar{w}_k \geq 0$$

for every $z \in \Omega$ and every $w \in \mathbb{C}^n$. In other words, f is psh on Ω iff the complex Hessian of f is positive semidefinite at each point of Ω. See [KRA4] for more on these matters.

Proposition 5.2.6 *If $f : \Omega \to \mathbb{R} \cup \{-\infty\}$ is psh and $\phi : \mathbb{R} \cup \{\infty\} \to \mathbb{R} \cup \{\infty\}$ is convex and monotonically increasing then $\phi \circ f$ is psh.*

Proof: Exercise. Use the chain rule.   □

Definition 5.2.7 A real-valued function $f \in C^2(\Omega)$, $\Omega \subseteq \mathbb{C}^n$, is said to be *strictly plurisubharmonic* if

$$\sum_{j,k=1}^{n} \frac{\partial^2 f}{\partial z_j \partial \bar{z}_k}(z) w_j \bar{w}_k > 0$$

for every $z \in \Omega$ and every $0 \neq w \in \mathbb{C}^n$ (see the preceding exercise for the reader for motivation).

Exercise for the Reader: With notation as in Definition 6.2.7, use the notion of homogeneity to see that if $K \subset\subset \Omega$, then there is a $C = C(K) > 0$ such that

$$\sum_{j,k=1}^{n} \frac{\partial^2 f}{\partial z_j \partial \bar{z}_k}(z) w_j \bar{w}_k \geq C|w|^2$$

for all $z \in K$, $w \in \mathbb{C}^n$.

Proposition 5.2.8 *Let $\Omega \subseteq \mathbb{C}^n$ and $f : \Omega \to \mathbb{R} \cup \{\infty\}$ a psh function. For $\epsilon > 0$ we set $\Omega_\epsilon = \{z \in \Omega : \mathrm{dist}(z, \partial\Omega) > \epsilon\}$. Then there is a family $f_\epsilon : \Omega_\epsilon \to \mathbb{R}$ for $\epsilon > 0$ such that $f_\epsilon \in C^\infty(\Omega_\epsilon)$, $f_\epsilon \searrow f$ as $\epsilon \to 0^+$, and each f_ϵ is psh on Ω_ϵ.*

Proof: Let $\phi \in C_c^\infty(\mathbb{C}^n)$ satisfy $\int \phi = 1$ and $\phi(z_1, \ldots, z_n) = \phi(|z_1|, \ldots, |z_n|)$ for all z. Assume that ϕ is supported in $B(0, 1)$.

   Define

$$f_\epsilon(z) = \int f(z - \epsilon\zeta)\phi(\zeta)dV(\zeta), \quad z \in \Omega_\epsilon.$$

Now a standard argument (see [KRA4]) shows that each f_ϵ is smooth and plurisubharmonic.   □

Continuous plurisubharmonic functions are often called *pseudoconvex functions*.

Exercise for the Reader: If $\Omega_1, \Omega_2 \subseteq \mathbb{C}^n$, Ω_1 is bounded, and $f : \Omega_2 \to \mathbb{R} \cup \{\infty\}$ is C^2, then f is psh if and only if $f \circ \phi$ is psh for every holomorphic map $\phi : \Omega_1 \to \Omega_2$. Now prove the result assuming only that f is u.s.c. Why must Ω_1 be bounded?

The deeper properties of psh functions are beyond the scope of this book. The potential theory of psh functions is a rather well developed subject, and is intimately connected with the theory of the complex Monge–Ampère equation. Good references for these matters are [CEG], [KLI]. See also the papers [BET1–BET3] and references therein. In earlier work ([BET1]), the theory of the Dirichlet problem for psh functions is developed.

We have laid the groundwork earlier for one aspect of the potential theory of psh functions, and we should like to say a bit about it now. Call a subset $\mathcal{P} \subseteq \mathbb{C}^n$ *pluripolar* if it is the $-\infty$ set of a plurisubharmonic function. Then zero sets of holomorphic functions are obviously pluripolar (because if f is holomorphic then $\log |f|$ is plurisubharmonic). It is a result of B. Josefson [JOS] that locally pluripolar sets are pluripolar. In [BET3], a capacity theory for pluripolar sets is developed that is a powerful tool for answering many natural questions about pluripolar sets. In particular, it gives another method for proving Josefson's theorem. Plurisubharmonic functions, which were first defined by Lelong, are treated in depth in the treatise by L. Gruman and P. Lelong [GRL].

Define the *Hartogs functions* on a domain in $\mathbb{C}^n$ to be the smallest class of functions on Ω that contains all $\log |f|$ for f holomorphic on Ω and that is closed under the operations **(1)–(6)** listed here:

(1) If $\phi_1, \phi_2 \in \mathcal{F}_\Omega$ then $\phi_1 + \phi_2 \in \mathcal{F}_\Omega$.
(2) If $\phi \in \mathcal{F}_\Omega$ and $a \geq 0$ then $a\phi \in \mathcal{F}_\Omega$.
(3) If $\{\phi_j\} \subseteq \mathcal{F}_\Omega, \phi_1 \geq \phi_2 \geq \cdots$ then $\lim_{j \to \infty} \phi_j \in \mathcal{F}_\Omega$.
(4) If $\{\phi_j\} \subseteq \mathcal{F}_\Omega, \phi_j$ uniformly bounded above on compacta, then $\sup_j \phi_j \in \mathcal{F}_\Omega$.
(5) If $\phi \in \mathcal{F}_\Omega$ then $\limsup_{z' \to z} \phi(z') \equiv \tilde{\phi}(z) \in \mathcal{F}_\Omega$.
(6) If $\phi \in \mathcal{F}_{\Omega'}$ for all $\Omega' \subset\subset \Omega$ then $\phi \in \mathcal{F}_\Omega$.

H. Bremerman [2] has shown that all psh functions are Hartogs (note that the converse is trivial) provided that Ω is a domain of holomorphy (to be defined later). He also showed that it is necessary for Ω to be a domain of holomorphy in order for this assertion to hold. This answered an old question of S. Bochner and W.T. Martin [BOM].

5.3 Basic Concepts of Convexity

Capsule: Convexity is one of the most elegant and important ideas in modern mathematics. Up to two thousand years old, the idea was first formalized in a book in 1934 [BOF]. It is now a prominent feature of geometry, functional analysis, tomography, and several complex variables (to name just a few subject areas). It turns out (thanks to the Riemann mapping theorem) that convexity is not preserved under holomorphic mappings. Thus one desires a version of convexity that will be invariant. This is pseudoconvexity. We exposit this point of view here.

The concept of convexity goes back to the work of Archimedes, who used the idea in his axiomatic treatment of arc length. The notion was treated sporadically, and in an ancillary fashion, by Fermat, Cauchy, Minkowski, and others. It was not until the 1930s, however, that the first treatise on convexity [BOF] appeared. An authoritative discussion of the history of convexity can be found in [FEN].

One of the most prevalent and classical definitions of convexity is as follows: a subset $S \subseteq \mathbb{R}^N$ is said to be convex if whenever $P, Q \in S$ and $0 \leq \lambda \leq 1$ then $(1 - \lambda)P + \lambda Q \in S$. In the remainder of this book we shall refer to a set, or domain, satisfying this condition as *geometrically convex*. From the point of view of analysis, this definition is of little use. We say this because the definition is *nonquantitative, nonlocal*, and *not formulated in the language of functions*. Put slightly differently, we have known since the time of Riemann that the most useful conditions in geometry are differential conditions. Thus we wish to find a differential characterization of convexity. We begin this chapter by relating classical, geometric notions of convexity to more analytic notions. All of these ideas are properly a part of *real analysis*, so we restrict attention to $\mathbb{R}^N$.

Let $\Omega \subseteq \mathbb{R}^N$ be a domain with C^1 boundary. Let $\rho : \mathbb{R}^N \to \mathbb{R}$ be a C^1 *defining function* for Ω. Such a function has these properties:

1. $\Omega = \{x \in \mathbb{R}^N : \rho(x) < 0\}$;
2. $^c\overline{\Omega} = \{x \in \mathbb{R}^N : \rho(x) > 0\}$;
3. $\operatorname{grad} \rho(x) \neq 0 \ \forall x \in \partial\Omega$.

If $k \geq 2$ and the boundary of Ω is a regularly embedded C^k manifold in the usual sense then it is straightforward to manufacture a C^1 (indeed a C^k) defining function for Ω by using the signed distance-to-the-boundary function $(\pm\delta_\Omega(x))$:

$$\rho(x) = \begin{cases} -\delta_\Omega(x) & \text{if } x \in \Omega, \\ 0 & \text{if } x \in \partial\Omega, \\ \delta_\Omega(x) & \text{if } x \in {}^c\overline{\Omega}. \end{cases}$$

See [KRA4] for the details.

Definition 5.3.1 Let $\Omega \subseteq \mathbb{R}^N$ have C^1 boundary and let ρ be a C^1 defining function. Let $P \in \partial\Omega$. An N-tuple $w = (w_1, \ldots, w_N)$ of real numbers is called a *tangent vector* to $\partial\Omega$ at P if

$$\sum_{j=1}^{N} (\partial\rho/\partial x_j)(P) \cdot w_j = 0.$$

We write $w \in T_P(\partial\Omega)$.

We will formulate our analytic definition of convexity, and later our analytic (Levi) definition of pseudoconvexity, in terms of the defining function and tangent vectors. A more detailed discussion of these basic tools of geometric analysis may be found in [KRA4].

5.3.1 The Analytic Definition of Convexity

For convenience, we restrict attention for this subsection to *bounded* domains. Many of our definitions would need to be modified, and extra arguments given in proofs, were we to consider unbounded domains as well. We continue, for the moment, to work in $\mathbb{R}^N$.

Definition 5.3.2 Let $\Omega \subset\subset \mathbb{R}^N$ be a domain with C^2 boundary and ρ a C^2 defining function for Ω. Fix a point $P \in \partial\Omega$. We say that $\partial\Omega$ is (weakly) *analytically convex* at P if

$$\sum_{j,k=1}^{N} \frac{\partial^2 \rho}{\partial x_j \partial x_k}(P) w_j w_k \geq 0, \quad \forall w \in T_P(\partial\Omega).$$

We say that $\partial\Omega$ is *strongly analytically convex* at P if the inequality is strict whenever $w \neq 0$.

If $\partial\Omega$ is analytically convex (resp. strongly analytically convex) at each boundary point then we say that Ω is analytically convex (resp. strongly analytically convex).

The quadratic form

$$\left(\frac{\partial^2 \rho}{\partial x_j \partial x_k}(P) \right)_{j,k=1}^{N}$$

is frequently called the "real Hessian" of the function ρ. This form carries considerable geometric information about the boundary of Ω. It is of course closely related to the second fundamental form of Riemannian geometry (see [ONE]).

Now we explore our analytic notions of convexity. The first lemma is a technical one:

Prelude: The point to see in the statement of this next lemma is that we get a strict inequality not just for tangent vectors but for *all* vectors. This is why *strong* convexity is so important—because it is stable under C^2 perturbations. Without this lemma the stability would not be at all clear.

Lemma 5.3.3 *Let $\Omega \subseteq \mathbb{R}^N$ be strongly convex. Then there is a constant $C > 0$ and a defining function $\widetilde{\rho}$ for Ω such that*

$$\sum_{j,k=1}^{N} \frac{\partial^2 \widetilde{\rho}}{\partial x_j \partial x_k}(P) w_j w_k \geq C|w|^2, \quad \forall P \in \partial\Omega, \, w \in \mathbb{R}^N. \tag{5.3.3.1}$$

Proof: Let ρ be some fixed C^2 defining function for Ω. For $\lambda > 0$ define

$$\rho_\lambda(x) = \frac{\exp(\lambda \rho(x)) - 1}{\lambda}.$$

We shall select λ large in a moment. Let $P \in \partial\Omega$ and set

$$X = X_P = \left\{ w \in \mathbb{R}^N : |w| = 1 \text{ and } \sum_{j,k} \frac{\partial^2 \rho}{\partial x_j \partial x_k}(P) w_j w_k \le 0 \right\}.$$

Then no element of X could be a tangent vector at P; hence $X \subseteq \{w : |w| = 1$ and $\sum_j \partial\rho/\partial x_j(P) w_j \ne 0\}$. Since X is defined by a nonstrict inequality, it is closed; it is of course also bounded. Hence X is compact and

$$\mu \equiv \min\left\{ \left| \sum_j \partial\rho/\partial x_j(P) w_j \right| : w \in X \right\}$$

is attained and is nonzero. Define

$$\lambda = \frac{- \min_{w \in X} \sum_{j,k} \frac{\partial^2 \rho}{\partial x_j \partial x_k}(P) w_j w_k}{\mu^2} + 1.$$

Set $\widetilde{\rho} = \rho_\lambda$. Then, for any $w \in \mathbb{R}^N$ with $|w| = 1$, we have (since $\exp(\rho(P)) = 1$) that

$$\sum_{j,k} \frac{\partial^2 \widetilde{\rho}}{\partial x_j \partial x_k}(P) w_j w_k = \sum_{j,k} \left\{ \frac{\partial^2 \rho}{\partial x_j \partial x_k}(P) + \lambda \frac{\partial\rho}{\partial x_j}(P) \frac{\partial\rho}{\partial x_k}(P) \right\} w_j w_k$$

$$= \sum_{j,k} \left\{ \frac{\partial^2 \rho}{\partial x_j \partial x_k} \right\} (P) w_j w_k + \lambda \left| \sum_j \frac{\partial\rho}{\partial x_j}(P) w_j \right|^2.$$

If $w \notin X$ then this expression is positive by definition (because the first sum is). If $w \in X$ then the expression is positive by the choice of λ. Since $\{w \in \mathbb{R}^N : |w| = 1\}$ is compact, there is thus a $C > 0$ such that

$$\sum_{j,k} \left\{ \frac{\partial^2 \widetilde{\rho}}{\partial x_j \partial x_k} \right\} (P) w_j w_k \ge C, \quad \forall w \in \mathbb{R}^N \text{ such that } |w| = 1.$$

This establishes our inequality (5.3.3.1) for $P \in \partial\Omega$ fixed and w in the unit sphere of $\mathbb{R}^N$. For arbitrary w, we set $w = |w|\widehat{w}$, with $\widehat{w}$ in the unit sphere. Then (5.3.3.1) holds for $\widehat{w}$. Multiplying both sides of the inequality for $\widehat{w}$ by $|w|^2$ and performing some algebraic manipulations gives the result for fixed P and all $w \in \mathbb{R}^N$. [In the future we shall refer to this type of argument as a "homogeneity argument."]

Finally, notice that our estimates—in particular the existence of C, hold uniformly over points in $\partial\Omega$ near P. Since $\partial\Omega$ is compact, we see that the constant C may be chosen uniformly over all boundary points of Ω. $\square$

Notice that the statement of the lemma has two important features: **(i)** the constant C may be selected uniformly over the boundary and **(ii)** the inequality (5.3.3.1) holds for all $w \in \mathbb{R}^N$ (not just tangent vectors). Our proof shows in fact that (5.3.3.1) is true not just for $P \in \partial\Omega$ but for P in a neighborhood of $\partial\Omega$. It is this sort of stability of the notion of strong analytic convexity that makes it a more useful device than ordinary (weak) analytic convexity.

Prelude: We certainly believe, and the results here bear out this belief, that the analytic approach to convexity is the most useful and most meaningful. However, it is comforting and heuristically appealing to relate that less familiar idea to the better known notion of convexity that one learns in grade school. This we now do.

Proposition 5.3.4 *If Ω is strongly analytically convex then Ω is geometrically convex.*

Proof: We use a connectedness argument.

Clearly $\Omega \times \Omega$ is connected. Set $S = \{(P_1, P_2) \in \Omega \times \Omega : (1 - \lambda)P_1 + \lambda P_2 \in \Omega$, all $0 \leq \lambda \leq 1\}$. Then S is plainly open and nonempty.

To see that S is closed, fix a defining function $\widetilde{\rho}$ for Ω as in the lemma. If S is not closed in $\Omega \times \Omega$ then there exist $P_1, P_2 \in \Omega$ such that the function

$$t \mapsto \widetilde{\rho}((1 - t)P_1 + t P_2)$$

assumes an interior maximum value of 0 on $[0, 1]$. But the positive definiteness of the real Hessian of $\widetilde{\rho}$ now contradicts that assertion. The proof is therefore complete. $\square$

We gave a special proof that strong convexity implies geometric convexity simply to illustrate the utility of the strong convexity concept. It is possible to prove that an arbitrary (weakly) convex domain is geometrically convex by showing that such a domain can be written as the increasing union of strongly convex domains. However, the proof is difficult and technical (the reader interested in these matters may wish to consider them after he or she has learned the techniques in the proof of Theorem 5.5.5). We thus give another proof of this fact:

Proposition 5.3.5 *If Ω is (weakly) analytically convex then Ω is geometrically convex.*

Proof: To simplify the proof we shall assume that Ω has at least C^3 boundary.

Assume without loss of generality that $N \geq 2$ and $0 \in \Omega$. Let $0 < M \in \mathbb{R}$ be large. Let f be a defining function for Ω. For $\epsilon > 0$, let $\rho_\epsilon(x) = \rho(x) + \epsilon |x|^{2M}/M$ and $\Omega_\epsilon = \{x : \rho_\epsilon(x) < 0\}$. Then $\Omega_\epsilon \subseteq \Omega_{\epsilon'}$ if $\epsilon' < \epsilon$ and $\cup_{\epsilon>0}\Omega_\epsilon = \Omega$. If $M \in \mathbb{N}$ is large and ϵ is small, then Ω_ϵ is strongly convex. By Proposition 4.1.5, each Ω_ϵ is geometrically convex, so Ω is convex. $\square$

We mention in passing that a nice treatment of convexity, from the classical point of view, appears in [VAL].

Proposition 5.3.6 *Let $\Omega \subset\subset \mathbb{R}^N$ have C^2 boundary and be geometrically convex. Then Ω is (weakly) analytically convex.*

Proof: Seeking a contradiction, we suppose that for some $P \in \partial\Omega$ and some $w \in T_P(\partial\Omega)$ we have

$$\sum_{j,k} \frac{\partial^2 \rho}{\partial x_j \partial x_k}(P) w_j w_k = -2K < 0. \tag{5.3.6.1}$$

Suppose without loss of generality that coordinates have been selected in $\mathbb{R}^N$ such that $P = 0$ and $(0, 0, \ldots, 0, 1)$ is the unit outward normal vector to $\partial\Omega$ at P. We may further normalize the defining function ρ so that $\partial\rho/\partial x_N(0) = 1$. Let $Q = Q^t = tw + \epsilon \cdot (0, 0, \ldots, 0, 1)$, where $\epsilon > 0$ and $t \in \mathbb{R}$. Then, by Taylor's expansion,

$$\rho(Q) = \rho(0) + \sum_{j=1}^N \frac{\partial \rho}{\partial x_j}(0) Q_j + \frac{1}{2} \sum_{j,k=1}^N \frac{\partial^2 \rho}{\partial x_j \partial x_k}(0) Q_j Q_k + o(|Q|^2)$$

$$= \epsilon \frac{\partial \rho}{\partial x_N}(0) + \frac{t^2}{2} \sum_{j,k=1}^N \frac{\partial^2 \rho}{\partial x_j \partial x_k}(0) w_j w_k + \mathcal{O}(\epsilon^2) + o(t^2)$$

$$= \epsilon - K t^2 + \mathcal{O}(\epsilon^2) + o(t^2).$$

Thus if $t = 0$ and $\epsilon > 0$ is small enough then $\rho(Q) > 0$. However, for that same value of ϵ, if $|t| > \sqrt{2\epsilon/K}$ then $\rho(Q) < 0$. This contradicts the definition of geometric convexity. $\qquad\square$

Remark: The reader can already see in the proof of the proposition how useful the quantitative version of convexity can be.

The assumption that $\partial\Omega$ be C^2 is not very restrictive, for convex functions of one variable are twice differentiable almost everywhere (see [ZYG] or [EVG]). On the other hand, C^2 smoothness of the boundary is essential for our approach to the subject.

Exercise for the Reader: If $\Omega \subseteq \mathbb{R}^N$ is a domain then the *closed convex hull* of Ω is defined to be the closure of the set $\{\sum_{j=1}^m \lambda_j s_j : s_j \in \Omega, m \in \mathbb{N}, \lambda_j \geq 0, \sum \lambda_j = 1\}$. Equivalently, Ω is the intersection of all closed, convex sets that contain Ω.

Assume in the following problems that $\overline{\Omega} \subseteq \mathbb{R}^N$ is closed, bounded, and convex. Assume that Ω has C^2 boundary.

A point p in the closure $\overline{\Omega}$ of our domain is called *extreme* if it cannot be written in the form $p = (1 - \lambda)x + \lambda y$ with $x, y \in \overline{\Omega}$, $x \neq y$, and $0 < \lambda < 1$.

(a) Prove that $\overline{\Omega}$ is the closed convex hull of its extreme points (this result is usually referred to as the *Krein–Milman theorem* and is true in much greater generality).
(b) Let $P \in \partial\Omega$ be extreme. Let $\mathbf{p} = P + T_P(\partial\Omega)$ be the geometric tangent affine hyperplane to the boundary of Ω that passes through P. Show by an example that it is not necessarily the case that $\mathbf{p} \cap \overline{\Omega} = \{P\}$.

(c) Prove that if Ω_0 is *any* bounded domain with C^2 boundary then there is a relatively open subset U of $\partial\Omega_0$ such that U is strongly convex. (Hint: Fix $x_0 \in \Omega_0$ and choose $P \in \partial\Omega_0$ that is as far as possible from x_0).

(d) If Ω is a convex domain then the Minkowski functional (see [LAY]) gives a convex defining function for Ω.

5.3.2 Convexity with Respect to a Family of Functions

Let $\Omega \subseteq \mathbb{R}^N$ be a domain and let $\mathcal{F}$ be a family of real-valued functions on Ω (we do not assume in advance that $\mathcal{F}$ is closed under any algebraic operations, although often in practice it will be). Let K be a compact subset of Ω. Then the *convex hull of K in Ω with respect to $\mathcal{F}$* is defined to be

$$\hat{K}_{\mathcal{F}} \equiv \left\{ x \in \Omega : f(x) \leq \sup_{t \in K} f(t) \text{ for all } f \in \mathcal{F} \right\}.$$

We sometimes denote this hull by $\hat{K}$ when the family $\mathcal{F}$ is understood or when no confusion is possible. We say that Ω is *convex* with respect to $\mathcal{F}$ provided $\hat{K}_{\mathcal{F}}$ is compact in Ω whenever K is. When the functions in $\mathcal{F}$ are complex-valued then $|f|$ replaces f in the definition of $\hat{K}_{\mathcal{F}}$.

Prelude: The next result relates geometric convexity to a flexible notion involving families of functions. This new idea can in turn be readily connected with defining functions and other geometric artifacts.

Proposition 5.3.7 *Let $\Omega \subset\subset \mathbb{R}^N$ and let $\mathcal{F}$ be the family of real linear functions. Then Ω is convex with respect to $\mathcal{F}$ if and only if Ω is geometrically convex.*

Proof: Exercise. Use the classical definition of geometric convexity at the beginning of the section. $\square$

Proposition 5.3.8 *Let $\Omega \subset\subset \mathbb{R}^N$ be any domain. Let $\mathcal{F}$ be the family of continuous functions on Ω. Then Ω is convex with respect to $\mathcal{F}$.*

Proof: If $K \subset\subset \Omega$ and $x \notin K$ then the function $F(t) = 1/(1+|x-t|)$ is continuous on Ω. Notice that $f(x) = 1$ and $|f(k)| < 1$ for all $k \in K$. Thus $x \notin \hat{K}_{\mathcal{F}}$. Therefore $\hat{K}_{\mathcal{F}} = K$ and Ω is convex with respect to $\mathcal{F}$. $\square$

Prelude: It is important to notice that the next result is stated and proved in complex dimension 1 only. In fact, in higher dimensions, a result like this can be true only for the special class of domains known as domains of holomorphy.

Proposition 5.3.9 *Let $\Omega \subseteq \mathbb{C}$ be an open set and let $\mathcal{F}$ be the family of all functions holomorphic on Ω. Then Ω is convex with respect to $\mathcal{F}$.*

Proof: First suppose that Ω is bounded. Let $K \subset\subset \Omega$. Let r be the Euclidean distance of K to the complement of Ω. Then $r > 0$. Suppose that $w \in \Omega$ is of distance less than r from $\partial\Omega$. Choose $w' \in \partial\Omega$ such that $|w - w'| = \text{dist}(w, {}^c\Omega)$. Then the function $f(\zeta) = 1/(\zeta - w')$ is holomorphic on Ω and $|f(w)| > \sup_{\zeta \in K} |f(\zeta)|$. Hence $w \notin \hat{K}_{\mathcal{F}}$ so $\hat{K}_{\mathcal{F}} \subset\subset \Omega$. Therefore Ω is convex with respect to $\mathcal{F}$.

In case Ω is unbounded we take a large disk $D(0, R)$ containing K and notice that $\hat{K}_{\mathcal{F}}$ with respect to Ω is equal to $\hat{K}_{\mathcal{F}}$ with respect to $\Omega \cap D(0, R)$, which by the first part of the proof is relatively compact. $\qquad\square$

5.3.3 A Complex Analogue of Convexity

Our goal now is to pass from convexity to a complex-analytic analogue of convexity. We shall first express the differential condition for convexity in complex notation. Then we shall isolate the portion of the complexified expression that is invariant under biholomorphic mappings. Because of its centrality we have gone to extra trouble to put all these ideas into context.

Now fix $\Omega \subset\subset \mathbb{C}^n$ with C^2 boundary and assume that $\partial\Omega$ is convex at $P \in \partial\Omega$. If $w \in \mathbb{C}^n$ then the complex coordinates for w are of course

$$w = (w_1, \ldots, w_n) = (\xi_1 + i\eta_1, \ldots, \xi_n + i\eta_n).$$

Then it is natural to (geometrically) identify $\mathbb{C}^n$ with $\mathbb{R}^{2n}$ via the map

$$(\xi_1 + i\eta_1, \ldots, \xi_n + i\eta_n) \longleftrightarrow (\xi_1, \eta_1, \ldots, \xi_n, \eta_n).$$

Similarly[1] we identify $x = (x_1, \ldots, x_n) = (x_1 + iy_1, \ldots, x_n + iy_n) \in \mathbb{C}^n$ with $(x_1, y_1, \ldots, x_n, y_n) \in \mathbb{R}^{2n}$. If ρ is a defining function for Ω that is C^2 near P then the condition that $w \in T_P(\partial\Omega)$ is

$$\sum_j \frac{\partial\rho}{\partial x_j}\xi_j + \sum_j \frac{\partial\rho}{\partial y_j}\eta_j = 0.$$

In complex notation we may write this equation as

$$\frac{1}{2}\sum_j \left[\left(\frac{\partial}{\partial z_j} + \frac{\partial}{\partial \overline{z}_j}\right)\rho(P)\right](w_j + \overline{w}_j)$$

$$+ \frac{1}{2}\sum_j \left[\left(\frac{1}{i}\right)\left(\frac{\partial}{\partial \overline{z}_j} - \frac{\partial}{\partial z_j}\right)\rho(P)\right]\left(\frac{1}{i}\right)(w_j - \overline{w}_j) = 0.$$

But this is the same as

$$2\,\mathrm{Re}\left(\sum_j \frac{\partial\rho}{\partial z_j}(P)w_j\right) = 0. \qquad (5.3.10)$$

Again, (5.3.10) is the equation for the *real tangent space* written in complex notation.

---

[1] Strictly speaking, $\mathbb{C}^n$ is $\mathbb{R}^n \otimes_{\mathbb{R}} \mathbb{C}$. Then one equips $\mathbb{C}^n$ with a linear map J, called the *complex structure tensor*, which mediates between the algebraic operation of multiplying by i and the geometric mapping $(\xi_1, \eta_1, \ldots, \xi_n, \eta_n) \mapsto (-\eta_1, \xi_1, \ldots, -\eta_n, \xi_n)$. In this book it would be both tedious and unnatural to indulge in these niceties. In other contexts they are essential. See [WEL] for a thorough treatment of this matter.

The space of vectors w that satisfy this last equation is not closed under multiplication by i, hence is not a natural object of study for our purposes. Instead, we restrict attention in the following discussion to vectors $w \in \mathbb{C}^n$ that satisfy

$$\sum_j \frac{\partial \rho}{\partial z_j}(P)w_j = 0. \tag{5.3.11}$$

The collection of all such vectors is termed the *complex tangent space* to $\partial\Omega$ at P and is denoted by $\mathcal{T}_P(\partial\Omega)$. Clearly $\mathcal{T}_P(\partial\Omega) \subset T_P(\partial\Omega)$ but the spaces are not equal; indeed the complex tangent space is a codimension-1 proper real subspace of the ordinary (real) tangent space. The reader should check that $\mathcal{T}_P(\partial\Omega)$ is the largest complex subspace of $T_P(\partial\Omega)$ in the following sense: first, $\mathcal{T}_P(\partial\Omega)$ *is* closed under multiplication by i; second, if S is a real linear subspace of $T_P(\partial\Omega)$ that is closed under multiplication by i then $S \subseteq \mathcal{T}_P(\partial\Omega)$. In particular, when $n = 1, \Omega \subseteq \mathbb{C}^n$, and $P \in \partial\Omega$ then $\mathcal{T}_P(\partial\Omega) = \{0\}$. At some level, this last fact explains many of the differences between the function theories of one and several complex variables. To wit, the complex tangent space is the part of the differential geometry of $\partial\Omega$ that behaves rigidly under biholomorphic mappings. Since the complex tangent space in dimension 1 is vacuous, this rigidity is gone. This is why the Riemann mapping theorem is possible. Now we turn our attention to the analytic convexity condition.

The convexity condition on tangent vectors is

$$0 \le \sum_{j,k=1}^n \frac{\partial^2\rho}{\partial x_j \partial x_k}(P)\xi_j\xi_k + 2\sum_{j,k=1}^n \frac{\partial^2\rho}{\partial x_j \partial y_k}(P)\xi_j\eta_k + \sum_{j,k=1}^n \frac{\partial^2\rho}{\partial y_j \partial y_k}(P)\eta_j\eta_k$$

$$= \frac{1}{4}\sum_{j,k=1}^n \left(\frac{\partial}{\partial z_j} + \frac{\partial}{\partial \bar{z}_j}\right)\left(\frac{\partial}{\partial z_k} + \frac{\partial}{\partial \bar{z}_k}\right)\rho(P)(w_j + \bar{w}_j)(w_k + \bar{w}_k)$$

$$+ 2 \cdot \frac{1}{4}\sum_{j,k=1}^n \left(\frac{\partial}{\partial z_j} + \frac{\partial}{\partial \bar{z}_j}\right)\left(\frac{1}{i}\right)\left(\frac{\partial}{\partial \bar{z}_k} - \frac{\partial}{\partial z_k}\right)\rho(P)$$

$$\times (w_j + \bar{w}_j)\left(\frac{1}{i}\right)(w_k - \bar{w}_k)$$

$$+ \frac{1}{4}\sum_{j,k=1}^n \left(\frac{1}{i}\right)\left(\frac{\partial}{\partial \bar{z}_j} - \frac{\partial}{\partial z_j}\right)\left(\frac{1}{i}\right)\left(\frac{\partial}{\partial \bar{z}_k} - \frac{\partial}{\partial z_k}\right)\rho(P)$$

$$\times \left(\frac{1}{i}\right)(w_j - \bar{w}_j)\left(\frac{1}{i}\right)(w_k - \bar{w}_k)$$

$$= \sum_{j,k=1}^{n} \frac{\partial^2 \rho}{\partial z_j \partial z_k}(P) w_j w_k + \sum_{j,k=1}^{n} \frac{\partial^2 \rho}{\partial \overline{z}_j \partial \overline{z}_k}(P) \overline{w}_j \overline{w}_k$$

$$+ 2 \sum_{j,k=1}^{n} \frac{\partial^2 \rho}{\partial z_j \partial \overline{z}_k}(P) w_j \overline{w}_k$$

$$= 2 \operatorname{Re} \left(\sum_{j,k=1}^{n} \frac{\partial^2 \rho}{\partial z_j \partial z_k}(P) w_j w_k \right) + 2 \sum_{j,k=1}^{n} \frac{\partial^2 \rho}{\partial z_j \partial \overline{z}_k}(P) w_j \overline{w}_k. \qquad (5.3.12)$$

[This formula could also have been derived by examining the complex form of Taylor's formula.] We see that the real Hessian, when written in complex coordinates, decomposes rather naturally into two Hessian-like expressions. Our next task is to see that the first of these does not transform canonically under biholomorphic mappings but the second one does. We shall thus dub the second quadratic expression the "complex Hessian" of ρ. It will also be called the "Levi form" of the domain Ω at the point P.

The Riemann mapping theorem tells us, in part, that the unit disk is biholomorphic to any simply connected smoothly bounded planar domain. Since many of these domains are not convex, we see easily that biholomorphic mappings do not preserve convexity (an explicit example of this phenomenon is the mapping $\phi : D \to \phi(D), \phi(\zeta) = (\zeta + 2)^4$). We wish now to understand analytically where the failure lies. So let $\Omega \subset\subset \mathbb{C}^n$ be a convex domain with C^2 boundary. Let U be a neighborhood of $\overline{\Omega}$ and $\rho : U \to \mathbb{R}$ a defining function for Ω. Assume that $\Phi : U \to \mathbb{C}^n$ is biholomorphic onto its image and define $\Omega' = \Phi(\Omega)$. We now use an old idea from partial differential equations—Hopf's lemma—to get our hands on a defining function for Ω'. Hopf's lemma has classical provenance, such as [COH]. A statement is this (see also [GRK12] for an elementary proof):

Prelude: The next result, Hopf's lemma, arose originally in the proof of the maximum principle for solutions of elliptic partial differential equations. Its utility in several complex variables is a fairly recent discovery. It has proved to have both practical and philosophical significance. Notice that it is a result about *real analysis*.

Lemma 5.3.13 *Let u be a real-valued harmonic function on a bounded domain $\Omega \subseteq \mathbb{R}^N$ with C^2 boundary. Assume that u extends continuously to a boundary point $p \in \partial\Omega$. We take it that $u(p) = 0$ and $u(x) < 0$ for all other points of the domain. Let v denote the unit outward normal vector at p. Then*

$$\frac{\partial}{\partial v} u(p) > 0.$$

What is interesting here is that the conclusion with $>$ replaced by $\geq$ is immediate by the definition of the derivative.

The proof shows that Hopf's lemma is valid for subharmonic, hence for plurisubharmonic, functions. The result guarantees that $\rho' \equiv \rho \circ \Phi^{-1}$ is a defining function for Ω'. Finally, fix a point $P \in \partial\Omega$ and corresponding point $P' \equiv \Phi(P) \in \partial\Omega'$. If $w \in T_P(\partial\Omega)$ then

$$w' = \left(\sum_{j=1}^{n} \frac{\partial\Phi_1(P)}{\partial z_j} w_j, \ldots, \sum_{j=1}^{n} \frac{\partial\Phi_n(P)}{\partial z_j} w_j \right) \in T_{P'}(\partial\Omega').$$

Let the complex coordinates on $\Phi(U)$ be $z'_1, \ldots, z'_n$. Our task is to write the expression (6.3.9) determining convexity,

$$2\,\mathrm{Re}\left(\sum_{j,k=1}^{n} \frac{\partial^2\rho}{\partial z_j \partial z_k}(P)w_j w_k \right) + 2\sum_{j,k=1}^{n} \frac{\partial^2\rho}{\partial z_j \partial \bar{z}_k}(P)w_j \bar{w}_k, \tag{5.3.14}$$

in terms of the z'_j and the w'_j. But

$$\frac{\partial^2\rho}{\partial z_j \partial z_k}(P) = \frac{\partial}{\partial z_j} \sum_{\ell=1}^{n} \frac{\partial\rho'}{\partial z'_\ell} \frac{\partial\Phi_\ell}{\partial z_k}$$

$$= \sum_{\ell,m=1}^{n} \left\{ \frac{\partial^2\rho'}{\partial z'_\ell \partial z'_m} \frac{\partial\Phi_\ell}{\partial z_k} \frac{\partial\Phi_m}{\partial z_j} \right\} + \sum_{\ell=1}^{n} \left\{ \frac{\partial\rho'}{\partial z'_\ell} \frac{\partial^2\Phi_\ell}{\partial z_j \partial z_k} \right\},$$

$$\frac{\partial^2\rho}{\partial z_j \partial \bar{z}_k}(P) = \frac{\partial}{\partial z_j} \sum_{\ell=1}^{n} \frac{\partial\rho'}{\partial \bar{z}'_\ell} \frac{\partial\overline{\Phi}_\ell}{\partial \bar{z}_k} = \sum_{\ell,m=1}^{n} \frac{\partial^2\rho'}{\partial z'_m \partial \bar{z}'_\ell} \frac{\partial\Phi_m}{\partial z_j} \frac{\partial\overline{\Phi}_\ell}{\partial \bar{z}_k}.$$

Therefore

$$(5.3.14) = 2\mathrm{Re}\underbrace{\left\{ \sum_{\ell,m=1}^{n} \frac{\partial^2\rho'}{\partial z'_\ell \partial z'_m} w'_\ell w'_m + \sum_{j,k=1}^{n} \sum_{\ell=1}^{n} \frac{\partial\rho'}{\partial z'_\ell} \frac{\partial^2\Phi_\ell}{\partial z_j \partial z_k} w_j w_k \right\}}_{\text{nonfunctorial}}$$

$$+ 2\underbrace{\sum_{\ell,m=1}^{n} \frac{\partial^2\rho'}{\partial z'_m \partial \bar{z}'_\ell} w'_m \bar{w}'_\ell}_{\text{functorial}}. \tag{5.3.15}$$

So we see that the part of the quadratic form characterizing convexity that is preserved under biholomorphic mappings (compare (5.3.15) with (5.3.12)) is

$$\sum_{j,k=1}^{n} \frac{\partial^2\rho}{\partial z_j \partial \bar{z}_k}(P)w_j \bar{w}_k.$$

The other part is definitely *not* preserved. Our calculations motivate the following definition.

Definition 5.3.16 Let $\Omega \subseteq \mathbb{C}^n$ be a domain with C^2 boundary and let $P \in \partial\Omega$. We say that $\partial\Omega$ is (weakly) *Levi pseudoconvex* at P if

$$\sum_{j,k=1}^{n} \frac{\partial^2 \rho}{\partial z_j \partial \overline{z}_k}(P) w_j \overline{w}_k \geq 0, \quad \forall w \in \mathcal{T}_P(\partial\Omega). \tag{5.3.16.1}$$

The expression on the left side of (5.3.16.1) is called the *Levi form*. The point P is said to be *strongly (or strictly) Levi pseudoconvex* if the expression on the left side of (5.3.16.1) is positive whenever $w \neq 0$. A domain is called *Levi pseudoconvex* (resp. *strongly Levi pseudoconvex*) if all of its boundary points are pseudoconvex (resp. strongly Levi pseudoconvex).

The reader may check that the definition of pseudoconvexity is independent of the choice of defining function for the domain in question.

The collection of Levi pseudoconvex domains is, in a local sense to be made precise later, the smallest class of domains that contains the convex domains and is closed under increasing union and biholomorphic mappings.

Prelude: Since pseudoconvexity is a complex-analytically invariant form of convexity, it is important to relate the two ideas. This next proposition gives a direct connection.

Proposition 5.3.17 *If $\Omega \subseteq \mathbb{C}^n$ is a domain with C^2 boundary and if $P \in \partial\Omega$ is a point of convexity then P is also a point of pseudoconvexity.*

Proof: Let ρ be a defining function for Ω. Let $w \in \mathcal{T}_P(\partial\Omega)$. Then iw is also in $\mathcal{T}_P(\partial\Omega)$. If we apply the convexity hypothesis to w at P we obtain

$$2\operatorname{Re}\left(\sum_{j,k=1}^{n} \frac{\partial^2 \rho}{\partial z_j \partial z_k}(P) w_j w_k\right) + 2\sum_{j,k=1}^{n} \frac{\partial^2 \rho}{\partial z_j \partial \overline{z}_k}(P) w_j \overline{w}_k \geq 0.$$

However, if we apply the convexity condition to iw at P we obtain

$$-2\operatorname{Re}\left(\sum_{j,k=1}^{n} \frac{\partial^2 \rho}{\partial z_j \partial z_k}(P) w_j w_k\right) + 2\sum_{j,k=1}^{n} \frac{\partial^2 \rho}{\partial z_j \partial \overline{z}_k}(P) w_j \overline{w}_k \geq 0.$$

Adding these two inequalities we find that

$$4\sum_{j,k=1}^{n} \frac{\partial^2 \rho}{\partial z_j \partial \overline{z}_k}(P) w_j \overline{w}_k \geq 0;$$

hence $\partial\Omega$ is Levi pseudoconvex at P. $\qquad\qquad\square$

The converse of this lemma is false. For instance, any product of smooth domains (take annuli, for example) $\Omega_1 \times \Omega_2 \subseteq \mathbb{C}^2$ is Levi pseudoconvex at boundary points that are smooth (for instance, off the distinguished boundary $\partial\Omega_1 \times \partial\Omega_2$).

From this observation a smooth example may be obtained simply by rounding off the product domain near its distinguished boundary. The reader should carry out the details of these remarks as an exercise.

There is no elementary geometric way to think about pseudoconvex domains. The collection of convex domains forms an important subclass, but by no means a representative subclass. As recently as 1972 it was conjectured that a pseudoconvex point $P \in \partial\Omega$ has the property that there is a holomorphic change of coordinates Φ on a neighborhood U of P such that $\Phi(U \cap \partial\Omega)$ is convex. This conjecture is false (see [KON2]). In fact, it is not known which pseudoconvex boundary points are "convexifiable."

The definition of Levi pseudoconvexity can be motivated by analogy with the real-variable definition of convexity. However, we feel that the calculations above, which we learned from J.J. Kohn, provide the most palpable means of establishing the importance of the Levi form.

We conclude this discussion by noting that pseudoconvexity is not an interesting condition in one complex dimension because the complex tangent space to the boundary of a domain is always trivial (i.e., just the zero vector). Any domain in the complex plane is vacuously pseudoconvex.

5.3.4 Further Remarks about Pseudoconvexity

The discussion thus far in this chapter has shown that convexity for domains and convexity for functions are closely related concepts. We now develop the latter notion a bit further.

Classically, a real-valued function f on a convex domain Ω is called *convex* if whenever $P, Q \in \Omega$ and $0 \leq \lambda \leq 1$, we have $f((1 - \lambda)P + \lambda Q) \leq (1 - \lambda)f(P) + \lambda f(Q)$. A C^2 function f is convex according to this definition if and only if the matrix $\left(\partial^2 f/\partial x_j \partial x_k\right)_{j,k=1}^{N}$ is positive semidefinite at each point of the domain of f. Put in other words, the function is convex at a point if this Hessian matrix is positive semidefinite at that point. It is *strongly* (or strictly) *convex* at a point of its domain if the matrix is strictly positive definite at that point.

Now let $\Omega \subseteq \mathbb{R}^N$ be any domain. A function $\phi : \Omega \to \mathbb{R}$ is called an *exhaustion function* for Ω if for any $c \in \mathbb{R}$, the set $\Omega_c \equiv \{x \in \Omega : \phi(x) \leq c\}$ is relatively compact in Ω. It is a fact (not easy to prove) that Ω is convex if and only if it possesses a (strictly) convex exhaustion function, and that is true if and only if it possesses a strictly convex exhaustion function. The reference [KRA4] provides some of the techniques for proving a result like this.

We now record several logical equivalences that are fundamental to the function theory of several complex variables. We stress that we shall *not* prove any of these— the proofs, generally speaking, are simply too complex and would take too much time and space. See [KRA4] for the details. But it is useful for us to have this information at our fingertips.

$$\Omega \text{ is a domain of holomorphy} \iff$$

Ω is Levi pseudoconvex $\Longleftrightarrow$

Ω has a C^∞ strictly psh exhaustion function $\Longleftrightarrow$

the equation $\overline{\partial} u = f$ can be solved on Ω for

every $\overline{\partial}$ − closed (p, q) form f on Ω.

The hardest part of these equivalences is that a Levi pseudoconvex domain is a domain of holomorphy. This implication is known as the *Levi problem*, and was solved completely for domains in $\mathbb{C}^n$, all n only in the mid-1950s. Some generalizations of the problem to complex manifolds remain open. An informative survey is [SIU].

The next section collects a number of geometric properties of pseudoconvex domains. Although some of these properties are not needed for a while, it is appropriate to treat these facts all in one place.

6

Pseudoconvexity and Domains of Holomorphy

Prologue: Pseudoconvexity and domains of holomorphy are the two fundamental ideas in the function theory of several complex variables. The first
is a differential geometric condition on the boundary. The second is an idea
that comes strictly from function theory. The big result in the subject—the
solution of the Levi problem—is that these two conditions are equivalent.

It requires considerable machinery to demonstrate the solution of the Levi
problem, and we cannot present all of it here. What we can do is to describe
all the key ingredients. And this is important, for these ingredients make up
the fundamental tools in the subject.

6.1 Comparing Convexity and Pseudoconvexity

Capsule: As indicated in the previous section, convexity and pseudoconvexity are closely related. Pseudoconvexity is, in a palpable sense, a biholomorphically invariant version of convexity. In this section we explore the
connections between convexity and pseudoconvexity.

A straightforward calculation (see [KRA4]) establishes the following
result:

Prelude: The next proposition demonstrates, just as we did above for strong convexity, that strong pseudoconvexity is a stable property under C^2 perturbations of the
boundary. This is one of the main reasons that strong pseudoconvexity is so useful.

Proposition 6.1.1 *If Ω is strongly pseudoconvex then Ω has a defining function $\widetilde{\rho}$
such that*

$$\sum_{j,k=1}^{n} \frac{\partial^2 \widetilde{\rho}}{\partial z_j \partial \overline{z}_k}(P) w_j \overline{w}_k \geq C|w|^2$$

for all $P \in \partial\Omega$, all $w \in \mathbb{C}^n$.

S.G. Krantz, *Explorations in Harmonic Analysis*, Applied and Numerical
Harmonic Analysis, DOI 10.1007/978-0-8176-4669-1_6,
© Birkhäuser Boston, a part of Springer Science+Business Media, LLC 2009

By continuity of the second derivatives of $\tilde{\rho}$, the inequality in the proposition must in fact persist for all z in a neighborhood of $\partial\Omega$. In particular, if $P \in \partial\Omega$ is a point of strong pseudoconvexity then so are all nearby boundary points.

Example 6.1.2 Let $\Omega = \{(z_1, z_2) \in \mathbb{C}^2 : |z_1|^2 + |z_2|^4 < 1\}$. Then $\rho(z_1, z_2) = -1 + |z_1|^2 + |z_2|^4$ is a defining function for Ω and the Levi form applied to (w_1, w_2) is $|w_1|^2 + 4|z_2|^2|w_2|^2$. Thus $\partial\Omega$ is strongly pseudoconvex except at boundary points where $|z_2|^2 = 0$ and the tangent vectors w satisfy $w_1 = 0$. Of course these are just the boundary points of the form $(e^{i\theta}, 0)$. The domain is (weakly) Levi pseudoconvex at these exceptional points.

Pseudoconvexity describes something more (and less) than classical geometric properties of a domain. However, it is important to realize that there is no simple geometric description of pseudoconvex points. Weakly pseudoconvex points are far from being well understood at this time. Matters are much clearer for strongly pseudoconvex points:

Prelude: One of the biggest unsolved problems in the function theory of several complex variables is to determine which pseudoconvex boundary points may be "convexified"—in the sense that there is a biholomorphic change of coordinates that makes the point convex in some sense. We see next that, for a strongly pseudoconvex point, matters are quite simple.

Lemma 6.1.3 (Narasimhan) *Let $\Omega \subset\subset \mathbb{C}^n$ be a domain with C^2 boundary. Let $P \in \partial\Omega$ be a point of strong pseudoconvexity. Then there is a neighborhood $U \subseteq \mathbb{C}^n$ of P and a biholomorphic mapping Φ on U such that $\Phi(U \cap \partial\Omega)$ is strongly convex.*

Proof: By Proposition 6.1.1 there is a defining function $\tilde{\rho}$ for Ω such that

$$\sum_{j,k} \frac{\partial^2 \tilde{\rho}}{\partial z_j \partial \bar{z}_k}(P) w_j \bar{w}_k \geq C|w|^2$$

for all $w \in \mathbb{C}^n$. By a rotation and translation of coordinates, we may assume that $P = 0$ and that $v = (1, 0, \ldots, 0)$ is the unit outward normal to $\partial\Omega$ at P. The second-order Taylor expansion of $\tilde{\rho}$ about $P = 0$ is given by

$$\tilde{\rho}(w) = \tilde{\rho}(0) + \sum_{j=1}^{n} \frac{\partial \tilde{\rho}}{\partial z_j}(P) w_j + \frac{1}{2} \sum_{j,k=1}^{n} \frac{\partial^2 \tilde{\rho}}{\partial z_j \partial z_k}(P) w_j w_k$$

$$+ \sum_{j=1}^{n} \frac{\partial \tilde{\rho}}{\partial \bar{z}_j}(P) \bar{w}_j + \frac{1}{2} \sum_{j,k=1}^{n} \frac{\partial^2 \tilde{\rho}}{\partial \bar{z}_j \partial \bar{z}_k}(P) \bar{w}_j \bar{w}_k$$

$$+ \sum_{j,k=1}^{n} \frac{\partial^2 \tilde{\rho}}{\partial z_j \partial \bar{z}_k}(P) w_j \bar{w}_k + o(|w|^2)$$

$$= 2\,\mathrm{Re}\left\{\sum_{j=1}^{n}\frac{\partial\widetilde{\rho}}{\partial z_j}(P)w_j + \frac{1}{2}\sum_{j,k=1}^{n}\frac{\partial^2\widetilde{\rho}}{\partial z_j\partial z_k}(P)w_j w_k\right\}$$

$$+\sum_{j,k=1}^{n}\frac{\partial^2\widetilde{\rho}}{\partial z_j\partial\overline{z}_k}(P)w_j\overline{w}_k + o(|w|^2)$$

$$= 2\,\mathrm{Re}\left\{w_1 + \frac{1}{2}\sum_{j,k=1}^{n}\frac{\partial^2\widetilde{\rho}}{\partial z_j\partial z_k}(P)w_j w_k\right\}$$

$$+\sum_{j,k=1}^{n}\frac{\partial^2\widetilde{\rho}}{\partial z_j\partial\overline{z}_k}(P)w_j\overline{w}_k + o(|w|^2) \tag{6.1.3.1}$$

by our normalization $v = (1, 0, \ldots, 0)$.

Define the mapping $w = (w_1, \ldots, w_n) \mapsto w' = (w'_1, \ldots, w'_n)$ by

$$w'_1 = \Phi_1(w) = w_1 + \frac{1}{2}\sum_{j,k=1}^{n}\frac{\partial^2\widetilde{\rho}}{\partial z_j\partial z_k}(P)w_j w_k,$$

$$w'_2 = \Phi_2(w) = w_2,$$

$$\cdot \qquad \cdot$$
$$\cdot \qquad \cdot$$
$$\cdot \qquad \cdot$$

$$w'_n = \Phi_n(w) = w_n.$$

By the implicit function theorem, we see that for w sufficiently small this is a well-defined invertible holomorphic mapping on a small neighborhood W of $P = 0$. Then equation (6.1.3.1) tells us that, in the coordinate w', the defining function becomes

$$\widehat{\rho}(w') = 2\,\mathrm{Re}\,w'_1 + \sum_{j,k=1}^{n}\frac{\partial^2\widetilde{\rho}}{\partial z'_j\partial\overline{z}'_k}(P)w_j\overline{w}'_k + o(|w'|^2).$$

[Notice how the canonical transformation property of the Levi form comes into play!] Thus the real Hessian at P of the defining function $\widehat{\rho}$ is precisely the Levi form; and the latter is positive definite by our hypothesis. Hence the boundary of $\Phi(W \cap \Omega)$ is strictly convex at $\Phi(P)$. By the continuity of the second derivatives of $\widehat{\rho}$, we may conclude that the boundary of $\Phi(W \cap \Omega)$ is strictly convex in a neighborhood V of $\Phi(P)$. We now select $U \subseteq W$ a neighborhood of P such that $\Phi(U) \subseteq V$ to complete the proof. $\qquad\square$

By a very ingenious (and complicated) argument, J.E. Fornæss [1] has refined Narasimhan's lemma in the following manner:

Prelude: The Fornæss embedding theorem is quite deep and difficult. But it makes good sense, and is a very natural generalization of Narasimhan's lemma.

Theorem 6.1.4 (Fornæss) *Let $\Omega \subseteq \mathbb{C}^n$ be a strongly pseudoconvex domain with C^2 boundary. Then there are an integer $n' > n$, a strongly convex domain $\Omega' \subseteq \mathbb{C}^{n'}$, a neighborhood $\widehat{\Omega}$ of $\overline{\Omega}$, and a one-to-one embedding $\Phi : \widehat{\Omega} \to \mathbb{C}^{n'}$ such that:*

1. $\Phi(\Omega) \subseteq \Omega'$;
2. $\Phi(\partial\Omega) \subseteq \partial\Omega'$;
3. $\Phi(\widehat{\Omega} \setminus \overline{\Omega}) \subseteq \mathbb{C}^{n'} \setminus \overline{\Omega}'$;
4. $\Phi(\widehat{\Omega})$ *is transversal to* $\partial\Omega'$.

Remark: In general, $n' \gg n$ in the theorem. Sharp estimates on the size of n', in terms of the Betti numbers of Ω and other analytic data, are not known. Figure 6.1 suggests, roughly, what Fornæss's theorem says.

It is known (see [YU]) that if Ω has real analytic boundary then the domain Ω' in the theorem can be taken to have real analytic boundary and the mapping Φ will extend real analytically across the boundary (see also [FOR1]). It is not known whether if Ω is described by a polynomial defining function, the mapping Φ can be taken to be a polynomial. Sibony has produced an example of a smooth weakly pseudoconvex domain that cannot be mapped properly into any weakly convex domain of any dimension (even if we discard the smoothness and transversality-at-the-boundary part of the conclusion). See [SIB] for details. It is not known which weakly pseudoconvex domains can be properly embedded in a convex domain of some dimension.

Definition 6.1.5 An *analytic disk* in $\mathbb{C}^n$ is a nonconstant holomorphic mapping $\phi : D \to \mathbb{C}^n$. We shall sometimes intentionally confuse the embedding with its image (the latter is denoted by $\mathbf{d}$ or $\mathbf{d}_\phi$). If ϕ extends continuously to $\overline{D}$ then we call $\phi(\overline{D})$ a *closed analytic disk* and $\phi(\partial D)$ the *boundary* of the analytic disk.

Example 6.1.6 The analytic disk $\phi(\zeta) = (1, \zeta)$ lies entirely in the boundary of the bidisk $D \times D$. By contrast, the boundary of the ball contains no nontrivial (i.e., nonconstant) analytic disks.

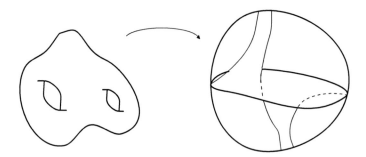

Figure 6.1. The Fornæss embedding theorem.

To see this last assertion, take the dimension to be 2. Assume that $\phi = (\phi_1, \phi_2)$: $D \to \partial B$ is an analytic disk. Let $\rho(x) = -1 + |z_1|^2 + |z_2|^2$ be a defining function for B. Then $\rho \circ \phi = -1 + |\phi_1|^2 + |\phi_2|^2$ is constantly equal to 0, or $|\phi_1(\zeta)|^2 + |\phi_2(\zeta)|^2 \equiv 1$. Each function on the left side of this identity is subharmonic. By the submean value property, if $\mathbf{d}$ is a small disk centered at $\zeta \in D$ with radius r, then

$$1 = |\phi_1(\zeta)|^2 + |\phi_2(\zeta)|^2 \leq \frac{1}{\pi r^2} \int_{\mathbf{d}} |\phi_1(\xi)|^2 + |\phi_2(\xi)|^2 \, dA(\xi) = 1.$$

Since the far left and far right of this string of inequalities are equal, we thus in fact have the equality

$$|\phi_1(\zeta)|^2 + |\phi_2(\zeta)|^2 = \frac{1}{\pi r^2} \int_{\mathbf{d}} |\phi_1(\xi)|^2 + |\phi_2(\xi)|^2 \, dA(\xi).$$

But also

$$|\phi_1(\zeta)|^2 \leq \frac{1}{\pi r^2} \int_{\mathbf{d}} |\phi_1(\xi)|^2 \, dA(\xi)$$

and

$$|\phi_2(\zeta)|^2 \leq \frac{1}{\pi r^2} \int_{\mathbf{d}} |\phi_2(\xi)|^2 \, dA(\xi).$$

It therefore must be that equality holds in each of these last two inequalities. But then, since ζ and r are arbitrary, $|\phi_1|^2$ and $|\phi_2|^2$ are harmonic. That can be true only if ϕ_1, ϕ_2 are constant.

Exercise for the Reader: Prove that the boundary of a strongly pseudoconvex domain cannot contain a nonconstant analytic disk.

In fact, more is true: If Ω is strongly pseudoconvex, $P \in \partial\Omega$, and $\phi : D \to \overline{\Omega}$ satisfies $\phi(0) = P$ then ϕ is identically equal to P.

Exercise for the Reader: There is a precise, quantitative version of the behavior of an analytic disk at a strongly pseudoconvex point. If $P \in \partial\Omega$ is a strongly pseudoconvex point then there is no analytic disk $\mathbf{d}$ with the property that

$$\lim_{\mathbf{d} \ni z \to P} \frac{\text{dist}(z, \partial\Omega)}{|z - P|^2} = 0.$$

This property distinguishes a weakly pseudoconvex boundary point from a strongly pseudoconvex boundary point. For example, the boundary point $(1, 0, 0)$ in the domain $\{z \in \mathbb{C}^3 : |z_1|^2 + |z_2|^2 + |z_3|^4 < 1\}$ has a zero eigenvalue of its Levi form in the direction $(0, 0, 1)$. Correspondingly, the analytic disk $\phi(\zeta) = (1, 0, \zeta)$ has order of contact greater than 2 (in point of fact the order of contact is 4) with the boundary at the point $(1, 0, 0)$.

6.1.1 Holomorphic Support Functions

Let $\Omega \subseteq \mathbb{C}^n$ be a domain and $P \in \partial\Omega$. We say that P possesses a *holomorphic support function* for the domain Ω provided that there is a neighborhood U_P of P

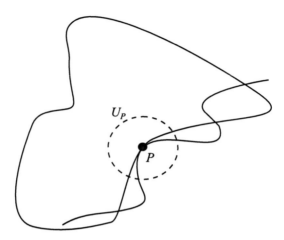

Figure 6.2. A holomorphic support function.

and a holomorphic function $f_P : U_P \to \mathbb{C}$ such that $\{z \in U_P : f_P(z) = 0\} \cap \overline{\Omega} = \{P\}$ (see Figure 6.2). Compare the notion of holomorphic support function with the classical notion of support line or support hypersurface for a convex body (see [VAL] or [LAY]).

Suppose now $\Omega \subseteq \mathbb{C}^n$ and that $P \in \partial\Omega$ is a point of strong convexity. Further assume that $T_P(\partial\Omega)$ denotes the ordinary, $(2n-1)$-dimensional, real tangent hyperplane to $\partial\Omega$ at P. Then there is a neighborhood U_P of P such that $T_P(\partial\Omega) \cap \overline{\Omega} \cap U_P = \{P\}$ (exercise). Identify $\mathbb{C}^n$ with $\mathbb{R}^{2n}$ in the usual way. Assume, for notational simplicity, that $P = 0$. Let $(a_1, b_1, \ldots, a_n, b_n) \simeq (a_1 + ib_1, \ldots, a_n + ib_n) \equiv (\alpha_1, \ldots, \alpha_n) = \alpha$ be the unit outward normal to $\partial\Omega$ at P. Then we may think of $T_P(\partial\Omega)$ as $\{(x_1, y_1, \ldots, x_n, y_n) : \sum_{j=1}^{n} a_j x_j + b_j y_j = 0\}$. Equivalently, identifying $(x_1, y_1, \ldots, x_n, y_n)$ with $(z_1, \ldots, z_n)$, we may identify $T_P(\partial\Omega)$ with

$$\left\{ (z_1, \ldots, z_n) : \mathrm{Re} \sum_{j=1}^{n} z_j \overline{\alpha}_j = 0 \right\}.$$

Let $f(z) = \sum z_j \overline{\alpha}_j = z \cdot \overline{\alpha}$. [The notation $\langle z, \alpha \rangle$ is used in some contexts in place of $z \cdot \overline{\alpha}$.] Then f is plainly holomorphic on $\mathbb{C}^n$ and f is a support function for Ω at P since the zero set of f lies in $T_P(\partial\Omega)$. The next proposition now follows from Narasimhan's lemma:

Prelude: In the 1970s there was great interest in support functions and peak functions. Here f is a peak function at $P \in \partial\Omega$ if **(i)** f is continuous on $\overline{\Omega}$, **(ii)** f is holomorphic on Ω, **(iii)** $f(P) = 1$, and **(iv)** $|f(z)| < 1$ for $z \in \overline{\Omega} \setminus \{P\}$. See [GAM] and Section 6.4.2 for more about peak functions. Often one can use a support function to manufacture a peak function. And peak functions are useful in function-algebraic considerations. Unfortunately there are no useful necessary and sufficient conditions for the existence of either support or peak functions; the problem seems to be quite difficult, and interest in the matter has waned.

Proposition 6.1.7 *If $\Omega \subseteq \mathbb{C}^n$ is a domain and $P \in \partial\Omega$ is a point of strong pseudo-convexity, then there exists a holomorphic support function for Ω at P.*

As already noted, the proposition follows immediately from Narasimhan's lemma. But the phenomenon of support functions turns out to be so important that we now provide a separate, self-contained proof:

Proof of the Proposition: Let ρ be a defining function for Ω with the property that

$$\sum_{j,k=1}^{n} \frac{\partial^2 \rho}{\partial z_j \bar{z}_k}(P) w_j \bar{w}_k \geq C|w|^2$$

for all $w \in \mathbb{C}^n$. Define

$$f(z) = \sum_{j=1}^{n} \frac{\partial \rho}{\partial z_j}(P)(z_j - P_j) + \frac{1}{2} \sum_{j,k=1}^{n} \frac{\partial^2 \rho}{\partial z_j \partial z_k}(P)(z_j - P_j)(z_k - P_k).$$

We call f the *Levi polynomial* at P. The function f is obviously holomorphic. We claim that f is a support function for Ω at P. To see this, we expand ρ in a Taylor expansion about P :

$$\rho(z) = 2\,\mathrm{Re}\left\{\sum_{j=1}^{n} \frac{\partial \rho}{\partial z_j}(P)(z_j - P_j) + \frac{1}{2} \sum_{j,k=1}^{n} \frac{\partial^2 \rho}{\partial z_j \partial z_k}(P)(z_j - P_j)(z_k - P_k)\right\}$$

$$+ \sum_{j,k=1}^{n} \frac{\partial^2 \rho}{\partial z_j \partial \bar{z}_k}(P)(z_j - P_j)(\overline{z_k - P_k}) + o(|z - P|^2)$$

$$= 2\,\mathrm{Re}\,f(z) + \sum_{j,k=1}^{n} \frac{\partial^2 \rho}{\partial z_j \partial \bar{z}_k}(P)(z_j - P_j)(\overline{z_k - P_k}) + o(|z - P|^2).$$

Note that $\rho(P) = 0$ so there is no constant term. Now let z be a point at which $f(z) = 0$. Then we find that

$$\rho(z) = \sum_{j,k=1}^{n} \frac{\partial^2 \rho}{\partial z_j \partial \bar{z}_k}(P)(z_j - P_j)(\overline{z_k - P_k}) + o(|z - P|^2)$$

$$\geq C|z - P|^2 + o(|z - P|^2).$$

Obviously if z is sufficiently closed to P then we find that

$$\rho(z) \geq \frac{C}{2}|z - P|^2.$$

Thus if z is near P and $f(z) = 0$ then either $\rho(z) > 0$, which means that z lies outside $\overline{\Omega}$, or $z = P$. But this means precisely that f is a holomorphic support function for Ω at P. $\qquad\square$

Example 6.1.8 Let $\Omega = D^3(0, 1) \subseteq \mathbb{C}^3$. Then

$$\partial\Omega = (\partial D \times D \times D) \cup (D \times \partial D \times D) \cup (D \times D \times \partial D).$$

In particular, $\partial\Omega$ contains the entire bidisk $\mathbf{d} = \{(\zeta_1, \zeta_2, 1) : |\zeta_1| < 1, |\zeta_2| < 1\}$. The point $P = (0, 0, 1)$ is the center of $\mathbf{d}$. If there were a support function f for Ω at P then $f|_{\mathbf{d}}$ would have an isolated zero at P. That is impossible for a function of two complex variables.[1]

On the other hand, the function $g(z_1, z_2, z_3) = z_3 - 1$ is a *weak support function* for Ω at P in the sense that $g(P) = 0$ and $\{z : g(z) = 0\} \cap \overline{\Omega} \subseteq \partial\Omega$.

If $\Omega \subseteq \mathbb{C}^n$ is any (weakly) convex domain and $P \in \partial\Omega$, then $T_P(\partial\Omega) \cap \overline{\Omega} \subseteq \partial\Omega$. As above, a weak support function for Ω at P can therefore be constructed.

As recently as 1972 it was hoped that a weakly pseudoconvex domain would have at least a weak support function at each point of its boundary. These hopes were dashed by the following theorem:

Prelude: As recently as 1972 the experts thought that any smooth Levi pseudoconvex point could be convexified. That is, in analogy with Narasim- han's lemma, it was supposed that a biholomorphic change of coordinates could be instituted that would make a pseudoconvex point convex. These hopes were dashed by the dramatic but simple example of Kohn and Nirenberg. This is a real analytic, pseudoconvex point in the boundary of a domain in $\mathbb{C}^2$ such that any complex variety passing through the point weaves in and out of the domain infinitely many times in any small neighborhood. The Kohn–Nirenberg example has proved to be one of the most important and influential artifacts of the subject.

Theorem 6.1.9 (Kohn and Nirenberg) *Let*

$$\Omega = \left\{ (z_1, z_2) \in \mathbb{C}^2 : \operatorname{Re} z_2 + |z_1 z_2|^2 + |z_1|^8 + \frac{15}{7}|z_1|^2 \operatorname{Re} z_1^6 < 0 \right\}.$$

Then Ω is strongly pseudoconvex at every point of $\partial\Omega$ except 0 (where it is weakly pseudoconvex). However, there is no weak holomorphic support function (and hence no support function) for Ω at 0. More precisely, if f is a function holomorphic in a neighborhood U of 0 such that $f(0) = 0$, then for every neighborhood V of zero, f vanishes both in $V \cap \Omega$ and in $V \cap {}^c\overline{\Omega}$.

Proof: The reader should verify the first assertion. For the second, consult [KON2]. $\qquad\square$

---

[1] Some explanation is required here. The celebrated *Hartogs extension phenomenon* says that if f is holomorphic on the domain $\Omega = B(0, 2) \setminus \overline{B}(0, 1)$ then f continues analytically to all of $B(0, 2)$. It follows that a holomorphic function h cannot have an isolated zero, because then $1/h$ would have an isolated singularity.

Necessary and sufficient conditions for the existence of holomorphic support functions, or of weak holomorphic support functions, are not known.

6.1.2 Peaking Functions

The ideas that we have been presenting are closely related to questions about the existence of peaking functions in various function algebras on pseudoconvex domains. We briefly summarize some of these. Let $\Omega \subseteq \mathbb{C}^n$ be a pseudoconvex domain.

If $\mathcal{A}$ is any algebra of functions on $\overline{\Omega}$, we say that $P \in \overline{\Omega}$ is a *peak point* for $\mathcal{A}$ if there is a function $f \in \mathcal{A}$ satisfying $f(P) = 1$ and $|f(z)| < 1$ for all $z \in \overline{\Omega} \setminus \{P\}$. Let $\mathcal{P}(\mathcal{A})$ denote the set of all peak points for the algebra $\mathcal{A}$.

Recall that $A(\Omega)$ is the subspace of $C(\overline{\Omega})$ consisting of those functions that are holomorphic on Ω. Let $A^j(\Omega) = A(\Omega) \cap C^j(\overline{\Omega})$. The maximum principle for holomorphic functions implies that any peak point for any subalgebra of $A(\Omega)$ must lie in $\partial\Omega$. If Ω is Levi pseudoconvex with C^∞ boundary then $\mathcal{P}(A(\Omega))$ is contained in the closure of the strongly pseudoconvex points (see [BASE]). Also the Šilov boundary for the algebra $A(\Omega)$ is equal to the closure of the set of peak points (this follows from classical function algebra theory—see [GAM]), which in turn equals the closure of the set of strongly pseudoconvex points (see [BASE]).

The Kohn–Nirenberg domain has no peak function at 0 that extends to be holomorphic in a neighborhood of 0 (for if f were a peak function then $f(z) - 1$ would be a holomorphic support function at 0). It is shown [HAS] that the same domain has no peak function at 0 for the algebra $A^8(\Omega)$. Fornæss [FOR2] has refined the example to exhibit a domain Ω that is strongly pseudoconvex except at one boundary point P but has no peak function at P for the algebra $A^1(\Omega)$. There is no known example of a smooth, pseudoconvex boundary point that is not a peak point for the algebra $A(\Omega)$. The deepest work to date on peaking functions is [BEDF1]. See also the more recent approaches in [FOSI] and in [FOM]. It is desirable to extend that work to a larger class of domains, and in higher dimensions, but that program seems to be intractable. See [YU] for some progress in the matter.

It is reasonable to hypothesize (see the remarks in the last paragraph about the Kohn–Nirenberg domain) that if $\Omega \subseteq \mathbb{C}^n$ has C^∞ boundary and a holomorphic support function at $P \in \partial\Omega$ then there is a neighborhood U_P of P and a holomorphic function $f : U_P \cap \Omega \to \mathbb{C}$ with a smooth extension to $\overline{U_P \cap \Omega}$ such that f peaks at P. This was proved false by [BLO]. However, this conjecture *is* true at a strongly pseudoconvex point.

The problem of the existence of peaking functions is still a matter of great interest, for peak functions can be used to study the boundary behavior of holomorphic mappings (see [BEDF2]). Recently [FOR3] modified the peaking function construction of [BEDF1] to construct reproducing formulas for holomorphic functions.

Exercise for the Reader: Show that if $\Omega \subseteq \mathbb{C}$ is a domain with boundary consisting of finitely many closed C^j Jordan curves (each of which closes up to be C^j)

then every $P \in \partial\Omega$ is a peak point for $A^{j-1}(\Omega)$. (Hint: The problem is local. The Riemann mapping of a simply connected Ω to D extends $C^{j-\epsilon}$ to the boundary. See [KEL], [TSU], [BEK].)

6.2 Pseudoconvexity by Way of Analytic Disks

Capsule: Certainly one of the most powerful and flexible tools in the analysis of several complex variables is the analytic disk. An analytic disk is simply a holomorphic mapping of the unit disk into $\mathbb{C}^n$. Analytic disks measure a holomorphically invariant version of convexity—namely pseudoconvexity—and detect other subtle phenomena such as tautness of a domain. An analytic disk may be used to detect pseudoconvexity in much the same way that a line segment may be used to detect convexity (see, for example [KRA14]). We explore these ideas in the present section.

At this point the various components of our investigations begin to converge to a single theme. In particular, we shall directly relate the notions of pseudoconvexity, plurisubharmonicity, and domains of holomorphy. On the one hand, the theorems that we now prove are fundamental. The techniques in the proofs are basic to the subject. On the other hand, the proofs are rather long and tedious. Because the main focus of the present book is harmonic analysis—*not* several complex variables—we shall place the proof of Theorems 6.2.5 and 6.3.6 at the ends of their respective sections. That way the reader may concentrate on the exposition and dip into the proofs as interest dictates.

In order to effect the desired unity—exhibiting the synthesis between pseudo-convexity and domains of holomorphy—we need a second notion of pseudoconvexity. This, in turn, requires some preliminary terminology:

Definition 6.2.1 A continuous function $\mu : \mathbb{C}^n \to \mathbb{R}$ is called a *distance function* if

1. $\mu \geq 0$;
2. $\mu(z) = 0$ if and only if $z = 0$;
3. $\mu(tz) = |t|\mu(z)$, $\forall t \in \mathbb{C}, z \in \mathbb{C}^n$.

Definition 6.2.2 Let $\Omega \subseteq \mathbb{C}^n$ be a domain and μ a distance function. For any $z \in \mathbb{C}^n$, define

$$\mu_\Omega(z) = \mu(z, {}^c\Omega) = \inf_{w \in {}^c\Omega} \mu(z - w).$$

If $X \subseteq \Omega$ is a set, we write

$$\mu_\Omega(z) = \inf_{z \in X} \mu_\Omega(z).$$

It is elementary to verify that the function μ_Ω is continuous. In the special case that $\mu(z) = |z|$, one checks that in fact μ_Ω must satisfy a classical Lipschitz condition with norm 1. Moreover, for this special μ, it turns out that when Ω has

C^j boundary, $j \geq 2$, then μ_Ω is a one-sided C^j function near $\partial\Omega$ (see [KRP1] or [GIL]). The assertion is false for $j = 1$.

Definition 6.2.3 Let $\Omega \subseteq \mathbb{C}^n$ be a (possibly unbounded) domain (with, possibly, unsmooth boundary). We say that Ω is *Hartogs pseudoconvex* if there is a distance function μ such that $-\log \mu_\Omega$ is plurisubharmonic on Ω.

This new definition is at first obscure. What do these different distance functions have to do with complex analysis? It turns out that they give us a flexibility that we shall need in characterizing domains of holomorphy. In practice we think of a Hartogs pseudoconvex domain as a domain that has a (strictly) plurisubharmonic exhaustion function. [The equivalence of this characterization with Definition 6.2.3 requires proof.] Theorem 6.2.5 below will clarify matters. In particular, we shall prove that a given domain Ω satisfies the definition of "Hartogs pseudoconvex" for one distance function if and only if it does so for all distance functions. The thing to note is that Hartogs pseudoconvexity makes sense on a domain regardless of the boundary smoothness; Levi pseudoconvexity requires C^2 boundary so that we can make sense of the Levi form.

In what follows we shall let $d_\Omega(z)$ denote the Euclidean distance of the point z to the boundary of the domain Ω.

Prelude: It is natural to wonder why "domain of holomorphy" and "pseudoconvex" are not discussed for domains in complex dimension one. The answer is that *every* domain in $\mathbb{C}^1$ is a domain of holomorphy, *every* domain is Hartogs pseudoconvex, and every C^2 domain is (vacuously) Levi pseudoconvex. So there is nothing to discuss.

Proposition 6.2.4 *Let $\Omega \subseteq \mathbb{C}$ be any planar domain. Then Ω is Hartogs pseudoconvex.*

Proof: We use the Euclidean distance $\delta(z) = |z|$. Let $\overline{D}(z_0, r) \subseteq \Omega$ and let h be real-valued and harmonic on a neighborhood of this closed disk. Assume that $h \geq -\log d_\Omega$ on $\partial D(z_0, r)$. Let $\widetilde{h}$ be a real harmonic conjugate for h on a neighborhood of $\overline{D}(z_0, r)$. So $h + i\widetilde{h}$ is holomorphic on $D(z_0, r)$ and continuous on $\overline{D}(z_0, r)$. Fix, for the moment, a point $P \in \partial\Omega$. Then

$$- \log d_\Omega(z) \leq h(z), \qquad\qquad z \in \partial D(z_0, r),$$

$$\Rightarrow \left| \exp\left(-h(z) - i\widetilde{h}(z)\right) \right| \leq d_\Omega(z), \qquad z \in \partial D(z_0, r),$$

$$\Rightarrow \left| \frac{\exp\left(-h(z) - i\widetilde{h}(z)\right)}{z - P} \right| \leq 1, \qquad z \in \partial D(z_0, r).$$

But the expression in absolute value signs is holomorphic on $D(z_0, r)$ and continuous on $\overline{D}(z_0, r)$. Hence

$$\left| \frac{\exp\left(-h(z) - i\widetilde{h}(z)\right)}{z - P} \right| \le 1, \quad \forall z \in \overline{D}(z_0, r).$$

Unwinding this inequality yields that

$$-\log|z - P| \le h(z), \quad \forall z \in \overline{D}(z_0, r).$$

Choosing for each $z \in \Omega$ a point $P = P_z \in \partial\Omega$ with $|z - P| = d_\Omega(z)$ yields now that

$$-\log d_\Omega(z) \le h(z).$$

It follows that $-\log d_\Omega$ is subharmonic; hence the domain Ω is Hartogs pseudo-convex. $\square$

Exercise for the Reader: Why does the proof of Proposition 6.2.4 break down when the dimension is two or greater?

Prelude: Pseudoconvexity is one of the fundamental ideas in the function theory of several complex variables. The next result, giving ten equivalent formulations of pseudoconvexity, is important **(a)** because these different formulations are actually used in practice and **(b)** because the *proofs* of the equivalences are some of the most fundamental techniques in the subject. In particular, the prominent role of plurisub-harmonic functions comes to the fore here.

Theorem 6.2.5 *Let $\Omega \subseteq \mathbb{C}^n$ be any connected open set. The following ten properties are then equivalent. [Note, however, that property* **(8)** *makes sense only when the boundary is C^2.]*

(1) $-\log \mu_\Omega$ *is plurisubharmonic on Ω for any distance function μ.*
(2) Ω *is Hartogs pseudoconvex.*
(3) *There exists a continuous plurisubharmonic (i.e., pseudoconvex) function Φ on Ω such that for every $c \in \mathbb{R}$, we have $\{z \in \Omega : \Phi(z) < c\} \subset\subset \Omega$.*
(4) *Same as* **(3)** *for a C^∞ strictly plurisubharmonic exhaustion function Φ.*
(5) *Let $\{d_a\}_{a \in A}$ be a family of closed analytic disks in Ω. If $\cup_{a \in A} \partial d_a \subset\subset \Omega$, then $\cup_{a \in A} d_a \subset\subset \Omega$. (This assertion is called the Kontinuitätssatz.)*
(6) *If μ is any distance function and if $\mathbf{d} \subseteq \Omega$ is any closed analytic disk then $\mu_\Omega(\partial\mathbf{d}) = \mu_\Omega(\mathbf{d})$.*
(7) *Same as* **(6)** *for just one particular distance function.*
(8) Ω *is Levi pseudoconvex.*
(9) $\Omega = \cup\Omega_j$, *where each Ω_j is Hartogs pseudoconvex and $\Omega_j \subset\subset \Omega_{j+1}$.*
(10) *Same as* **(9)** *except that each Ω_j is a bounded, strongly Levi pseudoconvex domain with C^∞ boundary.*

Remark: The strategy of the proof is shown in Figure 6.3.

Some parts of the proof are rather long. However, this proof (presented at the end of the section) contains many of the basic techniques of the theory of several complex variables. This is material that is worth mastering.

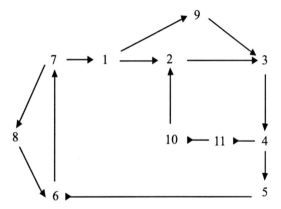

Figure 6.3. Scheme of the proof of Theorem 6.2.5.

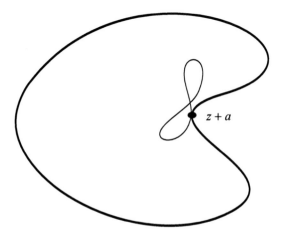

Figure 6.4. Levi pseudoconvexity.

Notice that the hypothesis of C^2 boundary is used only in the proof that $(\mathbf{1}) \Rightarrow$ $(\mathbf{8}) \Rightarrow (\mathbf{3})$. The implication $(\mathbf{1}) \Rightarrow (\mathbf{3})$ is immediate for any domain Ω.

Remark: The last half of the proof of $(\mathbf{8}) \Rightarrow (\mathbf{3})$ explains what is geometrically significant about Levi pseudoconvexity. For it is nothing other than a classical convexity condition along complex tangential directions. This convexity condition may be computed along analytic disks that are tangent to the boundary. With this in mind, we see that Figure 6.4 already suggests that Levi pseudoconvexity does *not* hold at a boundary point with a certain indicative geometry—think of an appropriate submanifold of the boundary as locally the graph of a function over the analytic disk.

We now formulate some useful consequences of the theorem, some of which are restatements of the theorem and some of which go beyond the theorem. Note that we may now use the word "pseudoconvex" to mean *either* Hartogs's notion or Levi's notion (at least for domains with C^2 boundary).

Proposition 6.2.6 *Let $\Omega_j \subseteq \mathbb{C}^n$ be pseudoconvex domains. If $\Omega \equiv \cap_{j=1}^{\infty} \Omega_j$ is open and connected, then Ω is pseudoconvex.*

Proof: Use part **(1)** of the theorem together with the Euclidean distance function. $\square$

Proposition 6.2.7 *If $\Omega_1 \subseteq \Omega_2 \subseteq \cdots$ are pseudoconvex domains then $\cup_j \Omega_j$ is pseudoconvex.*

Proof: Exercise. $\square$

Proposition 6.2.8 *Let $\Omega \subseteq \mathbb{C}^n$. Let $\Omega' \subseteq \mathbb{C}^{n'}$ be pseudoconvex, and assume that $\phi :$ $\Omega \rightarrow \Omega'$ is a surjective (but not necessarily injective) proper holomorphic mapping. Then Ω is pseudoconvex.*

Proof: Let $\Phi' : \Omega' \rightarrow \mathbb{R}$ be a pseudoconvex (continuous, plurisubharmonic) exhaustion function for Ω'. Let $\Phi \equiv \Phi' \circ \phi$. Then Φ is pseudoconvex. Let $\Omega'_c \equiv (\Phi')^{-1}\big((-\infty, c)\big)$. Then Ω'_c is relatively compact in Ω' since Φ' is an exhaustion function. Then $\Omega_c \equiv \Phi^{-1}\big((-\infty, c)\big) = \phi^{-1}(\Omega'_c)$. Thus Ω_c is relatively compact in Ω by the properness of ϕ. We conclude that Ω is pseudoconvex. $\square$

Check that Proposition 6.2.8 fails if the hypothesis of properness of ϕ is omitted.

Prelude: It is the *localness* of pseudoconvexity, more than anything, that makes it so useful. The property of being a domain of holomorphy is *not* obviously local—although it turns out in the end that it actually is local. One of the principal reasons that the solution of the Levi problem turns out to be so difficult is the dialectic about locality.

Proposition 6.2.9 *Hartogs pseudoconvexity is a local property. More precisely, if $\Omega \subseteq \mathbb{C}^n$ is a domain and each $P \in \partial\Omega$ has a neighborhood U_P such that $U_P \cap \Omega$ is Hartogs pseudoconvex, then Ω is Hartogs pseudoconvex.*

Proof: Since $U_P \cap \Omega$ is pseudoconvex, $-\log d_{U_P \cap \Omega}$ is psh on $U_P \cap \Omega$ (here δ is Euclidean distance). But for x sufficiently near P, $-\log d_{U_P \cap \Omega} = -\log d_\Omega$. It follows that $-\log d_\Omega$ is psh near $\partial\Omega$, say on $\Omega \setminus F$, where F is a closed subset of Ω. Let $\phi : \mathbb{C}^n \rightarrow \mathbb{R}$ be a convex increasing function of $|z|^2$ that satisfies $\phi(z) \rightarrow \infty$ when $|z| \rightarrow \infty$ and $\phi(z) > -\log d_\Omega(z)$ for all $x \in F$. Then the function

$$\Phi(z) = \max\{\phi(z), -\log d_\Omega(z)\}$$

is continuous and plurisubharmonic on Ω and is also an exhaustion function for Ω. Thus Ω is a pseudoconvex domain. $\square$

Remark: If $\partial\Omega$ is C^2 then there is an alternative (but essentially equivalent) proof of the last proposition as follows: Fix $P \in \partial\Omega$. Since $U_P \cap \Omega$ is pseudoconvex, the part of its boundary that it shares with $\partial\Omega$ is Levi pseudoconvex. Hence the Levi form at P is positive semidefinite. But P was arbitrary, hence each boundary point of Ω is Levi pseudoconvex. The result follows.

It is essentially tautological that Levi pseudoconvexity is a local property.

Proof of Theorem 6.2.5:

(2) $\Rightarrow$ **(3)** If Ω is unbounded then it is possible that $-\log\mu_\Omega(z)$ is not an exhaustion function (although, by hypothesis, it *is* psh). Thus we set $\Phi(z) = -\log\mu_\Omega(z) + |z|^2$, where μ is given by **(2)**. Then Φ will be a psh *exhaustion*.

(9) $\Rightarrow$ **(3)** We are assuming that Ω has C^2 boundary. If **(3)** is false, then the Euclidean distance function $d_\Omega(z) \equiv \text{dist}(z, {}^c\Omega)$ (which is C^2 on $U \cap \overline{\Omega}$, U a tubular neighborhood of $\partial\Omega$; see Exercise 4 at the end of the chapter) has the property that $-\log\delta_\Omega$ is *not* psh.

So plurisubharmonicity of $-\log d_\Omega$ fails at some $z \in \Omega$. Since Ω has a psh exhaustion function if and only if it has one defined near $\partial\Omega$ (exercise), we may as well suppose that $z \in U \cap \Omega$. So the complex Hessian of $-\log d_\Omega$ has a negative eigenvalue at z. Quantitatively, there is a direction w such that

$$\frac{\partial^2}{\partial\zeta\partial\bar\zeta}\log d_\Omega(z + \zeta w)\Big|_{\zeta=0} \equiv \lambda > 0.$$

To exploit this, we let $\phi(\zeta) = \log d_\Omega(z + \zeta w)$ and examine the Taylor expansion about $\zeta = 0$:

$$\log d_\Omega(z + \zeta w) = \phi(\zeta) = \phi(0) + 2\,\text{Re}\left\{\frac{\partial\phi}{\partial\zeta}(0)\cdot\zeta + \frac{\partial^2\phi}{\partial\zeta^2}(0)\cdot\frac{\zeta^2}{2}\right\}$$

$$+ \frac{\partial^2\phi}{\partial\zeta\partial\bar\zeta}(0)\cdot\zeta\cdot\bar\zeta + o(|\zeta|^2)$$

$$\equiv \log d_\Omega(z) + \text{Re}\{A\zeta + B\zeta^2\} + \lambda|\zeta|^2 + o(|\zeta|^2), \qquad (6.2.5.1)$$

where A, B are defined by the last equality.

Now choose $a \in \mathbb{C}^n$ such that $z + a \in \partial\Omega$ and $|a| = d_\Omega(z)$. Define the function

$$\psi(\zeta) = z + \zeta w + a\exp(A\zeta + B\zeta^2), \quad \zeta \in \mathbb{C} \text{ small,}$$

and notice that $\psi(0) = z + a \in \partial\Omega$. Also, by (6.2.5.1),

$$d_\Omega(\psi(\zeta)) \geq d_\Omega(z + \zeta w) - |a|\cdot|\exp(A\zeta + B\zeta^2)|$$

$$\geq d_\Omega(z)\cdot|\exp(A\zeta + B\zeta^2)|\exp\left(\lambda|\zeta|^2 + o(|\zeta|^2)\right)$$

$$- |a|\cdot|\exp(A\zeta + B\zeta^2)|$$

$$\geq |a| \cdot |\exp(A\zeta + B\zeta^2)| \left\{ \exp(\lambda|\zeta|^2/2) - 1 \right\}$$

$$\geq 0 \tag{6.2.5.2}$$

if ζ is small. These estimates also show that, up to reparametrization, ψ describes an analytic disk *that is contained in* $\overline{\Omega}$ and that is internally tangent to $\partial\Omega$. The disk intersects $\partial\Omega$ at the single point $z + a$ (see Figure 6.5).

On geometrical grounds, then, $(\partial/\partial\zeta)(d_\Omega \circ \psi)(0) = 0$. In order for (6.2.5.2) to hold, it must then be that $(\partial^2/\partial\zeta\partial\overline{\zeta})(d_\Omega \circ \psi)(0) > 0$ since the term $2\,\mathrm{Re}\,[(\partial^2/\partial\zeta^2) (d_\Omega \circ \psi)(0)\zeta^2]$ in the Taylor expansion of $d_\Omega \circ \psi$ is not of constant sign.

We have proved that the defining function

$$\rho(z) = \begin{cases} -d_\Omega(z) \text{ if } z \in \overline{\Omega} \cap U, \\ d_{c\Omega}(z) \text{ if } z \in {}^c\overline{\Omega} \cap U, \end{cases}$$

does not satisfy the Levi pseudoconvexity condition at $z + a = \psi(0) \in \partial\Omega$ in the tangent direction $\psi'(0)$, which is a contradiction. (See the important remark at the end of the proof of the theorem.)

(3) $\Rightarrow$ **(4)** Let Φ be the psh function whose existence is hypothesized in **(3)**. Let $\Omega_c \equiv \{z \in \Omega : \Phi(z) + |z|^2 < c\}, c \in \mathbb{R}$. Then each $\Omega_c \subset\subset \Omega$, by the definition of exhausting function, and $c' > c$ implies that $\Omega_c \subset\subset \Omega_{c'}$. Let $0 \leq \phi \in C_c^\infty(\mathbb{C}^n)$, $\int \phi = 1$, ϕ polyradial [i.e., $\phi(z_1, \ldots, z_n) = \phi(|z_1|, \ldots, |z_n|)$]. We may assume that ϕ is supported in $B(0, 1)$. Pick $\epsilon_j > 0$ such that $\epsilon_j < \mathrm{dist}(\Omega_{j+1}, \partial\Omega)$. For $z \in \Omega_{j+1}$, set

$$\Phi_j(z) = \int_\Omega [\Phi(\zeta) + |\zeta|^2]\epsilon_j^{-2n}\phi\left((z - \zeta)/\epsilon_j\right)dV(\zeta) + |z|^2 + 1.$$

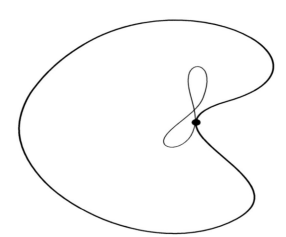

Figure 6.5. Analytic disks.

Then Φ_j is C^∞ and strictly psh on Ω_{j+1} (use Proposition 2.1.12). We know that $\Phi_j(\zeta) > \Phi(\zeta) + |\zeta|^2$ on $\overline{\Omega}_j$. Let $\chi \in C^\infty(\mathbb{R})$ be a convex function with $\chi(t) = 0$ when $t \le 0$ and $\chi'(t), \chi''(t) > 0$ when $t > 0$. Observe that $\Psi_j(z) \equiv \chi(\Phi_j(z) - (j-1))$ is positive and psh on $\Omega_j \setminus \overline{\Omega}_{j-1}$ and is, of course, C^∞. Now we inductively construct the desired function Φ'. First, $\Phi_0 > \Phi$ on Ω_0. If a_1 is large and positive, then $\Phi'_1 = \Phi_0 + a_1 \Psi_1 > \Phi$ on Ω_1. Inductively, if $a_1, \ldots, a_{\ell-1}$ have been chosen, select $a_\ell > 0$ such that $\Phi'_\ell = \Phi_0 + \sum_{j=1}^{\ell} a_j \Psi_j > \Phi$ on Ω_ℓ. Since $\Psi_{\ell+k} = 0$ on $\Omega_\ell, k > 0$, we see that $\Phi'_{\ell+k} = \Phi'_{\ell+k'}$ on Ω_ℓ for any $k, k' > 0$. So the sequence Φ'_ℓ stabilizes on compacta and $\Phi' \equiv \lim_{\ell \to \infty} \Phi'_\ell$ is a C^∞ strictly psh function that majorizes Φ. Hence Φ' is the smooth, strictly psh exhaustion function that we seek.

(4) $\Rightarrow$ (5) Immediate from the definition of convexity with respect to a family of functions.

(5) $\Rightarrow$ (6) Let $\mathbf{d} \subseteq \Omega$ be a closed analytic disk and let $u \in P(\Omega)$. Let $\phi : \overline{D} \to \mathbf{d}$ be a parametrization of $\mathbf{d}$. Then $u \circ \phi$ is subharmonic, so for any $\zeta \in \overline{D}$,

$$u \circ \phi(\zeta) \le \sup_{\zeta \in \partial D} u(\zeta).$$

It follows that for any $p \in \mathbf{d}$,

$$u(p) \le \sup_{\xi \in \partial \mathbf{d}} u(\xi).$$

Therefore $\mathbf{d} \subseteq \widehat{\partial \mathbf{d}}_{P(\Omega)}$. Thus if $\{\mathbf{d}_a\}_{a \in A}$ is a family of closed analytic disks in Ω, then $\cup \mathbf{d}_a \subseteq (\cup_a \partial \mathbf{d}_a)_{P(\Omega)}^{\widehat{}}$. Hence **(6)** holds.

(6) $\Rightarrow$ (7) If not, there is a closed analytic disk $\phi : \overline{D} \to \mathbf{d} \subseteq \Omega$ and a distance function μ such that $\mu_\Omega(\overset{\circ}{\mathbf{d}}) < \mu_\Omega(\partial \mathbf{d})$. Note that because the continuous image of a compact set is compact, $\mathbf{d}$ is both closed and bounded.

Let $p_0 \in \overset{\circ}{\mathbf{d}}$ be the μ-nearest point to $\partial\Omega$. We may assume that $\phi(0) = p_0$. Choose $z_0 \in \partial\Omega$ such that $\mu(p_0 - z_0) = \mu_\Omega(p_0)$. It follows that the disks $\mathbf{d}_j \equiv \mathbf{d} + (1 - (1/j))(z_0 - p_0)$ satisfy $\cup \partial \mathbf{d}_j \subset\subset \Omega$, whereas $\cup \mathbf{d}_j \supseteq \{(1 - (1/j))z_0 + (1/j)p_0\} \to z_0 \in \partial\Omega$. This contradicts **(6)**.

(7) $\Rightarrow$ (1) (This ingenious proof is due to Hartogs.)

It is enough to check plurisubharmonicity at a fixed point $z_0 \in \Omega$, and for any distance function μ. Fix a vector $a \in \mathbb{C}^n$: we must check the subharmonicity of $\psi : \zeta \mapsto -\log \mu_\Omega(z_0 + a\zeta), \zeta \in \mathbb{C}$ small. If $|a|$ is small enough, we may take $|\zeta| \le 1$. We then show that

$$\psi(0) \le \frac{1}{2\pi} \int_0^{2\pi} \psi(e^{i\theta}) d\theta.$$

Now $\psi|_{\partial D}$ is continuous. Let $\epsilon > 0$. By the Stone–Weierstrass theorem, there is a holomorphic polynomial p on $\mathbb{C}$ such that if $h = \operatorname{Re} p$ then

$$\sup_{\zeta \in \partial D} |\psi(\zeta) - h(\zeta)| < \epsilon.$$

We may assume that $h > \psi$ on ∂D. Let $b \in \mathbb{C}^n$ satisfy $\mu(b) \leq 1$. Define a closed analytic disk

$$\phi : \zeta \mapsto z_0 + \zeta a + b e^{-p(\zeta)}, \quad |\zeta| \leq 1.$$

Identifying ϕ with its image $\mathbf{d}$ as usual, our aim is to see that $\mathbf{d} \subseteq \Omega$. If we can prove this claim, then the proof of (1) is completed as follows: Since b was arbitrary, we conclude by setting $\zeta = 0$ that $z_0 + b e^{-p(0)} \in \Omega$ for every choice of b a vector of μ length not exceeding 1. It follows that the ball with μ-radius $|e^{-p(0)}|$ and center z_0 is contained in Ω. Thus

$$\mu_\Omega(z_0) \geq |e^{-p(0)}| = e^{-h(0)}.$$

Equivalently,

$$\psi(0) = -\log \mu_\Omega(z_0) \leq h(0) = \frac{1}{2\pi} \int_0^{2\pi} h(e^{i\theta})d\theta < \frac{1}{2\pi} \int_0^{2\pi} \psi(e^{i\theta})d\theta + \epsilon.$$

Letting $\epsilon \to 0^+$ then yields the result.

It remains to check that $\mathbf{d} \subseteq \Omega$. The proof we present will seem unnecessarily messy. For it is fairly easy to see that $\partial \mathbf{d}$ lies in Ω. The trouble is that while the spirit of (7) suggests that we may then conclude that $\mathbf{d}$ itself lies in Ω, this is not so. There are no complex analytic obstructions to this conclusion; but there *can* be topological obstructions. To show that $\mathbf{d}$ in its entirety lies in Ω, we must demonstrate that it is a continuous deformation of another disk that we know a priori lies in Ω. Thus there are some unpleasant details.

We define the family of disks

$$\mathbf{d}_\lambda : \zeta \mapsto z_0 + \zeta a + \lambda b e^{-p(\zeta)}, \quad 0 \leq \lambda \leq 1.$$

Let $S = \{\lambda : 0 \leq \lambda \leq 1 \text{ and } \mathbf{d} \subseteq \Omega\}$. We claim that $S = [0, 1]$. Of course $\mathbf{d}_1 = \mathbf{d}$, so that will complete the proof. We use a continuity method.

First notice that by the choice of a, $0 \in S$. Hence S is not empty.

Next, if $P \in \mathbf{d}_\lambda$ choose $\zeta_0 \in D$ such that $\mathbf{d}(\zeta_0) = P$. If $\lambda_j \to \lambda$ then $\mathbf{d}_{\lambda_j}(\zeta_0) \equiv P_j \to P$. So the disk $\mathbf{d}_\lambda$ is the limit of the disks $\mathbf{d}_{\lambda_j}$ in a natural way. Moreover, $\cup_{0 \leq \lambda \leq 1} \partial \mathbf{d}_\lambda \subset\subset \Omega$ because

$$\mu\left((z_0 + \zeta a) - (z_0 + \zeta a + \lambda b e^{-p(\zeta)})\right) = \mu(\lambda b e^{-p(\zeta)})$$

$$\leq e^{-h(\zeta)}$$

$$< e^{-\psi(\zeta)}$$

$$= \mu_\Omega(z_0 + \zeta a).$$

We may not now conclude from (7) *that* $\cup_{0 \leq \lambda \leq 1} \partial \mathbf{d}_\lambda \subset\subset \Omega$ *because the Kontinuitätssatz applies only to disks that are known a priori to lie in Ω. But we* *may* conclude from the Kontinuitätssatz and the remarks in the present paragraph that S is closed.

Since Ω is an open domain, it is also clear that S is open. We conclude that $S = [0, 1]$ and our proof is complete.

(1) $\Rightarrow$ **(2)** Trivial.

(4) $\Rightarrow$ **(11)** Let Φ be as in **(4)**. The level sets $\{z \in \Omega : \Phi(z) < c\}$ may not all have smooth boundary; at a boundary point where $\nabla\Phi$ vanishes, there could be a singularity. However, Sard's theorem [KRP1] guarantees that the set of c's for which this problem arises has measure zero in $\mathbb{R}$. Thus we let $\Omega_j = \{z \in \Omega : \Phi(z) < \lambda_j\}$, where $\lambda_j \to +\infty$ are such that each Ω_j is smooth. Since Φ is strictly psh, each Ω_j is strongly pseudoconvex.

(11) $\Rightarrow$ **(10)** It is enough to prove that a strongly pseudoconvex domain $\mathcal{D}$ with smooth boundary is Hartogs pseudoconvex. But this follows from **(8)** $\Rightarrow$ **(3)** $\Rightarrow$ **(4)** $\Rightarrow$ **(5)** $\Rightarrow$ **(6)** $\Rightarrow$ **(1)** $\Rightarrow$ **(2)** above (see Figure 6.3 to verify that we have avoided circular reasoning).

(10) $\Rightarrow$ **(2)** Let δ be the Euclidean distance. By **(2)** $\Rightarrow$ **(3)** $\Rightarrow$ **(4)** $\Rightarrow$ **(5)** $\Rightarrow$ **(6)** $\Rightarrow$ **(1)** above, $-\log d_{\Omega_j}$ is psh for each j. Hence $-\log d_\Omega$ is psh and Ω is Hartogs pseudoconvex.

(7) $\Rightarrow$ **(8)** Trivial.

(8) $\Rightarrow$ **(6)** Let μ be the distance function provided by **(8)**. If **(6)** fails then there is a sequence $\{\mathbf{d}_j\}$ of closed analytic disks lying in Ω with $\mu_\Omega(\partial\mathbf{d}_j) \geq \delta_0 > 0$, whereas $\mu_\Omega(\overset{\circ}{\mathbf{d}}_j) \to 0$. This is a contradiction.

(1) $\Rightarrow$ **(9)** Let δ be Euclidean distance. If $\partial\Omega$ is C^2, then $d_\Omega(\cdot)$ is C^2 at points P that are sufficiently near to $\partial\Omega$ (see [KRP1]). Consider such a $P \in \Omega$ and $w \in \mathbb{C}^n$. By **(1)**, $-\log d_\Omega$ is psh on Ω. Hence

$$\sum_{j,k=1}^n \left(-d_\Omega^{-1}(P)\frac{\partial^2 d_\Omega}{\partial z_j \partial\bar{z}_k}(P) + d_\Omega^{-2}(P)\left(\frac{\partial d_\Omega}{\partial z_j}(P)\right)\left(\frac{\partial d_\Omega}{\partial\bar{z}_k}(P)\right)\right) w_j\bar{w}_k \geq 0.$$

Multiply through by $d_\Omega(P)$, and restrict attention to w that satisfy

$$\sum(\partial d_\Omega/\partial z_j)(P)w_j = 0.$$

Letting $P \to \partial\Omega$, the inequality becomes

$$-\sum_{j,k=1}^n \frac{\partial^2 d_\Omega}{\partial z_j \partial\bar{z}_k}(P)w_j\bar{w}_k \geq 0, \tag{6.2.5.3}$$

for all $P \in \partial\Omega$, all $w \in \mathbb{C}^n$ satisfying

$$\sum_j \frac{\partial d_\Omega}{\partial z_j}(P)w_j = 0.$$

But the function

$$\rho(z) = \begin{cases} -d_\Omega(z) & \text{if } x \in \overline{\Omega}, \\ d_{c\Omega}(z) & \text{if } x \notin \overline{\Omega}, \end{cases}$$

is a C^2 defining function for x near $\partial\Omega$ (which may easily be extended to all of $\mathbb{C}^n$). Thus (6.2.5.3) simply says that Ω is Levi pseudoconvex. $\square$

6.3 Domains of Holomorphy

Capsule: A domain Ω in $\mathbb{C}$ or $\mathbb{C}^n$ is a domain of holomorphy if it is the natural domain of definition of some holomorphic function. In other words, there is an f holomorphic on Ω that cannot be analytically continued to any larger domain. It is natural to want to give an extrinsic geometric charac- terization of domains of holomorphy. In one complex variable, any domain whatever is a domain of holomorphy. But in several-variables the question is genuine: for some domains *are* domains of holomorphy and some not. The concept of pseudoconvexity turns out to be the right differential-geometric measure for distinguishing these important domains.

We begin this section with a statement and proof of the historical result that began investigations of domains of holomorphy. Note the striking contrast with the situation in complex dimension 1.

Theorem 6.3.1 *Let*

$$\Omega \equiv D^2(0, 3) \setminus \overline{D}^2(0, 1) \subseteq \mathbb{C}^2.$$

Then every holomorphic function on Ω analytically continues to the domain $\hat{\Omega} \equiv D^2(0, 3)$.

Proof: Let h be holomorphic on Ω. For z_1 fixed, $|z_1| < 3$, we write

$$h_{z_1}(z_2) = h(z_1, z_2) = \sum_{j=-\infty}^{\infty} a_j(z_1) z_2^j, \qquad (6.3.1.1)$$

where the coefficients of this Laurent expansion are given by

$$a_j(z_1) = \frac{1}{2\pi i} \oint_{|z_2|=2} \frac{h(z_1, \zeta)}{\zeta^{j+1}} d\zeta.$$

In particular, $a_j(z_1)$ depends holomorphically on z_1 (by Morera's theorem, for instance). But $a_j(z_1) = 0$ for $j < 0$ and $1 < |z_1| < 3$. Therefore, by analytic continuation, $a_j(z_1) \equiv 0$ for $j < 0$. But then the series expansion (6.3.1.1) becomes

$$\sum_{j=0}^{\infty} a_j(z_1) z_2^j$$

and this series *defines* a holomorphic function $\hat{h}$ on all of $D^2(0, 3)$ that agrees with the original function h on Ω. Thus Ω is *not* a domain of holomorphy—all holomorphic functions on Ω continue to the larger domain $D^2(0, 3)$. This completes the proof of the "Hartogs extension phenomenon." □

We now direct the machinery that has been developed to derive several characterizations of domains of holomorphy. These are presented in Theorem 6.3.6. As an immediate consequence, we shall see that every domain of holomorphy is pseudoconvex. This leads to a solution of the Levi problem, that is, to see that every pseudoconvex domain is a domain of holomorphy. Thus Theorems 6.2.5 and 6.3.6, together with the solution of the Levi problem, give a total of 21 equivalent characterizations of the principal domains that are studied in the theory of several complex variables.

Recall that a domain of holomorphy is a domain in $\mathbb{C}^n$ with the property that there is a holomorphic function f defined on Ω such that f cannot be analytically continued to any larger domain. There are some technicalities connected with this definition that we cannot treat here; see [KRA4] for the details. It raises many questions because other reasonable definitions of "domain of holomorphy" are not manifestly equivalent to it. For instance, suppose that $\Omega \subseteq \mathbb{C}^n$ has the property that $\partial\Omega$ may be covered by finitely many open sets $\{U_j\}_{j=1}^M$ such that $\Omega \cap U_j$ is a domain of holomorphy, $j = 1, \ldots, M$. Is Ω then a domain of holomorphy? Suppose instead that to each $P \in \partial\Omega$ we may associate a neighborhood U_P and a holomorphic function $f_P : U_P \cap \Omega \to \mathbb{C}$ such that f_P cannot be continued analytically past P. Is Ω then a domain of holomorphy?

Fortunately, all these definitions are equivalent to the original definition of domain of holomorphy, as we shall soon see. We will ultimately learn that the property of being a domain of holomorphy is purely a local one.

In what follows, it will occasionally prove useful to allow our domains of holomorphy to be disconnected (contrary to our customary use of the word "domain"). We leave it to the reader to sort out this detail when appropriate. Throughout this section, the family of functions $\mathcal{O} = \mathcal{O}(\Omega)$ will denote the holomorphic functions on Ω.

If K is a compact set in Ω then we define

$$\widehat{K}_{\mathcal{O}} \equiv \left\{ z \in \Omega : |f(z)| \leq \sup_{w\in\Omega} |f(w)| \text{ for all } f \in \mathcal{O}(\Omega) \right\}.$$

We call $\widehat{K}_{\mathcal{O}}$ the *hull* of K with respect to $\mathcal{O}$. We say that the domain Ω is *convex* with respect to $\mathcal{O}$ if whenever $K \subset\subset \Omega$, then $\widehat{K}_{\mathcal{O}} \subset\subset \Omega$.

By way of warming up to our task, let us prove a few simple assertions about $\widehat{K}_{\mathcal{O}}$ when $K \subset\subset \Omega$ (note that we are assuming, in particular, that K is bounded).

Lemma 6.3.2 *The set $\widehat{K}_{\mathcal{O}}$ is bounded (even if Ω is not).*

Proof: The set K is bounded if and only if the holomorphic functions $f_1(z) = z_1, \ldots, f_n(z) = z_n$ are bounded on K. This, in turn, is true if and only if $f_1, \ldots, f_n$ are bounded on $\widehat{K}_{\mathcal{O}}$; and this last is true if and only if $\widehat{K}_{\mathcal{O}}$ is bounded. □

Lemma 6.3.3 *The set $\widehat{K}_{\mathcal{O}}$ is contained in the closed convex hull of K.*

Proof: Apply the definition of $\widehat{K}_{\mathcal{O}}$ to the real parts of all complex linear functionals on $\mathbb{C}^n$ (which, of course, are elements of $\mathcal{O}$). Then use the fact that $|\exp(f)| = \exp(\mathrm{Re}\, f)$. □

Prelude: Analytic disks turn out to be one of the most useful devices for thinking about domains of holomorphy. This next lemma captures what is most interesting about an analytic disk from our current perspective.

Lemma 6.3.4 *Let $\mathbf{d} \subseteq \Omega$ be a closed analytic disk. Then $\mathbf{d} \subseteq \widehat{\partial \mathbf{d}}_{\mathcal{O}}$.*

Proof: Let $f \in \mathcal{O}$. Let $\phi : \overline{D} \to \mathbf{d}$ be a parametrization of $\mathbf{d}$. Then $f \circ \phi$ is holomorphic on D and continuous on $\overline{D}$. Therefore it assumes its maximum modulus on ∂D. It follows that

$$\sup_{z \in \mathbf{d}} |f(z)| = \sup_{z \in \partial \mathbf{d}} |f(z)|.$$ □

Exercise for the Reader: Consider the Hartogs domain $\Omega = D^2(0, 3) \setminus \overline{D}^2(0, 1)$. Let $K = \{(0, 2e^{i\theta}) : 0 \leq \theta < 2\pi\} \subset\subset \Omega$. Verify that $\widehat{K}_{\mathcal{O}} = \{(0, re^{i\theta}) : 0 \leq \theta < 2\pi : 1 < r < 2\}$, which is *not* compact in Ω.

It is sometimes convenient to calculate the hull of a compact set K with respect to an arbitrary family $\mathcal{F}$ of functions. The definition is the same:

$$\widehat{K}_{\mathcal{F}} \equiv \left\{ z \in \Omega : |f(z)| \leq \sup_{w \in K} |f(w)| \text{ for all } f \in \mathcal{F}(\Omega) \right\}.$$

Sometimes, for instance, we shall find it useful to take $\mathcal{F}$ to be the family of plurisubharmonic functions.

Now we turn our attention to the main results of this section. First, a definition is needed. This is a localized version of the definition of domain of holomorphy.

Definition 6.3.5 Let $U \subseteq \mathbb{C}^n$ be an open set. We say that $P \in \partial U$ is *essential* if there is a holomorphic function h on U such that for no connected neighborhood U_2 of P and nonempty $U_1 \subseteq U_2 \cap U$ is there an h_2 holomorphic on U_2 with $h = h_2$ on U_1.

Prelude: Just as for pseudoconvexity, it is useful to have many different ways to think about domains of holomorphy. And the proofs of their equivalence are some of the most basic arguments in the subject.

Theorem 6.3.6 *Let $\Omega \subseteq \mathbb{C}^n$ be an open set (no smoothness of $\partial \Omega$ or boundedness of Ω need be assumed). Let $\mathcal{O} = \mathcal{O}(\Omega)$ be the family of holomorphic functions on Ω. Then the following are equivalent.*

(1) *Ω is convex with respect to $\mathcal{O}$.*
(2) *There is an $h \in \mathcal{O}$ that cannot be holomorphically continued past any $P \in \partial \Omega$ (i.e., in the definition of essential point the same function h can be used for every boundary point P).*

(3) *Each $P \in \partial\Omega$ is essential (Ω is a domain of holomorphy).*

(4) *Each $P \in \partial\Omega$ has a neighborhood U_P such that $U_P \cap \Omega$ is a domain of holomorphy.*

(5) *Each $P \in \partial\Omega$ has a neighborhood U_P such that $U_P \cap \Omega$ is convex with respect to $\mathcal{O}_P \equiv$ {holomorphic functions on $U_P \cap \Omega$}.*

(6) *For any $f \in \mathcal{O}$, any $K \subset\subset \Omega$, any distance function μ, the inequality*

$$|f(z)| \leq \mu_\Omega(z), \quad \forall z \in K,$$

implies that

$$|f(z)| \leq \mu_\Omega(z), \quad \forall z \in \widehat{K}_\mathcal{O}.$$

(7) *For any $f \in \mathcal{O}$, any $K \subset\subset \Omega$, and any distance function μ, we have*

$$\sup_{z \in K}\left\{\frac{|f(z)|}{\mu_\Omega(z)}\right\} = \sup_{z \in \widehat{K}_\mathcal{O}}\left\{\frac{|f(z)|}{\mu_\Omega(z)}\right\}.$$

(8) *If $K \subset\subset \Omega$, then for any distance function μ,*

$$\mu_\Omega(K) = \mu_\Omega(\widehat{K}_\mathcal{O}).$$

(9) *Same as* **(6)** *for just one distance function μ.*

(10) *Same as* **(7)** *for just one distance function μ.*

(11) *Same as* **(8)** *for just one distance function μ.*

Remark: The scheme of the proof is shown in Figure 6.6. Observe that **(4)** and **(5)** are omitted. They follow easily once the Levi problem has been solved.

We shall defer the proof of Theorem 6.3.6 until the end of the section; meanwhile, we discuss the theorem's many consequences.

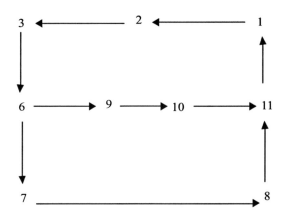

Figure 6.6. Scheme of the proof of Theorem 6.3.5.

6.3.1 Consequences of Theorems 6.2.5 and 6.3.6

Corollary 6.3.7 *If $\Omega \subseteq \mathbb{C}^n$ is a domain of holomorphy, then Ω is pseudoconvex.*

Proof: By part **(1)** of Theorem 6.3.6, Ω is holomorphically convex. It follows a fortiori that Ω is convex with respect to the family $P(\Omega)$ of all psh functions on Ω since $|f|$ is psh whenever f is holomorphic on Ω. Thus, by part **(5)** of Theorem 6.2.5, Ω is pseudoconvex. □

Corollary 6.3.8 *Let $\{\Omega_a\}_{a \in A}$ be domains of holomorphy in $\mathbb{C}^n$. If $\Omega \equiv \cap_{a \in A}\Omega_a$ is open, then Ω is a domain of holomorphy.*

Proof: Use part **(8)** of Theorem 6.3.6. □

Corollary 6.3.9 *If Ω is geometrically convex, then Ω is a domain of holomorphy.*

Proof: Let $P \in \partial\Omega$. Let $(a_1, \ldots, a_n) \in \mathbb{C}^n$ be any unit outward normal to $\partial\Omega$ at P. Then the real tangent hyperplane to $\partial\Omega$ at P is $\{z : \text{Re}\left[\sum_{j=1}^{n}(z_j - P_j)\bar{a}_j\right] = 0\}$. But then the function

$$f_P(z) = \frac{1}{\left(\sum_{j=1}^{n}(z_j - P_j)\bar{a}_j\right)}$$

is holomorphic on Ω and shows that P is essential. By part **(3)** of Theorem 6.3.6, Ω is a domain of holomorphy. □

Exercise for the Reader: Construct another proof of Corollary 6.3.9 using part **(1)** of Theorem 6.3.6.

Remark: Corollary 6.3.9 is a poor man's version of the Levi problem. The reader should consider why this proof fails on, say, strongly pseudoconvex domains: Let $\Omega \subseteq \mathbb{C}^n$ be strongly pseudoconvex, ρ the defining function for Ω. For $P \in \partial\Omega$,

$$\rho(z) = \rho(P) + 2\text{Re}\left\{\sum_{j=1}^{n}\frac{\partial\rho}{\partial z_j}(P)(z_j - P_j)\right.$$
$$\left. + \frac{1}{2}\sum_{j,k=1}^{n}\frac{\partial^2\rho}{\partial z_j \partial z_k}(P)(z_j - P_j)(z_k - P_k)\right\}$$
$$+ \sum_{j,k=1}^{n}\frac{\partial^2\rho}{\partial z_j \partial \bar{z}_k}(P)(z_j - P_j)(\bar{z}_k - \overline{P}_k) + o(|z - P|^2).$$

Define

$$L_P(t) = \sum_{j=1}^{n}\frac{\partial\rho}{\partial z_j}(P)(z_j - P_j) + \frac{1}{2}\sum_{j,k=1}^{n}\frac{\partial^2\rho}{\partial z_j \partial z_k}(P)(z_j - P_j)(z_k - P_k).$$

The function $L_P(z)$ is called the *Levi polynomial* at P. For $|z - P|$ sufficiently small, $z \in \overline{\Omega}$, we have that $L_P(z) = 0$ if and only if $z = P$. For $L_P(z) = 0$ means that $\rho(z) \geq C|z - P|^2 + o(|z - P|^2)$. Hence L_P is an Ersatz for f_P in the proof of Corollary 6.3.9 *near* P. In short, P is "locally essential." It requires powerful additional machinery to conclude from this that P is (globally) essential.

6.3.2 Consequences of the Levi Problem

Assume for the moment that we have proved that pseudoconvex domains are domains of holomorphy (we *did* prove the converse of this statement in Corollary 6.3.7.). Then we may quickly dispatch an interesting question. In fact, historically, this result played a crucial role in the solution of the Levi problem.

Prelude: In classical studies of the Levi problem, the Behnke–Stein theorem was basic. It gave a method of reducing the question to the study of *strongly* pseudoconvex domains.

Theorem 6.3.10 (Behnke–Stein) *Let* $\Omega_1 \subseteq \Omega_2 \subseteq \cdots$ *be domains of holomorphy. Then* $\Omega \equiv \cup_j \Omega_j$ *is a domain of holomorphy.*

Proof: Each Ω_j is pseudoconvex (by Corollary 6.3.7); hence, by Proposition 6.2.7, Ω is pseudoconvex. By the Levi problem, Ω is a domain of holomorphy.

□

Remark: It is possible, but rather difficult, to prove the Behnke–Stein theorem directly, without any reference to the Levi problem. Classically, the Levi problem was solved for strongly pseudoconvex domains and then the fact that any weakly pseudoconvex domain is the increasing union of strongly pseudoconvex domains, together with Behnke–Stein, was used to complete the argument. See [BER] for a treatment of this approach.

Proof of Theorem 6.3.6:

(2) $\Rightarrow$ **(3)** Trivial.

(11) $\Rightarrow$ **(1)** Trivial.

(1) $\Rightarrow$ **(2)** Choose a dense sequence $\{w_j\}_{j=1}^{\infty} \subseteq \Omega$ that repeats every point infinitely often. For each j, let D_j be the largest polydisk $D^n(w_j, r)$ contained in Ω. Choose a sequence $K_1 \subset\subset K_2 \subset\subset K_3 \subset\subset \cdots$ with $\cup_{j=1}^{\infty} K_j = \Omega$. For each j, $(\widehat{K}_j)_{\mathcal{O}} \subset\subset \Omega$ by hypothesis. Thus there is a point $z_j \in D_j \setminus (\widehat{K}_j)_{\mathcal{O}}$. This means that we may choose an $h_j \in \mathcal{O}$ such that $h_j(z_j) = 1, |.h_j|_{K_j}| < 1$. By replacing h_j by $h_j^{M_j}$, M_j large, we may assume that $|h_j|_{K_j}| < 2^{-j}$. Write

$$h(z) = \prod_{j=1}^{\infty}(1 - h_j)^j.$$

Then the product converges uniformly on each K_j, hence normally on Ω, and the limit function h is not identically zero (this is just standard one variable theory—see, for instance, L. Ahlfors [AHL]). By the choice of the points w_j, every D_j contains infinitely many of the z_ℓ and hence contains points at which h vanishes to arbitrarily high order. Any analytic continuation of h to a neighborhood of a point $P \in \partial\Omega$ is a continuation to a neighborhood of some $\overline{D}_j$ and hence would necessitate that h vanish to infinite order at some point. This would imply that $h \equiv 0$, which is a clear contradiction.

(3) $\Rightarrow$ **(6)** Fix $r = (r_1, \ldots, r_n) > 0$. Define a distance function $\mu^r(z) = \max_{1 \le j \le n}\{|z_j|/r_j\}$. We first prove **(6)** for this distance function. Let $f \in \mathcal{O}$, $K \subset\subset \Omega$ satisfy $|f(z)| \le \mu_\Omega^r(z)$, $z \in K$. We claim that for all $g \in \mathcal{O}$, all $p \in \widehat{K}_\mathcal{O}$, it holds that g has a normally convergent power series expansion on

$$\{z \in \mathbb{C}^n : \mu^r(z-p) < |f(p)|\} = D^1(p_1, |f(p)| \cdot r_1) \times \cdots \times D^1(p_n, |f(p)| \cdot r_n).$$

This implies **(6)** for this particular distance function. For if $|f(p)| > \mu_\Omega^r(p)$ for some $p \in \widehat{K}_\mathcal{O}$, then $D^1(p_1, |f(p)| \cdot r_1) \times \cdots \times D^1(p_n, |f(p)| \cdot r_n)$ has points in it that lie outside $\overline{\Omega}$ (to which every $g \in \mathcal{O}$ extends analytically!). That would contradict **(3)**. To prove the claim, let $0 < t < 1$, let $g \in \mathcal{O}$, and let

$$S_t = \bigcup_{k \in K}\{z \in \mathbb{C}^n : \mu^r(z-k) \le t|f(k)|\}.$$

Since $S_t \subset\subset \Omega$ by the hypothesis on f, there is an $M > 0$ such that $|g| \le M$ on S_t. By Cauchy's inequalities,

$$\left|\left(\frac{\partial}{\partial z}\right)^\alpha g(k)\right| \le \frac{\alpha!M}{t^{|\alpha|}r^\alpha|f(k)|^{|\alpha|}}, \qquad \forall k \in K, \text{ all multi-indices } \alpha.$$

But then the same estimate holds on $\widehat{K}_\mathcal{O}$. So the power series of g about $p \in \widehat{K}_\mathcal{O}$ converges on $D^1(p_1, t|f(p)| \cdot r_1) \times \cdots \times D^1(p_n, t|f(p)| \cdot r_n)$. Since $0 < t < 1$ was arbitrary, the claim is proved. So, in the special case of distance function μ^r, the implication **(3)** $\Rightarrow$ **(6)** is proved.

Now fix any distance function μ. Define, for any $w \in \mathbb{C}^n$,

$$S_\Omega^w(z) = \sup\{r \in \mathbb{R} : z + \tau w \in \Omega, \forall |\tau| < r, \tau \in \mathbb{C}\}.$$

Then, trivially,

$$\mu_\Omega(z) = \inf_{\mu(w)=1} S_\Omega^w(z). \qquad (6.3.6.1)$$

If we prove **(6)** for S_Ω^w instead of μ_Ω, w fixed, then the full result follows from (6.3.6.1).

After a rotation and dilation, we may suppose that $w = (1, 0, \ldots, 0)$. If $k \in \mathbb{N}$, we apply the special case of **(6)** to the n-tuple $r^k = (1, 1/k, \ldots, 1/k)$. Notice that $\mu_\Omega^{r^k} \nearrow S_\Omega^w$ as $k \to +\infty$. Let $K \subset\subset \Omega$. Assume that $|f(z)| \le S_\Omega^w(z)$ for $z \in K$. Let $\epsilon > 0$. Define

$$A_k = \{z : |f(z)| < (1+\epsilon)\mu_\Omega^{r^k}(z)\}.$$

Then $\{A_k\}$ is an increasing sequence of open sets whose union contains K. Thus one of the sets A_{k_0} covers K. In other words,

$$|f(z)| \le (1+\epsilon)\mu_\Omega^{r^{k_0}}(z), \quad z \in K.$$

By what we have already proved for the μ^r,

$$|f(z)| \le (1+\epsilon)\mu_\Omega^{r^{k_0}}(z) \le (1+\epsilon)S_\Omega^w(z), \quad \forall z \in \widehat{K}_\mathcal{O}.$$

Letting $\epsilon \to 0^+$ yields $|f(z)| \le S_\Omega^w(z), z \in \widehat{K}_\mathcal{O}$, as desired.

(6) $\Rightarrow$ **(9)** Trivial.

(9) $\Rightarrow$ **(10)** Trivial.

(10) $\Rightarrow$ **(11)** Apply **(10)** with $f \equiv 1$.

(7) $\Rightarrow$ **(8)** Apply **(7)** with $f \equiv 1$.

(8) $\Rightarrow$ **(11)** Trivial.

(6) $\Rightarrow$ **(7)** Trivial. $\square$

7

Canonical Complex Integral Operators

Prologue: In the function theory of one complex variable the obvious "canonical" integral kernels are the Cauchy kernel and the Poisson kernel. The Cauchy kernel may be discovered naturally by way of power series considerations, or partial differential equations considerations, or conformality considerations. The Poisson kernel is the real part of the Cauchy kernel. It also arises naturally as the solution operator for the Dirichlet problem. It is rather more difficult to get one's hands on integral reproducing kernels in several complex variables.

There are in fact a variety of formal mechanisms for generating canonical kernels. These include those due to Bergman, Szegő, Poisson, and others. Thus arises a plethora of canonical kernels, both in one and several complex variables. In one complex variable, all these different kernels boil down to the Cauchy kernel or the Poisson kernel. In several-variables they tend to be different.

There are also noncanonical methods—due to Henkin, Ramirez, Kerzman, Grauert, Lieb, and others—for creating reproducing kernels for holomorphic functions. Thus there exist, in principle, infinitely many kernels worthy of study on any domain (satisfying reasonable geometric conditions) in $\mathbb{C}^n$. In general (in several-variables), these kernels will be distinct.

It is a matter of considerable interest to relate the canonical kernels described above to the noncanonical ones. The canonical kernels have many wonderful properties, but tend to be difficult to compute. The noncanonical kernels are more readily computed, but of course do not behave canonically.

There are also kernels that can be constructed by abstract methods of function algebras. It is nearly impossible to say anything concrete about these kernels. In particular, nothing is known about the nature of their singularities. But it can be shown abstractly that *every* domain has a kernel holomorphic in the free variable (such as the Cauchy kernel) that reproduces holomorphic functions.

S.G. Krantz, *Explorations in Harmonic Analysis*, Applied and Numerical Harmonic Analysis, DOI 10.1007/978-0-8176-4669-1_7,
© Birkhäuser Boston, a part of Springer Science+Business Media, LLC 2009

It is difficult to create an explicit integral formula, with holomorphic reproducing kernel, for holomorphic functions on a domain in $\mathbb{C}^n$. This difficulty can be traced back to the Hartogs extension phenomenon: the zeros of a holomorphic function of several-variables are never isolated, and thus the singularities of the putative kernel are difficult to control (contrast the Cauchy kernel $1/(\zeta - z)$ of one variable, which has an isolated singularity at $\zeta = z$). Early work on integral formulas in several-variables concentrated on the polydisk and on bounded symmetric domains—see [WEI].

There are special classes of domains—such as strongly pseudoconvex domains—on which the creation of integral formulas may be carried out quite constructively. We cannot treat the details here, but see [KRA4, Ch. 5]. We now examine one of several nonconstructive (but canonical) approaches to the matter.[1] One of the main points of the present book is to study some of these canonical kernels on the Siegel upper half-space and the Heisenberg group.

It may be noted that the explicitly constructible kernels and the canonical kernels on any given domain (in dimension greater than 1) are in general different. *On the disk in $\mathbb{C}^1$ they all coincide.* This is one of the remarkable facts about the function theory of several complex variables, and one that is not yet fully understood. The papers [KST1] and [KST2] explore the connection between the two types of kernels on strongly pseudoconvex domains.

The circle of ideas regarding canonical kernels, which we shall explore in the present chapter, are due to S. Bergman ([BERG] and references therein) and to G. Szegő (see [SxE]; some of the ideas presented here were anticipated by the thesis of S. Bochner [BOC1]). They have profound applications to the boundary regularity of holomorphic mappings. We shall say a few words about those at the end.

7.1 Elementary Concepts of the Bergman Kernel

Capsule: Certainly the Bergman kernel construction was one of the great ideas of modern complex function theory. It not only gives rise to a useful and important canonical reproducing kernel, but also to the Bergman metric. The Bergman kernel and metric are expedient in the study of biholomorphic mappings (again because of their invariance) and also in the study of elliptic partial differential equations. In the hands of Charles Fefferman [FEF8], the Bergman theory has given rise to important biholomorphic invariants. The Bergman kernel is closely related philosophically, and sometimes computationally, to the Szegő kernel. The Bergman kernel deals with integration over the solid region Ω, while the Szegő kernel deals with integration over the boundary $\partial\Omega$. Each kernel looks at a different function space. We close

---

[1] One cannot avoid observing that in one complex variable, there is an explicit reproducing formula—the Cauchy formula—on *every* domain. And it is the *same* on every domain—that is, the kernel itself is the same explicitly given function. Such is not the case in several complex variables.

by noting (see [FEF7]) that Bergman metric geodesics carry important information about biholomorphic mappings and about complex function theory and harmonic analysis.

In this section we shall construct the Bergman kernel from first principles and prove some of its basic properties. We shall also see some of the invariance properties of the Bergman kernel. This will lead to the definition of the Bergman metric (in which all biholomorphic mappings become isometries). The Bergman kernel has certain extremal properties that make it a powerful tool in the theory of partial differential equations (see S. Bergman and M. Schiffer [BERS]). Also the form of the singularity of the Bergman kernel (calculable for some interesting classes of domains) explains many phenomena of the theory of several complex variables.

Let $\Omega \subseteq \mathbb{C}^n$ be a domain. Define the *Bergman space*

$$A^2(\Omega) = \left\{ f \text{ holomorphic on } \Omega : \int_\Omega |f(z)|^2 \, dV(z)^{1/2} \equiv \|f\|_{A^2(\Omega)} < \infty \right\}.$$

Observe that A^2 is quite a different space from the space H^2 that is considered in our Chapter 8 below. The former is defined with integration over the $2n$-dimensional *interior* of the domain, while the latter is defined with respect to integration over the $(2n - 1)$-dimensional *boundary*.

Prelude: The entire Bergman theory hinges on the next very simple lemma. In fact, Aronsajn's more general theory of Bergman space with reproducing kernel (see [ARO]) also hinges on this result.

Lemma 7.1.1 *Let $K \subseteq \Omega \subseteq \mathbb{C}^n$ be compact. There is a constant $C_K > 0$, depending on K and on n, such that*

$$\sup_{z \in K} |f(z)| \leq C_K \|f\|_{A^2(\Omega)}, \quad \text{for all } f \in A^2(\Omega).$$

Proof: Since K is compact, there is an $r(K) = r > 0$ such that we have, for any $z \in K$, that $B(z, r) \subseteq \Omega$. Therefore, for each $z \in K$ and $f \in A^2(\Omega)$,

$$|f(z)| = \frac{1}{V(B(z, r))} \left| \int_{B(z,r)} f(t) \, dV(t) \right|$$

$$\leq (V(B(z, r)))^{-1/2} \|f\|_{L^2(B(z,r))}$$

$$\leq C(n) r^{-n} \|f\|_{A^2(\Omega)}$$

$$\equiv C_K \|f\|_{A^2(\Omega)}. \qquad \square$$

Lemma 7.1.2 *The space $A^2(\Omega)$ is a Hilbert space with the inner product $\langle f, g \rangle \equiv \int_\Omega f(z)\overline{g(z)} \, dV(z)$.*

Proof: Everything is clear except for completeness. Let $\{f_j\} \subseteq A^2$ be a sequence that is Cauchy in norm. Since L^2 is complete there is an L^2 limit function f. We need to see that f is holomorphic. But Lemma 7.1.1 yields that norm convergence implies normal convergence (uniform convergence on compact sets). And holomorphic functions are closed under normal limits (exercise). Therefore f is holomorphic and $A^2(\Omega)$ is complete. $\square$

Remark 7.1.3 It fact, it can be shown that in case Ω is a bounded domain, the space $A^2(\Omega)$ is separable (of course it is a subspace of the separable space $L^2(\Omega)$). The argument is nontrivial, and the reader may take it as an exercise (or see [GRK12]).

Lemma 7.1.4 *For each fixed $z \in \Omega$, the functional*

$$\Phi_z : f \mapsto f(z), \quad f \in A^2(\Omega),$$

is a continuous linear functional on $A^2(\Omega)$.

Proof: This is immediate from Lemma 7.1.1 if we take K to be the singleton $\{z\}$. $\square$

We may now apply the Riesz representation theorem to see that there is an element $k_z \in A^2(\Omega)$ such that the linear functional Φ_z is represented by inner product with k_z: if $f \in A^2(\Omega)$ then for all $z \in \Omega$ we have

$$f(z) = \Phi_z(f) = \langle f, k_z \rangle.$$

Definition 7.1.5 The Bergman kernel is the function $K(z, \zeta) = \overline{k_z(\zeta)}$, $z, \zeta \in \Omega$. It has the reproducing property

$$f(z) = \int K(z, \zeta) f(\zeta) \, dV(\zeta), \quad \forall f \in A^2(\Omega).$$

Prelude: There are not many domains on which we can calculate the Bergman kernel explicitly. In any presentation of the subject, the disk and the ball are the two primary examples. Also the polydisk can be handled with the same calculations. In the seminal work [HUA], the kernel on the bounded symmetric domains of Cartan is considered. Absent a transitive group of holomorphic automorphisms, it is virtually impossible to get a formula for the kernel. What one can sometimes do instead is to derive an *asymptotic expansion*—see for example [FEF7] or [KRPE].

Proposition 7.1.6 *The Bergman kernel $K(z, \zeta)$ is conjugate symmetric: $K(z, \zeta) = \overline{K(\zeta, z)}$.*

Proof: By its very definition, $\overline{K(\zeta, \cdot)} \in A^2(\Omega)$ for each fixed ζ. Therefore the reproducing property of the Bergman kernel gives

$$\int_\Omega K(z, t) \overline{K(\zeta, t)} \, dV(t) = \overline{K(\zeta, z)}.$$

On the other hand,

$$\int_\Omega K(z,t)\overline{K(\zeta,t)}\,dV(t) = \overline{\int K(\zeta,t)\overline{K(z,t)}\,dV(t)}$$

$$= \overline{K(z,\zeta)} = K(z,\zeta). \qquad \square$$

Prelude: Part of the appeal of the Bergman theory is that these simple and elegant preliminary results have beautiful, soft proofs. The next result is important in giving us a number of means to construct the Bergman kernel.

Proposition 7.1.7 *The Bergman kernel is uniquely determined by the properties that it is an element of $A^2(\Omega)$ in z, is conjugate symmetric, and reproduces $A^2(\Omega)$.*

Proof: Let $K'(z,\zeta)$ be another such kernel. Then

$$K(z,\zeta) = \overline{K(\zeta,z)} = \overline{\int K'(z,t)\overline{K(\zeta,t)}\,dV(t)}$$

$$= \int K(\zeta,t)\overline{K'(z,t)}\,dV(t)$$

$$= \overline{K'(z,\zeta)} = K'(z,\zeta). \qquad \square$$

Since $A^2(\Omega)$ is separable, there is a complete orthonormal basis $\{\phi_j\}_{j=1}^\infty$ for $A^2(\Omega)$.

Prelude: The next method of constructing the Bergman kernel is particularly appealing. It is rarely useful, however, because it is not often that we can write down an orthonormal basis for the Bergman space. Even in cases in which we can (such as the annulus), it is generally impossible to sum the series.

Proposition 7.1.8 *The series*

$$\sum_{j=1}^\infty \phi_j(z)\overline{\phi_j(\zeta)}$$

sums uniformly on $E \times E$ to the Bergman kernel $K(z,\zeta)$ for any compact subset $E \subseteq \Omega$.

Proof: By the Riesz–Fischer and Riesz representation theorems, we obtain

$$\sup_{z\in E}\left(\sum_{j=1}^\infty |\phi_j(z)|^2\right)^{1/2} = \sup_{z\in E}\left\|\{\phi_j(z)\}_{j=1}^\infty\right\|_{\ell^2}$$

$$= \sup_{\substack{\|(a_j)\|_{\ell^2}=1 \\ z \in E}} \left| \sum_{j=1}^{\infty} a_j \phi_j(z) \right|$$

$$= \sup_{\substack{\|f\|_{A^2}=1 \\ z \in E}} |f(z)| \le C_E. \tag{7.1.8.1}$$

In the last inequality we have used Lemma 7.1.1 and the Hahn–Banach theorem. Therefore

$$\sum_{j=1}^{\infty} \left| \phi_j(z)\overline{\phi_j(\zeta)} \right| \le \left(\sum_{j=1}^{\infty} |\phi_j(z)|^2 \right)^{1/2} \left(\sum_{j=1}^{\infty} |\phi_j(\zeta)|^2 \right)^{1/2}$$

and the convergence is uniform over $z, \zeta \in E$. For fixed $z \in \Omega$, (7.1.8.1) shows that $\{\phi_j(z)\}_{j=1}^{\infty} \in \ell^2$. Hence we have that $\sum \phi_j(z)\overline{\phi_j(\zeta)} \in A^2(\Omega)$ as a function of ζ. Let the sum of the series be denoted by $K'(z, \zeta)$. Notice that K' is conjugate symmetric by its very definition. Also, for $f \in A^2(\Omega)$, we have

$$\int K'(\cdot, \zeta)f(\zeta)\,dV(\zeta) = \sum \widehat{f}(j)\phi_j(\cdot) = f(\cdot),$$

where convergence is in the Hilbert space topology. [Here $\widehat{f}(j)$ is the jth Fourier coefficient of f with respect to the basis $\{\phi_j\}$.] But Hilbert space convergence dominates pointwise convergence (Lemma 7.1.1), so

$$f(z) = \int K'(z, \zeta)f(\zeta)\,dV(\zeta), \quad \text{for all } f \in A^2(\Omega).$$

Therefore, by Proposition 7.1.7, K' is the Bergman kernel. $\quad\square$

Remark: It is worth noting explicitly that the proof of Proposition 7.1.8 shows that

$$\sum \phi_j(z)\overline{\phi_j(\zeta)}$$

equals the Bergman kernel $K(z, \zeta)$ *no matter what the choice* of complete orthonormal basis $\{\phi_j\}$ for $A^2(\Omega)$.

Proposition 7.1.9 *If Ω is a bounded domain in $\mathbb{C}^n$ then the mapping*

$$P : f \mapsto \int_{\Omega} K(\cdot, \zeta)f(\zeta)\,dV(\zeta)$$

is the Hilbert space orthogonal projection of $L^2(\Omega, dV)$ onto $A^2(\Omega)$.

Proof: Notice that P is idempotent and self-adjoint and that $A^2(\Omega)$ is precisely the set of elements of L^2 that are fixed by P. $\quad\square$

Definition 7.1.10 Let $\Omega \subseteq \mathbb{C}^n$ be a domain and let $f : \Omega \to \mathbb{C}^n$ be a *holomorphic mapping*, that is, $f(z) = (f_1(z), \ldots, f_n(z))$ with $f_1, \ldots, f_n$ holomorphic on Ω. Let $w_j = f_j(z)$, $j = 1, \ldots, n$. Then the *holomorphic (or complex) Jacobian matrix* of f is the matrix

$$J_{\mathbb{C}} f = \frac{\partial(w_1, \ldots, w_n)}{\partial(z_1, \ldots, z_n)}.$$

Write $z_j = x_j + iy_j$, $w_k = \xi_k + i\eta_k$, $j, k = 1, \ldots, n$. Then the *real Jacobian matrix* of f is the matrix

$$J_{\mathbb{R}} f = \frac{\partial(\xi_1, \eta_1, \ldots, \xi_n, \eta_n)}{\partial(x_1, y_1, \ldots, x_n, y_n)}.$$

Prelude: The next simple fact goes in other contexts under the guise of the Lusin area integral. It is central to the quadratic theory, and has many manifestations and many consequences.

Proposition 7.1.11 *With notation as in the definition, we have*

$$\det J_{\mathbb{R}} f = |\det J_{\mathbb{C}} f|^2$$

whenever f is a holomorphic mapping.

Proof: We exploit the functoriality of the Jacobian. Let $w = (w_1, \ldots, w_n) = f(z) = (f_1(z), \ldots, f_n(z))$. Write $z_j = x_j + iy_j$, $w_j = \xi_j + i\eta_j$, $j = 1, \ldots, n$. Then

$$d\xi_1 \wedge d\eta_1 \wedge \cdots \wedge d\xi_n \wedge d\eta_n = (\det J_{\mathbb{R}} f(x, y)) dx_1 \wedge dy_1 \wedge \cdots \wedge dx_n \wedge dy_n.$$

$$(7.1.11.1)$$

On the other hand, using the fact that f is holomorphic,

$$d\xi_1 \wedge d\eta_1 \wedge \cdots \wedge d\xi_n \wedge d\eta_n \qquad\qquad (7.1.1)$$

$$= \frac{1}{(2i)^n} d\overline{w}_1 \wedge dw_1 \wedge \cdots \wedge d\overline{w}_n \wedge dw_n$$

$$= \frac{1}{(2i)^n} \overline{(\det J_{\mathbb{C}} f(z))}(\det J_{\mathbb{C}} f(z)) d\overline{z}_1 \wedge dz_1 \wedge \cdots \wedge d\overline{z}_n \wedge dz_n$$

$$= |\det J_{\mathbb{C}} f(z)|^2 dx_1 \wedge dy_1 \wedge \cdots \wedge dx_n \wedge dy_n. \qquad (7.1.11.2)$$

Equating (7.1.11.1) and (7.1.11.2) gives the result. $\qquad\square$

Exercise for the Reader: Prove Proposition 7.1.11 using only matrix theory (no differential forms). This will give rise to a great appreciation for the theory of differential forms (see [BER] for help).

Now we can prove the holomorphic implicit function theorem:

Theorem 7.1.12 *Let $f_j(w, z)$, $j = 1, \ldots, m$, be holomorphic functions of $(w, z) = ((w_1, \ldots, w_m), (z_1, \ldots, z_n))$ near a point $(w^0, z^0) \in \mathbb{C}^m \times \mathbb{C}^n$. Assume that*

$$f_j(w^0, z^0) = 0, \quad j = 1, \ldots, m,$$

and that

$$\det \left(\frac{\partial f_j}{\partial w_k} \right)^m_{j,k=1} \neq 0 \text{ at } (w^0, z^0).$$

Then the system of equations

$$f_j(w, z) = 0, \quad j = 1, \ldots, m,$$

has a unique holomorphic solution $w(z)$ in a neighborhood of z^0 that satisfies $w(z^0) = w^0$.

Proof: We rewrite the system of equations as

$$\operatorname{Re} f_j(w, z) = 0, \quad \operatorname{Im} f_j(w, z) = 0,$$

for the $2m$ real-variables $\operatorname{Re} w_k, \operatorname{Im} w_k, k = 1, \ldots, m$. By Proposition 7.1.11 the determinant of the Jacobian over $\mathbb{R}$ of this new system is the modulus squared of the determinant of the Jacobian over $\mathbb{C}$ of the old system. By our hypothesis, this number is nonvanishing at the point (w^0, z^0). Therefore the classical implicit function theorem (see [KRP2]) implies that there exist C^1 functions $w_k(z)$, $k = 1, \ldots, m$, with $w(z^0) = w^0$ that solve the system. Our job is to show that these functions are in fact holomorphic. When properly viewed, this is purely a problem of geometric algebra:

Applying exterior differentiation to the equations

$$0 = f_j(w(z), z), \quad j = 1, \ldots, m,$$

yields that

$$0 = df_j = \sum^m_{k=1} \frac{\partial f_j}{\partial w_k} dw_k + \sum^m_{k=1} \frac{\partial f_j}{\partial z_k} dz_k.$$

There are no $d\bar{z}_j$'s and no $d\bar{w}_k$'s because the f_j's are holomorphic.

The result now follows from linear algebra only: the hypothesis on the determinant of the matrix $(\partial f_j / \partial w_k)$ implies that we can solve for dw_k in terms of dz_j. Therefore w is a holomorphic function of z. □

A holomorphic mapping $f : \Omega_1 \rightarrow \Omega_2$ of domains $\Omega_1 \subseteq \mathbb{C}^n$, $\Omega_2 \subseteq \mathbb{C}^m$ is said to be *biholomorphic* if it is one-to-one, onto, and $\det J_{\mathbb{C}} f(z) \neq 0$ for every $z \in \Omega_1$. Of course these conditions entail that f will have a holomorphic inverse.

Exercise for the Reader: Use Theorem 7.1.11 to prove that a biholomorphic mapping has a holomorphic inverse (hence the name).

Remark: It is true, but not at all obvious, that the nonvanishing of the Jacobian determinant is a superfluous condition in the definition of "biholomorphic mapping"; that is, the nonvanishing of the Jacobian follows from the univalence of the mapping (see [KRA4]). There is great interest in proving an analogous result for holomorphic mappings in infinite dimensions; the problem remains wide open.

In what follows we denote the Bergman kernel for a given domain Ω by K_Ω.

Prelude: Next is the fundamental transformation formula for the Bergman kernel. The Bergman metric is an outgrowth of this formula. The invariance of the kernel is also key to many of Bergman's applications of the kernel to partial differential equations.

Proposition 7.1.13 *Let Ω_1, Ω_2 be domains in $\mathbb{C}^n$. Let $f : \Omega_1 \to \Omega_2$ be biholomorphic. Then*

$$\det J_\mathbb{C} f(z) K_{\Omega_2}(f(z), f(\zeta)) \det \overline{J_\mathbb{C} f(\zeta)} = K_{\Omega_1}(z, \zeta).$$

Proof: Let $\phi \in A^2(\Omega_1)$. Then, by change of variable,

$$\int_{\Omega_1} \det J_\mathbb{C} f(z) K_{\Omega_2}(f(z), f(\zeta)) \det \overline{J_\mathbb{C} f(\zeta)} \phi(\zeta) \, dV(\zeta)$$

$$= \int_{\Omega_2} \det J_\mathbb{C} f(z) K_{\Omega_2}(f(z), \widetilde{\zeta}) \det \overline{J_\mathbb{C} f(f^{-1}(\widetilde{\zeta}))} \phi(f^{-1}(\widetilde{\zeta}))$$

$$\times \det J_\mathbb{R} f^{-1}(\widetilde{\zeta}) \, dV(\widetilde{\zeta}).$$

By Proposition 7.1.11 this simplifies to

$$\det J_\mathbb{C} f(z) \int_{\Omega_2} K_{\Omega_2}(f(z), \widetilde{\zeta}) \left\{ \left(\det J_\mathbb{C} f(f^{-1}(\widetilde{\zeta})) \right)^{-1} \phi\left(f^{-1}(\widetilde{\zeta}) \right) \right\} dV(\widetilde{\zeta}).$$

By change of variables, the expression in braces { } is an element of $A^2(\Omega_2)$. So the reproducing property of K_{Ω_2} applies and the last line equals

$$\det J_\mathbb{C} f(z) \, (\det J_\mathbb{C} f(z))^{-1} \phi\left(f^{-1}(f(z)) \right) = \phi(z).$$

By the uniqueness of the Bergman kernel, the proposition follows. □

Proposition 7.1.14 *For $z \in \Omega \subset\subset \mathbb{C}^n$ it holds that $K_\Omega(z, z) > 0$.*

Proof: Let $\{\phi_j\}$ be a complete orthonormal basis for $A^2(\Omega)$. Now

$$K_\Omega(z, z) = \sum_{j=1}^{\infty} |\phi_j(z)|^2 \geq 0.$$

If in fact $K(z, z) = 0$ for some z then $\phi_j(z) = 0$ for all j; hence $f(z) = \sum_j a_j \phi_j(z) = 0$ for every $f \in A^2(\Omega)$. This is absurd. □

It follows from this last proposition that $\log K(z, z)$ makes sense on Ω and is smooth.

Definition 7.1.15 For any $\Omega \subseteq \mathbb{C}^n$ we define a Hermitian metric on Ω by

$$g_{ij}(z) = \frac{\partial^2}{\partial z_i \partial \bar{z}_j} \log K(z, z), \quad z \in \Omega.$$

This means that the square of the length of a tangent vector $\xi = (\xi_1, \ldots, \xi_n)$ at a point $z \in \Omega$ is given by

$$|\xi|_{B,z}^2 = \sum_{i,j} g_{ij}(z)\xi_i \bar{\xi}_j. \tag{7.1.15.1}$$

The metric that we have defined is called the *Bergman metric*.

In a Hermitian metric $\{g_{ij}\}$, the *length* of a C^1 curve $\gamma : [0, 1] \to \Omega$ is given by

$$\ell(\gamma) = \int_0^1 |\gamma_j'(t)|_{B, \gamma(t)}dt = \int_0^1 \left(\sum_{i,j} g_{i,j}(\gamma(t))\gamma_i'(t)\overline{\gamma_j'(t)} \right)^{1/2} dt.$$

If P, Q are points of Ω then their distance $d_\Omega(P, Q)$ in the metric is defined to be the infimum of the lengths of all piecewise C^1 curves connecting the two points.

It is not a priori obvious that the Bergman metric (7.1.15.1) for a bounded domain Ω is given by a positive definite matrix at each point. See [KRA4] for a sketch of the proof of that assertion.

Prelude: Here we see that the Bergman theory gives rise to an entirely new way to construct holomorphically invariant metrics. This is the foundation of Kähler geometry,[2] and of many other key ideas of modern complex differential geometry.

Proposition 7.1.16 *Let $\Omega_1, \Omega_2 \subseteq \mathbb{C}^n$ be domains and let $f : \Omega_1 \to \Omega_2$ be a biholomorphic mapping. Then f induces an isometry of Bergman metrics:*

$$|\xi|_{B,z} = |(J_\mathbb{C}f)\xi|_{B,f(z)}$$

for all $z \in \Omega_1, \xi \in \mathbb{C}^n$. Equivalently, f induces an isometry of Bergman distances in the sense that

$$d_{\Omega_2}(f(P), f(Q)) = d_{\Omega_1}(P, Q).$$

Proof: This is a formal exercise but we include it for completeness:

From the definitions, it suffices to check that

$$\sum g_{i,j}^{\Omega_2}(f(z))(J_\mathbb{C}f(z)w)_i \overline{(J_\mathbb{C}f(z)w)}_j = \sum_{i,j} g_{ij}^{\Omega_1}(z)w_i\bar{w}_j \tag{7.1.16.1}$$

for all $z \in \Omega, w = (w_1, \ldots, w_n) \in \mathbb{C}^n$. But by Proposition 7.1.13,

$$g_{ij}^{\Omega_1}(z) = \frac{\partial^2}{\partial z_i \bar{x}_j} \log K_{\Omega_1}(z, z)$$

---

[2] A Kähler metric is one in which the complex structure tensor is invariant under parallel translation in the metric.

$$= \frac{\partial^2}{\partial z_i \bar{x}_j} \log \left\{ |\det J_{\mathbb{C}} f(z)|^2 K_{\Omega_2}(f(z), f(z)) \right\}$$

$$= \frac{\partial^2}{\partial z_i \bar{x}_j} \log K_{\Omega_2}(f(z), f(z)) \tag{7.1.16.2}$$

since $\log |\det J_{\mathbb{C}} f(z)|^2$ is locally

$$\log (\det J_{\mathbb{C}} f) + \log \overline{(\det J_{\mathbb{C}} f)} + C$$

hence is annihilated by the mixed second derivative. But line (7.1.16.2) is nothing other than

$$\sum_{\ell, m} g_{\ell, m}^{\Omega_2}(f(z)) \frac{\partial f_\ell(z)}{\partial z_i} \frac{\partial \overline{f_m(z)}}{\partial \bar{x}_j}$$

and (7.1.16.1) follows.   □

Proposition 7.1.17 Let $\Omega \subset\subset \mathbb{C}^n$ be a domain. Let $z \in \Omega$. Then

$$K(z, z) = \sup_{f \in A^2(\Omega)} \frac{|f(z)|^2}{\|f\|_{A^2}^2} = \sup_{\|f\|_{A^2(\Omega)} = 1} |f(z)|^2.$$

Proof: Now

$$K(z, z) = \sum |\phi_j(z)|^2$$

$$= \left(\sup_{\|\{a_j\}\|_{\ell^2} = 1} \left| \sum \phi_j(z) a_j \right| \right)^2$$

$$= \sup_{\|f\|_{A^2} = 1} |f(z)|^2,$$

which by the Riesz–Fischer theorem is equal to

$$\sup_{f \in A^2} \frac{|f(z)|^2}{\|f\|_{A^2}^2}. \qquad □$$

7.1.1 Smoothness to the Boundary of K_Ω

It is of interest to know whether K_Ω is smooth on $\overline{\Omega} \times \overline{\Omega}$. We shall see below that the Bergman kernel of the disk D is given by

$$K_D(z, \zeta) = \frac{1}{\pi} \frac{1}{(1 - z \cdot \bar{\zeta})^2}.$$

It follows that $K_D(z, z)$ is smooth on $\overline{D} \times \overline{D} \setminus$ (boundary diagonal) and blows up as $z \to 1^-$. In fact, this property of blowing up along the diagonal at the boundary prevails at any boundary point of a domain at which there is a peaking function (apply Proposition 7.1.17 to a high power of the peaking function). The reference [GAM] contains background information on peaking functions.

However, there is strong evidence that—as long as Ω is smoothly bounded—on compact subsets of

$$\overline{\Omega} \times \overline{\Omega} \setminus ((\partial\Omega \times \partial\Omega) \cap \{z = \zeta\}),$$

the Bergman kernel will be smooth. For strongly pseudoconvex domains, this statement is true; its proof (see [KER]) uses deep and powerful methods of partial differential equations. It is now known that this property fails for the Diederich–Fornæss worm domain (see [DIF] and [LIG] as well as [CHE] and [KRPE]).

Perhaps the most central open problem in the function theory of several complex variables is to prove that a biholomorphic mapping of two smoothly bounded domains extends to a diffeomorphism of the closures. Fefferman [FEF7] proved that if Ω_1, Ω_2 are both strongly pseudoconvex and $\Phi : \Omega_1 \to \Omega_2$ is biholomorphic then Φ does indeed extend to a diffeomorphsm of $\overline{\Omega}_1$ to $\overline{\Omega}_2$. Bell and Ligocka [BELL] and [BEL], using important technical results of Catlin [CAT1], [CAT3], proved the same result for finite type domains in any dimension. The basic problem is still open. There are no smoothly bounded domains in $\mathbb{C}^n$ for which the result is known to be false.

It is known (see [BEB]) that a sufficient condition for this mapping problem to have an affirmative answer on a smoothly bounded domain $\Omega \subseteq \mathbb{C}^n$ is that for any multi-index α there are constants $C = C_\alpha$ and $m = m_\alpha$ such that the Bergman kernel K_Ω satisfies

$$\sup_{z \in \Omega} \left| \frac{\partial^\alpha}{\partial x^\alpha} K_\Omega(z, \zeta) \right| \leq C \cdot d_\Omega(\zeta)^{-m}$$

for all $\zeta \in \Omega$. Here $d_\Omega(w)$ denotes the distance of the point $w \in \Omega$ to the boundary of the domain.

7.1.2 Calculating the Bergman Kernel

The Bergman kernel can almost never be calculated explicitly; unless the domain Ω has a great deal of symmetry—so that a useful orthonormal basis for $A^2(\Omega)$ can be determined (or else Proposition 7.1.13 can be used)—there are few techniques for determining K_Ω.

In 1974, C. Fefferman [FEF7] introduced a new technique for obtaining an asymptotic expansion for the Bergman kernel on a large class of domains. (For an alternative approach see L. Boutet de Monvel and J. Sjöstrand [BMS].) This work enabled rather explicit estimations of the Bergman metric and opened up an entire branch of analysis on domains in $\mathbb{C}^n$ (see [FEF8], [CHM], [KLE], [GRK1]–[GRK11] for example).

The Bergman theory that we have presented here would be a bit hollow if we did not at least calculate the kernel in a few instances. We complete the section by addressing that task.

Restrict attention to the ball $B \subseteq \mathbb{C}^n$. The functions z^α, α a multi-index, are each in $A^2(B)$ and are pairwise orthogonal by the symmetry of the ball. By the uniqueness of the power series expansion for an element of $A^2(B)$, the elements z^α form a complete orthogonal system on B (their closed linear span is $A^2(B)$). Setting

$$\gamma_\alpha = \int_B |z^\alpha|^2 \, dV(z),$$

we see that $\{z^\alpha / \sqrt{\gamma_\alpha}\}$ is a complete orthonornal system in $A^2(B)$. Thus, by Proposition 7.1.8,

$$K_B(z, \zeta) = \sum_\alpha \frac{z^\alpha \overline{\zeta}^\alpha}{\gamma_\alpha}.$$

If we want to calculate the Bergman kernel for the ball in closed form, we need to calculate the γ_α's. This requires some lemmas from real analysis. These lemmas will be formulated and proved on $\mathbb{R}^N$ and $B_N = \{z \in \mathbb{R}^N : |z| < 1\}$.

Prelude: The next simple fact has many different proofs. It is important to know the integral of the Gaussian, since it arises fundamentally in probability theory and harmonic analysis.

Lemma 7.1.18 *We have that*

$$\int_{\mathbb{R}^N} e^{-\pi |x|^2} dx = 1.$$

Proof: The case $N = 1$ is familiar from calculus (or see [STG1]). For the N-dimensional case, write

$$\int_{\mathbb{R}^N} e^{-\pi |x|^2} dx = \int_{\mathbb{R}} e^{-\pi x_1^2} dx_1 \cdots \int_{\mathbb{R}} e^{-\pi x_N^2} dx_N$$

and apply the one-dimensional result.   □

Let σ be the unique rotationally invariant area measure on $S_{N-1} = \partial B_N$ and let $\omega_{N-1} = \sigma(\partial B)$.

Lemma 7.1.19 *We have*

$$\omega_{N-1} = \frac{2\pi^{N/2}}{\Gamma(N/2)},$$

where

$$\Gamma(z) = \int_0^\infty t^{z-1} e^{-t} dt$$

is Euler's gamma function.

Proof: Introducing polar coordinates we have

$$1 = \int_{\mathbb{R}^N} e^{-\pi |x|^2} dx = \int_{S^{N-1}} d\sigma \int_0^\infty e^{-\pi r^2} r^{N-1} dr,$$

or

$$\frac{1}{\omega_{N-1}} = \int_0^\infty e^{-\pi r^2} r^N \frac{dr}{r}.$$

Letting $s = r^2$ in this last integral and doing some obvious manipulations yields the result. $\qquad\square$

Now we return to $B \subseteq \mathbb{C}^n$. We set

$$\eta(k) = \int_{\partial B} |z_1|^{2k} d\sigma(z), \quad N(k) = \int_B |z_1|^{2k} dV(z), \quad k = 0, 1, \ldots.$$

Lemma 7.1.20 *We have*

$$\eta(k) = \pi^n \frac{2(k!)}{(k+n-1)!}, \quad N(k) = \pi^n \frac{k!}{(k+n)!}.$$

Proof: Polar coordinates show easily that $\eta(k) = 2(k+n)N(k)$. So it is enough to calculate $N(k)$. Let $z = (z_1, z_2, \ldots, z_n) = (z_1, z')$. We write

$$N(k) = \int_{|z|<1} |z_1|^{2k} dV(z)$$

$$= \int_{|z'|<1} \left(\int_{|z_1| \le \sqrt{1-|z'|^2}} |z_1|^{2k} dV(z_1) \right) dV(z')$$

$$= 2\pi \int_{|z'|<1} \int_0^{\sqrt{1-|z'|^2}} r^{2k} r\, dr\, dV(z')$$

$$= 2\pi \int_{|z'|<1} \frac{(1-|z'|^2)^{k+1}}{2k+2} dV(z')$$

$$= \frac{\pi}{k+1} \omega_{2n-3} \int_0^1 (1-r^2)^{k+1} r^{2n-3} dr$$

$$= \frac{\pi}{k+1} \omega_{2n-3} \int_0^1 (1-s)^{k+1} s^{n-1} \frac{ds}{2s}$$

$$= \frac{\pi}{2(k+1)} \omega_{2n-3} \beta(n-1, k+2),$$

where β is the classical beta function of special function theory (see [CCP] or [WHW]). By a standard identity for the beta function we then have

$$N(k) = \frac{\pi}{2(k+1)} \omega_{2n-3} \frac{\Gamma(n-1)\Gamma(k+2)}{\Gamma(n+k+1)}.$$

$$= \frac{\pi}{2(k+1)\,\Gamma(n-1)} \frac{2\pi^{n-1}}{\Gamma(n+k+1)} \frac{\Gamma(n-1)\,\Gamma(k+2)}{\Gamma(n+k+1)}$$

$$= \frac{\pi^n k!}{(k+n)!}.$$

This is the desired result.   □

Lemma 7.1.21 Let $z \in B \subseteq \mathbb{C}^n$ and $0 < r < 1$. The symbol **1** denotes the point $(1, 0, \ldots, 0)$. Then

$$K_B(z, r\mathbf{1}) = \frac{n!}{\pi^n} \frac{1}{(1 - rz_1)^{n+1}}.$$

Proof: Refer to the formula preceding Lemma 7.1.18. Then

$$K_B(z, r\mathbf{1}) = \sum_{\alpha} \frac{z^\alpha (r\mathbf{1})^\alpha}{\gamma_\alpha} = \sum_{k=0}^{\infty} \frac{z_1^k r^k}{N(k)}$$

$$= \frac{1}{\pi^n} \sum_{k=0}^{\infty} (rz_1)^k \cdot \frac{(k+n)!}{k!}$$

$$= \frac{n!}{\pi^n} \sum_{k=0}^{\infty} (rz_1)^k \binom{k+n}{n}$$

$$= \frac{n!}{\pi^n} \cdot \frac{1}{(1 - rz_1)^{n+1}}.$$

This is the desired result.   □

Theorem 7.1.22 If $z, \zeta \in B$ then

$$K_B(z, \zeta) = \frac{n!}{\pi^n} \frac{1}{(1 - z \cdot \bar{\zeta})^{n+1}},$$

where $z \cdot \bar{\zeta} = z_1 \bar{\zeta}_1 + z_2 \bar{\zeta}_2 + \cdots + z_n \bar{\zeta}_n$.

Proof: Let $z = r\tilde{z} \in B$, where $r = |z|$ and $|\tilde{z}| = 1$. Also, fix $\zeta \in B$. Choose a unitary rotation ρ such that $\rho\tilde{z} = \mathbf{1}$. Then, by Lemmas 7.1.13 and 7.1.21 we have

$$K_B(z, \zeta) = K_B(r\tilde{z}, \zeta) = K(r\rho^{-1}\mathbf{1}, \zeta)$$

$$= K(r\mathbf{1}, \rho\zeta) = \overline{K(\rho\zeta, r\mathbf{1})}$$

$$= \frac{n!}{\pi^n} \cdot \frac{1}{\left(1 - r\overline{(\rho\zeta)_1}\right)^{n+1}}$$

$$= \frac{n!}{\pi^n} \cdot \frac{1}{\left(1 - (r\mathbf{1}) \cdot \overline{(\rho\zeta)}\right)^{n+1}}$$

$$= \frac{n!}{\pi^n} \cdot \frac{1}{\left(1 - (r\rho^{-1}\mathbf{1}) \cdot \overline{\zeta}\right)^{n+1}}$$

$$= \frac{n!}{\pi^n} \cdot \frac{1}{(1 - z \cdot \overline{\zeta})^{n+1}}. \qquad \Box$$

Corollary 7.1.23 *The Bergman kernel for the disk is given by*

$$K_D(z, \zeta) = \frac{1}{\pi} \cdot \frac{1}{(1 - z \cdot \overline{\zeta})^2}.$$

Prelude: As with the Bergman kernel, the Bergman metric is generally quite difficult (or impossible) to compute. The paper [FEF7] gives a very useful asymptotic formula for the Bergman metric on a strongly pseudoconvex domain.

Proposition 7.1.24 *The Bergman metric for the ball $B = B(0, 1) \subseteq \mathbb{C}^n$ is given by*

$$g_{ij}(z) = \frac{n+1}{(1 - |z|^2)^2}[(1 - |z|^2)\delta_{ij} + \overline{z}_i z_j].$$

Proof: Since $K(z, z) = n!/(\pi^n(1 - |z|^2)^{n+1})$, this is a routine computation that we leave to the reader. $\qquad \Box$

Corollary 7.1.25 *The Bergman metric for the disk (i.e., the ball in dimension one) is*

$$g_{ij}(z) = \frac{2}{(1 - |z|^2)^2}, \quad i = j = 1.$$

This is the well-known Poincaré, or Poincaré–Bergman, metric.

Proposition 7.1.26 *The Bergman kernel for the polydisk $D^n(0, 1) \subseteq \mathbb{C}^n$ is the product*

$$K(z, \zeta) = \frac{1}{\pi^n} \prod_{j=1}^{n} \frac{1}{(1 - z_j \overline{\zeta}_j)^2}.$$

Proof: Exercise: Use the uniqueness property of the Bergman kernel. $\qquad \Box$

Exercise for the Reader: Calculate the Bergman metric for the polydisk.

It is a matter of great interest to calculate the Bergman kernel on various domains. This is quite difficult, even for domains in the plane. Just as an instance, an explicit formula for the Bergman kernel on the annulus involves elliptic functions (see [BERG]). In several complex variables matters are considerably more complicated. It was a watershed event when, in 1974, Charles Fefferman [FEF7] was able to calculate an *asymptotic expansion* for the Bergman kernel on a strongly pseudoconvex domain. Thus Fefferman was able to write

$$K(z, \zeta) = P(z, \zeta) + E(z, \zeta),$$

where P is an explicitly given *principal term* and E is an error term that is measurably smaller, or more tame, than the principal term. Our material on pseudodifferential operators (see Appendix 3) should give the reader some context in which to interpret these remarks. Fefferman used this asymptotic expansion to obtain information about the boundary behavior of geodesics in the Bergman metric. This in turn enabled him to prove the stunning result (for which he won the Fields Medal) that a biholomorphic mapping of smoothly bounded, strongly pseudoconvex domains will extend to a diffeomorphism of the closures.

Work continues on understanding Bergman kernels on a variety of domains in different settings. One recent result of some interest is that Krantz and Peloso [KRPE] have found an asymptotic expansion for the Bergman kernel on the worm domain of Kiselman [KIS] and Diederich–Fornæss [DIF].[3]

7.2 The Szegő Kernel

Capsule: The Szegő kernel is constructed with a mechanism similar to that used for the Bergman kernel, but focused on a different function space (H^2 instead of A^2). The Szegő kernel does not have all the invariance properties of the Bergman kernel; it does not give rise to a new invariant metric. But it is still a powerful tool in function theory. It also gives rise to the not-very-well-known Poisson–Szegő kernel, a positive reproducing kernel that is of considerable utility.

The basic theory of the Szegő kernel is similar to that for the Bergman kernel—they are both special cases of a general theory of "Hilbert spaces with reproducing kernel" (see N. Aronszajn [ARO]). Thus we only outline the basic steps here, leaving details to the reader.

Let $\Omega \subseteq \mathbb{C}^n$ be a bounded domain with C^2 boundary. Let $A(\Omega)$ be those functions continuous on $\overline{\Omega}$ that are holomorphic on Ω. Let $H^2(\partial\Omega)$ be the space consisting of the closure in the $L^2(\partial\Omega, d\sigma)$ topology of the restrictions to $\partial\Omega$ of elements of $A(\Omega)$ (see also our treatment in Chapter 8). Then $H^2(\partial\Omega)$ is a proper Hilbert subspace of $L^2(\partial\Omega, \partial\sigma)$. Here $d\sigma$ is the $(2n-1)$-dimensional area measure (i.e., Hausdorff measure—see Section 9.9.3) on $\partial\Omega$. Each element $f \in H^2(\partial\Omega)$ has a natural holomorphic extension to Ω given by its Poisson integral Pf. We prove in the next chapter that for σ-almost every $\zeta \in \partial\Omega$, it holds that

$$\lim_{\epsilon \to 0^+} f(\zeta - \epsilon v_\zeta) = f(\zeta).$$

Here, as usual, v_ζ is the unit outward normal to $\partial\Omega$ at the point ζ.

---

[3] This was an important example from 1976 of a pseudoconvex domain with special geometric properties—namely the closure of the worm does not have a Stein neighborhood basis. In recent years the worm has proved to be important in studies of the $\overline{\partial}$ operator and the Bergman theory. The book [CHS] provides an excellent exposition of some of these ideas.

For each fixed $z \in \Omega$ the functional

$$\psi_z : H^2(\Omega) \ni f \mapsto Pf(z)$$

is continuous (why?). Let $k_z(\zeta)$ be the Hilbert space representative (given by the Riesz representation theorem) for the functional ψ_z. Define the Szegő kernel $S(z, \zeta)$ by the formula

$$S(z, \zeta) = \overline{k_z(\zeta)}, \quad z \in \Omega, \ \zeta \in \partial\Omega.$$

If $f \in H^2(\partial\Omega)$ then

$$Pf(z) = \int_{\partial\Omega} S(z, \zeta) f(\zeta) d\sigma(\zeta)$$

for all $z \in \Omega$. We shall not explicitly formulate and verify the various uniqueness and extremal properties for the Szegő kernel. The reader is invited to consider these topics, referring to Section 7.1 for statements.

Let $\{\phi_j\}_{j=1}^{\infty}$ be an orthonormal basis for $H^2(\partial\Omega)$. Define

$$S'(z, \zeta) = \sum_{j=1}^{\infty} \phi_j(z) \overline{\phi_j(\zeta)}, \quad z, \ \zeta \in \Omega.$$

For convenience we tacitly identify here each function with its Poisson extension to the interior of the domain. Then, for $K \subseteq \Omega$ compact, the series defining S' converges uniformly on $K \times K$. By a Riesz–Fischer argument, $S'(\cdot, \zeta)$ is the Poisson integral of an element of $H^2(\partial\Omega)$, and $S'(z, \cdot)$ is the conjugate of the Poisson integral of an element of $H^2(\partial\Omega)$. So S' extends to $(\overline{\Omega} \times \Omega) \cup (\Omega \times \overline{\Omega})$, where it is understood that all functions on the boundary are defined only almost everywhere. The kernel S' is conjugate symmetric. Also, by Riesz–Fischer theory, S' reproduces $H^2(\partial\Omega)$. Since the Szegő kernel is unique, it follows that $S = S'$.

The Szegő kernel may be thought of as representing a map

$$S : f \mapsto \int_{\partial\Omega} f(\zeta) S(\cdot, \zeta) d\sigma(\zeta)$$

from $L^2(\partial\Omega)$ to $H^2(\partial\Omega)$. Since S is self-adjoint and idempotent, it is the Hilbert space projection of $L^2(\partial\Omega)$ to $H^2(\partial\Omega)$.

Prelude: The Poisson–Szegő kernel is obtained by a formal procedure from the Szegő kernel: this procedure manufactures a *positive* reproducing kernel from one that is not in general positive. Note in passing that just as we argued for the Bergman kernel in the last section, $S(z, z)$ is always positive when $z \in \Omega$. The history of the Poisson–Szegő kernel is obscure, but seeds of the idea seem to have occurred in [HUA] and in later work of Koranyi [KOR3].

Proposition 7.2.1 *Define*

$$\mathcal{P}(z, \zeta) = \frac{|S(z, \zeta)|^2}{S(z, z)}, \quad z \in \Omega, \ \zeta \in \partial\Omega.$$

Then $\mathcal{P}$ is positive and, for any $f \in A(\Omega)$ and $z \in \Omega$, it holds that

$$f(z) = \int_{\partial\Omega} f(\zeta)\mathcal{P}(z,\zeta)d\sigma(\zeta).$$

Proof: Fix $z \in \Omega$ and $f \in A(\Omega)$ and define

$$u(\zeta) = f(\zeta)\frac{\overline{S(z,\zeta)}}{S(z,z)}, \quad \zeta \in \partial\Omega.$$

Then $u \in H^2(\partial\Omega)$; hence

$$f(z) = u(z)$$

$$= \int_{\partial\Omega} S(z,\zeta)u(\zeta)d\sigma(\zeta)$$

$$= \int_{\partial\Omega} \mathcal{P}(z,\zeta)f(\zeta)d\sigma(\zeta).$$

This is the desired formula.   □

Remark: In passing to the Poisson–Szegő kernel we gain the advantage of positivity of the kernel (for more on this circle of ideas, see Chapter 1 of [KAT]). However, we lose something in that $\mathcal{P}(z,\zeta)$ is no longer holomorphic in the z variable nor conjugate holomorphic in the ζ variable. The literature on this kernel is rather sparse and there are many unresolved questions.

As an exercise, use the paradigm of Proposition 7.2.1 to construct a positive kernel from the Cauchy kernel on the disk (be sure to first change notation in the usual Cauchy formula so that it is written in terms of arc length measure on the boundary). What familiar kernel results?

Like the Bergman kernel, the Szegő and Poisson–Szegő kernels can almost never be explicitly computed. They can be calculated asymptotically in a number of important instances, however (see [FEF7], [BMS]). We shall give explicit formulas for these kernels on the ball. The computations are similar in spirit to those in Section 7.1; fortunately, we may capitalize on much of the work done there.

Lemma 7.2.2 *The functions $\{z^\alpha\}$, where α ranges over multi-indices, are pairwise orthogonal and span $H^2(\partial B)$.*

Proof: The orthogonality follows from symmetry considerations. For the completeness, notice that it suffices to see that the span of $\{z^\alpha\}$ is dense in $A(B)$ in the uniform topology on the boundary. By the Stone–Weierstrass theorem, the closed uniform algebra generated by $\{z^\alpha\}$ and $\{\bar{z}^\alpha\}$ is all of $C(\partial B)$. But the monomials $\bar{z}^\alpha$, $\alpha \neq 0$, are orthogonal to $A(B)$ (use the power series expansion about the origin to see this). The claimed density follows.   □

Lemma 7.2.3 *Let* $\mathbf{1} = (1, 0, \ldots, 0)$. *Then*

$$S(z, \mathbf{1}) = \frac{(n-1)!}{2\pi^n} \frac{1}{(1-z_1)^n}.$$

Proof: We have that

$$S(z, \mathbf{1}) = \sum_\alpha \frac{z^\alpha \cdot \mathbf{1}^\alpha}{\|z_1^\alpha\|^2_{L^2(\partial B)}}$$

$$= \sum_{k=0}^\infty \frac{z_1^k}{\eta(k)}$$

$$= \frac{1}{2\pi^n} \sum_{k=0}^\infty \frac{z_1^k (k+n-1)!}{k!}$$

$$= \frac{(n-1)!}{2\pi^n} \sum_{k=0}^\infty \binom{k+n-1}{n-1} z_1^k$$

$$= \frac{(n-1)!}{2\pi^n} \frac{1}{(1-z_1)^n}. \qquad \square$$

Lemma 7.2.4 *Let* ρ *be a unitary rotation on* $\mathbb{C}^n$. *For any* $z \in \overline{B}, \zeta \in \partial B$, *we have that* $S(z, \zeta) = S(\rho z, \rho\zeta)$.

Proof: This is a standard change of variables argument and we omit it. $\qquad \square$

Theorem 7.2.5 *The Szegő kernel for the ball is*

$$S(z, \zeta) = \frac{(n-1)!}{2\pi^n} \frac{1}{(1-z \cdot \overline{\zeta})^n}.$$

Proof: Let $z \in B$ be arbitrary. Let ρ be the unique unitary rotation such that ρz is a multiple of $\mathbf{1}$. Then, by Lemma 7.2.4,

$$S(z, \zeta) = S(\rho^{-1}\mathbf{1}, \zeta)$$

$$= S(\mathbf{1}, \rho\zeta)$$

$$= \overline{S(\rho\zeta, \mathbf{1})}$$

$$= \frac{(n-1)!}{2\pi^n} \frac{1}{\left(1 - (\overline{\rho\zeta}) \cdot \mathbf{1}\right)^n}$$

$$= \frac{(n-1)!}{2\pi^n} \frac{1}{\left(1 - \bar{\zeta} \cdot (\rho^{-1}\mathbf{1})\right)^n}$$

$$= \frac{(n-1)!}{2\pi^n} \frac{1}{(1 - z \cdot \bar{\zeta})^n}. \qquad \square$$

Corollary 7.2.6 *The Poisson–Szegő kernel for the ball is*

$$\mathcal{P}(z, \zeta) = \frac{(n-1)!}{2\pi^n} \frac{(1 - |z|^2)^n}{|1 - z \cdot \bar{\zeta}|^{2n}}.$$

Exercise for the Reader: Calculate the Szegő and Poisson–Szegő kernels for the polydisk.

Hardy Spaces Old and New

Prologue: The theory of Hardy spaces dates back to G.H. Hardy and M. Riesz in the early twentieth century. Part of the inspiration here is the celebrated theorem of P. Fatou that a *bounded* holomorphic function on the unit disk D has radial (indeed nontangential) boundary limits almost everywhere. Hardy and Riesz wished to expand the space of holomorphic functions for which such results could be obtained.

The motivation, and the ultimate payoff, for this work on the disk is obvious. For theorems like those described in the last paragraph link the holomorphic function theory of the disk to the Fourier analysis of the boundary. The result is a deep and powerful theory that continues even today to fascinate and to yield new and profound ideas.

In several-variables the picture is a bit reversed. By dint of a huge effort by A. Koranyi, E.M. Stein, and many others, there has arisen a theory of Hardy spaces on domains in $\mathbb{C}^n$. But this was done in the absence of a preexisting Fourier analysis on the boundary. That was developed somewhat later, using independent techniques. There is still a fruitful symbiosis between the boundary theory and the holomorphic function theory on the interior, but this is still in the developmental stages. Work continues on this exciting path.

The work described in the last paragraph was facilitated on the unit ball $B \subseteq \mathbb{C}^n$ because the boundary of the ball may be canonically identified with the Heisenberg group. Of course the Heisenberg group is a natural venue (as we now understand, and as is explained in the last two chapters of this book) for harmonic analysis. On the boundary of a general strongly pseudoconvex domain there is generally no group action, so no natural Fourier analysis is possible. There are, however, various approximation procedures that can serve as a substitute (see [FOST1]). Also the work [NAS] provides a calculus of pseudodifferential operators that can serve on the boundaries of strongly pseudoconvex domains and also certain finite type domains.

There are now several different approaches to the study of the boundary behavior of holomorphic functions on domains in $\mathbb{C}^n$. Certainly Koranyi [KOR1], [KOR2] and E.M. Stein [STE4] were the pioneers of the modern

S.G. Krantz, *Explorations in Harmonic Analysis,* Applied and Numerical
Harmonic Analysis, DOI 10.1007/978-0-8176-4669-1_8,
© Birkhäuser Boston, a part of Springer Science+Business Media, LLC 2009

theory (there were earlier works by Zygmund and others on special domains). Barker [BAR] and Krantz [KRA7] and Lempert [LEM] have other approaches to the matter. Di Biase [DIB] has made some remarkable contributions using graph theory. The ideas continue to develop.

The study of the boundary behavior of holomorphic functions on the unit disk is a venerable one. Beginning with the thesis of P. Fatou in 1906—in which the boundary behavior of *bounded* holomorphic functions was considered—the field has blossomed and grown to consider other classes of holomorphic functions, and a variety of means of deriving the boundary limits. This work has both aesthetic appeal and manifold applications to various parts of function theory, harmonic analysis, and partial differential equations. This chapter provides an introduction to the key ideas in both one complex variable and several complex variables.

8.1 Hardy Spaces on the Unit Disk

Capsule: The Hardy space (H^p) theory was born on the unit disk in the complex plane. In that context the Cauchy kernel and the Poisson kernel are familiar tools, and they both play a decisive role in the development of the theory. The set of ideas transfers rather naturally to the upper half-plane by way of the Cayley transform. In $\mathbb{R}^N$ on the unit ball there is certainly a theory of boundary values for *harmonic* functions; but the range of p for which the theory is valid is restricted to $p > 1$. Certainly, with suitable estimates on the Poisson kernel (see [KRA4, Ch. 8]), these ideas can be transferred to any domain in $\mathbb{R}^N$ with smooth boundary. For domains in $\mathbb{C}^n$, Koranyi began his investigations in 1969 (see also [KOR3] from 1965). Stein made his decisive contribution in 1972. Barker's contribution was in 1978 and the work of Lempert and Krantz was later still. Di Biase's thesis was published in 1998. And the book of Di Biase and Krantz [DIK] is forthcoming.

Throughout this section we let $D \subseteq \mathbb{C}$ denote the unit disk. Let $0 < p < \infty$. We define

$$H^p(D) = \left\{ f \text{ holomorphic on } D : \sup_{0<r<1} \left[\frac{1}{2\pi} \int_0^{2\pi} |f(re^{i\theta})|^p d\theta \right]^{1/p} \right.$$

$$\left. \equiv \|f\|_{H^p} < \infty \right\}$$

Also define

$$H^\infty(D) = \left\{ f \text{ holomorphic on } D : \sup_D |f| \equiv \|f\|_{H^p} < \infty \right\}.$$

The fundamental result in the subject of H^p, or *Hardy*, spaces (and also one of the fundamental results of this section) is that if $f \in H^p(D)$ then the limit

$$\lim_{r \to 1^-} f(re^{i\theta}) \equiv \widetilde{f}(e^{i\theta})$$

exists for almost every $\theta \in [0, 2\pi)$. For $1 \leq p \leq \infty$, the function f can be recovered from $\widetilde{f}$ by way of the Cauchy or Poisson (or even the Szegő) integral formulas; for $p < 1$ this "recovery" process is more subtle and must proceed by way of distributions. Once this pointwise boundary limit result is established, then an enormous and rich mathematical structure unfolds (see [KAT], [HOF], [GAR]).

Recall from Chapter 1 that the Poisson kernel for the disk is

$$P_r(e^{i\theta}) = \frac{1}{2\pi} \frac{1 - r^2}{1 - 2r\cos\theta + r^2}.$$

Our studies will be facilitated by first considering the boundary behavior of *harmonic* functions.

Let

$$\mathbf{h}^p(D) = \left\{ f \text{ harmonic on } D : \sup_{0 < r < 1} \left[\frac{1}{2\pi} \int_0^{2\pi} |f(re^{i\theta})|^p d\theta \right]^{1/p} \right.$$

$$\equiv \|f\|_{\mathbf{h}^p} < \infty \right\}$$

and

$$\mathbf{h}^\infty(D) = \left\{ f \text{ harmonic on } D : \sup_D |f| \equiv \|f\|_{\mathbf{h}^\infty} < \infty \right\}.$$

Throughout this section, arithmetic and measure theory on $[0, 2\pi)$ (equivalently on ∂D by way of the map $\theta \mapsto e^{i\theta}$) is done by identifying $[0, 2\pi)$ with $\mathbb{R}/2\pi\mathbb{Z}$. See [KAT] for more on this identification procedure.

Prelude: It is important here to get the logic sorted out. Our primary interest is in the boundary behavior of holomorphic (i.e., H^p) functions. But it turns out to be useful as a tool to study at first the boundary limits of harmonic functions. And that topic has its own interest as well.

The harmonic functions (in $\mathbf{h}^p$) are well behaved when $1 < p \leq \infty$. But the holomorphic functions (in H^p) work out nicely for all $0 < p \leq \infty$. The details of these assertions are fascinating, for they give a glimpse of a number of important techniques. And they also lead to important new ideas, such as the real-variable theory of Hardy spaces (for which see [KRA5]).

Proposition 8.1.1 *Let $1 < p \leq \infty$ and $f \in \mathbf{h}^p(D)$. Then there is an $\widetilde{f} \in L^p(\partial D)$ such that*

$$f(re^{i\theta}) = \int_0^{2\pi} \widetilde{f}(e^{i\psi}) P_r(e^{i(\theta - \psi)}) d\psi.$$

Proof: Define $f_r(e^{i\theta}) = f(re^{i\theta})$, $0 < r < 1$. Then $\{f_r\}_{0 < r < 1}$ is a bounded subset of $(L^{p'}(\partial D))^*$, $p' = p/(p-1)$. By the Banach–Alaoglu theorem (see [RUD3]), there is a subsequence f_{r_j} that converges weak-$*$ to some $\widetilde{f}$ in $L^p(\partial D)$. For any $0 < r < 1$, let $r < r_j < 1$. Then

$$f(re^{i\theta}) = f_{r_j}((r/r_j)e^{i\theta})) = \int_0^{2\pi} f_{r_j}(e^{i\psi})P_{r/r_j}(e^{i(\theta-\psi)})d\psi$$

because $f_{r_j} \in C(\overline{D})$. Now $P_{r/r_j} \in C(\partial D) \subseteq L^{p'}(\partial D)$. Thus the right-hand side of the last equation is

$$\int_0^{2\pi} f_{r_j}(e^{i\psi})P_r(e^{i(\theta-\psi)})d\psi + \int_0^{2\pi} f_{r_j}(e^{i\psi})[P_{r/r_j}(e^{i(\theta-\psi)}) - P_r(e^{i(\theta-\psi)})]d\psi.$$

As $j \to \infty$, the second integral vanishes (because the expression in brackets converges uniformly to 0) and the first integral tends to

$$\int_0^{2\pi} \widetilde{f}(e^{i\psi})P_r(e^{i(\theta-\psi)})d\psi$$

by weak-$*$ convergence. This is the desired result. $\qquad\qquad\square$

Remark: It is easy to see that the proof breaks down for $p = 1$ since L^1 is not the dual of any Banach space. This breakdown is not merely ostensible: the harmonic function

$$f(re^{i\theta}) = P_r(e^{i\theta})$$

satisfies

$$\sup_{0<r<1} \int_0^{2\pi} |f(re^{i\theta})|d\theta < \infty,$$

but the Dirac δ mass is the only measure of which f is the Poisson integral.

Exercise for the Reader: If $f \in h^1$ then there is a Borel measure μ_f on ∂D such that $f(re^{i\theta}) = P_r(\mu_f)(e^{i\theta})$.

Proposition 8.1.2 Let $f \in L^p(\partial D)$, $1 \le p < \infty$. Then $\lim_{r\to 1^-} P_r f = f$ in the L^p norm.

Remark: The result is false for $p = \infty$ if f is discontinuous. The correct analogue in the uniform case is that if $f \in C(\partial D)$ then $P_r f \to f$ uniformly.

As an exercise, consider a Borel measure μ on ∂D. Show that its Poisson integral converges (in a suitable sense) in the weak-$*$ topology to μ.

Note that we considered results of this kind, from another point of view (i.e., that of Fourier series), in Chapter 2.

Proof of Proposition 8.1.2: If $f \in C(\partial D)$, then the result is clear by the solution of the Dirichlet problem. If $f \in L^p(\partial D)$ is arbitrary, let $\epsilon > 0$ and choose $g \in C(\partial D)$ such that $\|f - g\|_{L^p} < \epsilon$. Then

$$\|P_r f - f\|_{L^p} \le \|P_r(f - g)\|_{L^p} + \|P_r g - g\|_{L^p} + \|g - f\|_{L^p}$$

$$\leq \|P_r\|_{L^1} \|f - g\|_{L^p} + \|P_r g - g\|_{L^p} + \epsilon$$

$$\leq \epsilon + o(1) + \epsilon$$

as $r \to 1^-$. $\square$

We remind the reader of the following result, which we saw earlier in another guise in Chapter 2. In that context we were not considering harmonic extensions, but were rather concerned with pointwise summability of Fourier series. We will see that all these different points of view lead us to the same place.

First we recall the definition of the nontangential approach regions.

Definition 8.1.3 Let $\alpha > 1$ and $P \in \partial D$. The *nontangential approach region* or *Stolz region* in D at the point P with aperture α is given by

$$\Gamma_\alpha(P) = \{z \in D : |z - P| < \alpha(1 - |z|)\}.$$

See Figure 8.1.

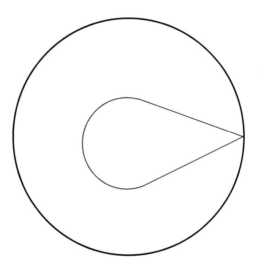

Figure 8.1. A nontangential approach region.

Prelude: Even in the original and seminal paper [FAT], nontangential approach regions are considered. For a variety of reasons—from the point of view of both boundary uniqueness and boundary regularity—nontangential approach is much more powerful and much more useful than radial approach. See [GAR] and [KOO] for some of the details.

Theorem 8.1.4 Let $f \in \mathbf{h}^p(D)$ and $1 < p \leq \infty$. Let $\widetilde{f}$ be as in Proposition 8.1.1 and $1 < \alpha < \infty$. Then

$$\lim_{\Gamma_\alpha(e^{i\theta}) \ni z \to e^{i\theta}} f(z) = \widetilde{f}(e^{i\theta}), \quad a.e. \ e^{i\theta} \in \partial D.$$

The informal statement of Theorem 8.1.4 is that f has nontangential boundary limits almost everywhere. We shall prove this result as the section develops. Recall from Chapter 2 that we can pointwise majorize the Poisson integral by the Hardy–Littlewood maximal function. That result serves us, together with Functional Analysis Principle II, in good stead here to prove the last result.

We shall need a basic covering lemma in order to control the relevant maximal function. See Lemma A1.6.2.

Definition 8.1.5 If $f \in L^1(\partial D)$, let

$$Mf(\theta) = \sup_{R>0} \frac{1}{2R} \int_{-R}^{R} |f(e^{i(\theta-\psi)})| d\psi.$$

The function Mf is called the *Hardy–Littlewood maximal function* of f.

Definition 8.1.6 Let (X, μ) be a measure space and let $f : X \to \mathbb{C}$ be measurable. We say that f is of *weak-type* p, $0 < p < \infty$, if $\mu\{x : |f(x)| > \lambda\} \leq C/\lambda^p$, for all $0 < \lambda < \infty$. The space weak-type ∞ is defined to be L^∞.

Lemma 8.1.7 (Chebyshev's inequality) *If $f \in L^p(X, d\mu)$, then f is weak-type p, $1 \leq p < \infty$.*

Proof: Let $\lambda > 0$. Then

$$\mu\{x : |f(x)| > \lambda\} \leq \int_{\{x:|f(x)|>\lambda\}} |f(x)|^p/\lambda^p \, d\mu(x) \leq \lambda^{-p} \|f\|_{L^p}^p. \qquad \square$$

Exercise for the Reader: There exist functions that are of weak-type p but not in L^p, $1 \leq p < \infty$.

Definition 8.1.8 An operator $T : L^p(X, d\mu) \to$ {measurable functions} is said to be of *weak-type* (p, p), $0 < p < \infty$, if

$$\mu\{x : |Tf(x)| > \lambda\} \leq C\|f\|_{L^p}^p/\lambda^p, \qquad \text{for all} \quad f \in L^p, \quad \lambda > 0.$$

Proposition 8.1.9 *Let $K \subseteq \partial D$ be a compact set and let $\{I_\alpha\}_{\alpha \in A}$ be a covering of K by open intervals. There is a subcollection $I_{\alpha_1}, I_{\alpha_2}, \ldots, I_{\alpha_p}$ that still covers K in such a way that no point is covered more than twice. We say that the subcovering has valence at most 2.*

Proof: This is a special case of a theorem of Lebesgue that characterizes dimension. See [HUW] for the details. $\qquad \square$

Proposition 8.1.10 *The operator M is of weak-type $(1, 1)$.*

Proof: Let $\lambda > 0$. Set $S_\lambda = \{\theta : |Mf(e^{i\theta})| > \lambda\}$. Let $K \subseteq S_\lambda$ be a compact subset with $2m(K) \geq m(S_\lambda)$. For each $k \in K$, there is an interval $I_k \ni k$

with $|I_k|^{-1} \int_{I_k} |f(e^{i\psi})| d\psi > \lambda$. Then $\{I_k\}_{k \in K}$ is an open cover of K. By Proposition 8.1.9, there is a subcover $\{I_{k_j}\}_{j=1}^M$ of K of valence not exceeding 2. Then

$$m(S_\lambda) \leq 2m(K) \leq 2m \left(\bigcup_{j=1}^M I_{k_j} \right) \leq 2 \sum_{j=1}^M m(I_{k_j})$$

$$\leq \sum_{j=1}^M \frac{2}{\lambda} \int_{I_{k_j}} |f(e^{i\psi})| d\psi$$

$$\leq \frac{4}{\lambda} \|f\|_{L^1}. \tag{8.1.10.1}$$

$\square$

Proposition 8.1.11 *If $e^{i\theta} \in \partial D$, $1 < \alpha < \infty$, then there is a constant $C_\alpha > 0$ such that if $f \in L^1(\partial D)$, then*

$$\sup_{re^{i\phi} \in \Gamma_\alpha(e^{i\theta})} |P_r f(e^{i\phi})| \leq C_\alpha M f(e^{i\theta}).$$

Proof: This result is treated in detail in Appendix 1. See particularly Proposition A1.6.4 and its proof at the end of the appendix. $\square$

Prelude: Now we turn our attention to extending Theorem 8.1.4 to Hardy spaces H^p when $0 < p < 1$. In the classical setting, the primary device for establishing such a result is the Blaschke products.

Definition 8.1.12 *If $a \in \mathbb{C}$, $|a| < 1$, then the Blaschke factor at a is*

$$B_a(z) = \frac{z - a}{1 - \bar{a}z}.$$

It is elementary to verify that B_a is holomorphic on a neighborhood of $\overline{D}$ and that $|B_a(e^{i\theta})| = 1$ for all θ.

Lemma 8.1.13 *If $0 < r < 1$ and f is holomorphic on a neighborhood of $\overline{D}(0, r)$, let $p_1, \ldots, p_k$ be the zeros of f (listed with multiplicity) in $D(0, r)$. Assume that $f(0) \neq 0$ and that $f(re^{it}) \neq 0$, for all t. Then*

$$\log |f(0)| + \log \prod_{j=1}^k r|p_j|^{-1} = \frac{1}{2\pi} \int_0^{2\pi} \log |f(re^{it})| dt.$$

Proof: Omitted. See [KRA4, Ch. 8]. $\square$

Notice that, by the continuity of the integral, Lemma 8.1.13 holds even if f has zeros on $\{re^{it}\}$.

Corollary 8.1.14 *If f is holomorphic in a neighborhood of $\overline{D}(0,r)$ then*

$$\log |f(0)| \le \frac{1}{2\pi} \int_0^{2\pi} \log |f(re^{it})| dt.$$

Proof: Omitted. See [KRA4, Ch. 8]. □

Corollary 8.1.15 *If f is holomorphic on D, $f(0) \ne 0$, and $\{p_1, p_2, \dots\}$ are the zeros of f counting multiplicities, then*

$$\log |f(0)| + \log \prod_{j=1}^{\infty} \frac{1}{|p_j|} \le \sup_{0<r<1} \frac{1}{2\pi} \int_0^{2\pi} \log^+ |f(re^{it})| dt.$$

Proof: Omitted. See [KRA4, Ch. 8]. □

Prelude: The characterization of the zero sets of H^p functions on the disk is elegant and incisive. And it is extremely useful. Note particularly that the condition is the *same* for every H^p space, $0 < p < \infty$. Such is not the case in several complex variables. First, necessary and sufficient conditions for a variety to be the zero set of an H^p function are not known. Some sufficient conditions are known, but they are so imprecise that we can say only that the variety is the zero set of some H^p function for some p—but we cannot say which p.

Corollary 8.1.16 *If $f \in H^p(D)$, $0 < p \le \infty$, and $\{p_1, p_2, \dots\}$ are the zeros of f counting multiplicities, then $\sum_{j=1}^{\infty}(1 - |p_j|) < \infty$.*

Proof: Since f vanishes to some finite order k at 0, we may replace f by $f(z)/z^k$ and assume that $f(0) \ne 0$. It follows from Corollary 8.1.15 that

$$\log \prod_{j=1}^{\infty} \left\{ \frac{1}{|p_j|} \right\} < \infty$$

or $\prod(1/|p_j|)$ converges; hence $\prod |p_j|$ converges. So $\sum_j (1 - |p_j|) < \infty$. □

Proposition 8.1.17 *If $\{p_1, p_2, \dots\} \subseteq D$ satisfy $\sum_j (1 - |p_j|) < \infty$, $p_j \ne 0$ for all j, then*

$$\prod_{j=1}^{\infty} \frac{-\overline{p}_j}{|p_j|} B_{p_j}(z)$$

converges normally on D.

Proof: Restrict attention to $|z| \le r < 1$. Then the assertion that the infinite product converges uniformly on this disk is equivalent to the assertion that

$$\sum_j \left| 1 + \frac{\overline{p}_j}{|p_j|} B_{p_j}(z) \right|$$

converges uniformly. But

$$\left| 1 + \frac{\overline{p}_j}{|p_j|} B_{p_j}(z) \right| = \left| \frac{|p_j| - |p_j|\overline{p}_j z + \overline{p}_j z - |p_j|^2}{|p_j|(1 - z\overline{p}_j)} \right|$$

$$= \left| \frac{(|p_j| + z\overline{p}_j)(1 - |p_j|)}{|p_j|(1 - z\overline{p}_j)} \right|$$

$$\leq \frac{(1 + r)(1 - |p_j|)}{1 - r},$$

so the convergence is uniform.   □

Corollary 8.1.18 *Fix* $0 < p \leq \infty$. *If* a_j *are complex constants with* $\sum_j (1 - |a_j|) < \infty$ *then there is an* $f \in H^p(D)$ *with zero set* $\{a_j\}$.

Prelude: There is nothing like a theory of Blaschke products in several complex variables. In fact, it is only a recent development that we know that there are *inner functions* in several complex variables—that is, holomorphic functions with unimodular boundary values almost everywhere. Although these inner functions can be used to effect a number of remarkable constructions, they have not proved to be nearly as useful as the inner functions and the corresponding canonical factorization in one complex variable. See [KRA4] for the former and [HOF] for the latter.

Definition 8.1.19 Let $0 < p \leq \infty$ and $f \in H^p(D)$. Let $\{p_1, p_2, \dots\}$ be the zeros of f counted according to multiplicities. Let

$$B(z) = \prod_{j=1}^{\infty} \frac{-\overline{p}_j}{|p_j|} B_{p_j}(z)$$

(where each $p_j = 0$ is understood to give rise to a factor of z only). Then B is a well-defined holomorphic function on D by Proposition 8.1.10. Let $F(z) = f(z)/B(z)$. By the Riemann removable singularities theorem, F is a well-defined, nonvanishing holomorphic function on D. The representation $f = F \cdot B$ is called the *canonical factorization* of f.

Exercise for the Reader: All the assertions of Definition 8.1.19 hold for $f \in N(D)$, the Nevanlinna class (see [GAR] or [HOF] for details of this class of holomorphic functions—these are functions that satisfy a logarithmic integrability condition).

Proposition 8.1.20 *Let* $f \in H^p(D)$, $0 < p \leq \infty$, *and let* $f = F \cdot B$ *be its canonical factorization. Then* $F \in H^p(D)$ *and* $\|F\|_{H^p(D)} = \|f\|_{H^p(D)}$.

Proof: Trivially, $|F| = |f/B| \geq |f|$, so $\|F\|_{H^p} \geq \|f\|_{H^p}$. If $N = 1, 2, \dots$, let

$$B_N(z) = \prod_{j=1}^{N} \frac{-\overline{p}_j}{|p_j|} B_{p_j}(z)$$

(where the factors corresponding to $p_j = 0$ are just z).

Let $F_N = f/B_N$. Since $|B_N(e^{it})| = 1$, for all t, it holds that $\|F_N\|_{H^p} = \|f\|_{H^p}$ (use Proposition 8.1.2 and the fact that $B_N(re^{it}) \to B_N(e^{it})$ uniformly in t as $r \to 1^-$). If $0 < r < 1$ then

$$\int_0^{2\pi} |F(re^{it})|^p dt^{1/p} = \lim_{N\to\infty} \int_0^{2\pi} |F_N(re^{it})|^p dt^{1/p}$$

$$\leq \lim_{N\to\infty} \|F_N\|_{H^p} = \|f\|_{H^p}.$$

Therefore $\|F\|_{H^p} \leq \|f\|_{H^p}$. $\square$

Corollary 8.1.21 *If $\{p_1, p_2, \ldots\}$ is a sequence of points in D satisfying $\sum_j (1 - |p_j|) < \infty$ and if $B(z) = \prod_j (-\overline{p}_j/|p_j|) B_{p_j}(z)$ is the corresponding Blaschke product, then*

$$\lim_{r\to 1^-} B(re^{it})$$

exists and has modulus 1 almost everywhere.

Proof: The conclusion that the limit exists follows from Theorem 8.1.4 and the fact that $B \in H^\infty$. For the other assertion, note that the canonical factorization for B is $B = 1 \cdot B$. Therefore, by Proposition 8.1.20,

$$\int |\widetilde{B}(e^{it})|^2 dt^{1/2} = \|B\|_{H^2} = \|1\|_{H^2} = 1;$$

hence $|\widetilde{B}(e^{it})| = 1$ almost everywhere. $\square$

Theorem 8.1.22 *If $f \in H^p(D)$, $0 < p \leq \infty$, and $1 < \alpha < \infty$, then*

$$\lim_{\Gamma_\alpha(D)\ni z\to e^{i\theta}} f(z)$$

exists for almost every $e^{i\theta} \in \partial D$ and equals $\widetilde{f}(e^{i\theta})$. Also, $\widetilde{f} \in L^p(\partial D)$ and

$$\|\widetilde{f}\|_{L^p} = \|f\|_{H^p} \equiv \sup_{0<r<1} \int_0^{2\pi} |f(re^{i\theta})|^p d\theta^{1/p}.$$

Proof: By Definition 8.1.19, write $f = B \cdot F$ where F has no zeros and B is a Blaschke product. Then $F^{p/2}$ is a well-defined H^2 function and thus has the appropriate boundary values almost everywhere. A fortiori, F has nontangential boundary limits almost everywhere. Since $B \in H^\infty$, B has nontangential boundary limits almost everywhere. It follows that f does as well. The final assertion follows from the corresponding fact for H^2 functions (exercise). $\square$

8.2 Key Properties of the Poisson Kernel

Capsule: According to the way that Hardy space theory is presented here, estimates on the Poisson kernel are critical to the arguments. On the disk, the upper half-plane, and the ball, there are explicit formulas for the Poisson kernel. This makes the estimation process straightforward. On more general domains some new set of ideas is required to gain workable estimates for this important kernel—an explicit formula is essentially impossible. The work [KRA4] presents one technique, due to N. Kerzman, for estimating the Poisson kernel. The work [KRA10] gives a more natural and more broadly applicable approach to the matter.

The crux of our arguments in Section 8.1 was the fact that the Poisson integral is majorized by the Hardy–Littlewood maximal function. In this estimation, the explicit form of the Poisson kernel for the disk was exploited. If we wish to use a similar program to study the boundary behavior of harmonic and holomorphic functions on general domains in $\mathbb{R}^N$ and $\mathbb{C}^n$, then we must again estimate the Poisson integral by a maximal function.

However, there is no hope of obtaining an explicit formula for the Poisson kernel of an arbitrary smoothly bounded domain. In this section we shall instead obtain some rather sharp estimates that will suffice for our purposes. The proofs of these results that appear in [KRA4] are rather classical, and depend on harmonic majorization. It may be noted that there are modern methods for deriving these results rather quickly. One is to use the theory of Fourier integral operators, for which see [TRE]. Another is to use scaling (see [KRA10]).

The Poisson kernel for a C^2 domain $\Omega \subseteq \mathbb{R}^N$ is given by $P(x, y) = -\nu_y G(x, y)$, $x \in \Omega$, $y \in \partial\Omega$. Here ν_y is the unit outward normal vector field to $\partial\Omega$ at y, and $G(x, y)$ is the Green's function for Ω (see [KRA4] for full details of these assertions). Recall that for $N > 2$, we have $G(x, y) = c_n|x - y|^{-N+2} - F_x(y)$, where F depends in a $C^{2-\epsilon}$ fashion on x and y jointly and F is harmonic in y. It is known (again see [KRA4]) that G is $C^{2-\epsilon}$ on $\overline{\Omega} \times \overline{\Omega} \setminus \{\text{diagonal}\}$ and $G(x, y) = G(y, x)$. It follows that $P(x, y)$ behaves qualitatively like $|x - y|^{-N+1}$. [These observations persist in $\mathbb{R}^2$ by a slightly different argument.] The results enunciated in the present section will refine these rather crude estimates.

We begin with a geometric fact:

Geometric Fact: Let $\Omega \subset\subset \mathbb{R}^N$ have C^2 boundary. There are numbers $r, \widetilde{r} > 0$ such that for each $y \in \partial\Omega$ there are balls $B(c_y, r) \equiv B_y \subseteq \Omega$ and $B(\widetilde{c}_y, \widetilde{r}) \equiv \widetilde{B}_y \subseteq {}^c\overline{\Omega}$ that satisfy

(i) $\overline{B}(\widetilde{c}_y, \widetilde{r}) \cap \overline{\Omega} = \{y\}$;
(ii) $\overline{B}(c_y, r) \cap {}^c\overline{\Omega} = \{y\}$.

See Figure 8.2.

Let us indicate why these balls exist. At each point $\zeta \in \partial\Omega$, let ν_ζ denote the unit outward normal vector. Fix $P \in \partial\Omega$. Applying the implicit function theorem to the mapping

$$\partial\Omega \times (-1, 1) \to \mathbb{R}^N,$$

$$(\zeta, t) \mapsto \zeta + t\nu_\zeta,$$

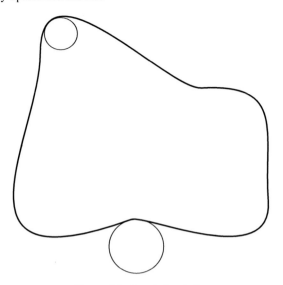

Figure 8.2. Osculating balls.

at the point $(P, 0)$, we find a neighborhood U_P of the point P on which the mapping is one-to-one. By the compactness of the boundary, there is thus a neighborhood U of $\partial\Omega$ such that each $x \in U$ has a unique nearest point in $\partial\Omega$. It further follows that there is an $\epsilon > 0$ such that if ζ_1, ζ_2 are distinct points of $\partial\Omega$ then $I_1 = \{\zeta_1 + t\nu_{\zeta_1} : |t| < 2\epsilon\}$ and $I_2 = \{\zeta_2 + t\nu_{\zeta_2} : |t| < 2\epsilon\}$ are disjoint sets (that is, the normal bundle is locally trivial in a natural way). From this it follows that if $y \in \partial\Omega$, then we may take $c_y = y - \epsilon\nu_y$, $\tilde{c}_y = y + \epsilon\nu_y$, and $r = \tilde{r} = \epsilon$.

We may assume in what follows that $r = \tilde{r} < \text{diam } \Omega/2$. See Figure 8.3.

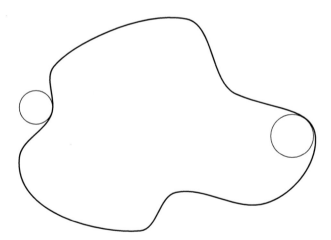

Figure 8.3. Osculating balls radius smaller than diam $\Omega/2$.

We now consider estimates for $P_\Omega(x, y)$. The proofs of these results are technical and tedious, and we cannot treat them here. See [KRA4, Ch. 8] for the details. In what follows, $\delta(x) = \delta_\Omega(x)$ denotes the (nonnegative) Euclidean distance of the point x to the boundary of Ω.

Prelude: Just as with the Bergman kernel and other canonical kernels, we have little hope of explicitly calculating the Poisson kernel on most domains. In the case of a domain like the ball or the upper half-space we can do it. But generally we cannot (see the paper [KRA10] for a consideration of some of these issues). The next result gives a quantitative estimate for the size of the Poisson kernel that proves to be of great utility in many circumstances. For a really precise and detailed asymptotic expansion, something like Fourier integral operators is needed.

Proposition 8.2.1 *Let $\Omega \subseteq \mathbb{R}^N$ be a domain with C^2 boundary. Let $P = P_\Omega : \Omega \times \partial\Omega \to \mathbb{R}$ be its Poisson kernel. Then for each $x \in \Omega$ there is a positive constant C_x such that*

$$0 < C_x \le P(x, y) \le \frac{C}{|x - y|^{N-1}} \le \frac{C}{\delta(x)^{N-1}}.$$

Here $\delta(x) = \text{dist}\,(x, \partial\Omega)$.

Prelude: This next is the most precise size estimate for the Poisson kernel that is of general utility. Its proof is rather delicate, and details may be found in [KRA4]. An alternative proof is in [KRA10]. Fourier integral operators give a more high-level, but in some ways more natural, proof. The reader may find it enlightening to think about what this estimate says on the disk and on the upper half-plane.

Proposition 8.2.2 *If $\Omega \subset\subset \mathbb{R}^N$ is a domain with C^2 boundary, then there are constants $0 < c < C < \infty$ such that*

$$c \cdot \frac{\delta(x)}{|x - y|^N} \le P_\Omega(x, y) \le C \cdot \frac{\delta(x)}{|x - y|^N}.$$

8.3 The Centrality of Subharmonicity

Capsule: One of E.M. Stein's many contributions to this subject is to teach us that we may free ourselves from an artificial dependence on Blaschke products (which really work only on the disk and the half-plane in one complex dimension) by exploiting subharmonicity and harmonic majorization. What is nice about this new approach is that it applies in the classical setting but it also applies to domains in $\mathbb{R}^N$ (for the harmonic function theory) and to domains in $\mathbb{C}^n$ (for the holomorphic function theory). It is a flexible methodology that can be adapted to a variety of situations.

Let $\Omega \subseteq \mathbb{C}^N$ be a domain and $f : \Omega \to \mathbb{R}$ a function. The function f is said to have a *harmonic majorant* if there is a (necessarily) nonnegative harmonic u on Ω

with $|f| \leq u$. We are interested in harmonic majorants for subharmonic functions. As usual, B denotes the unit ball in $\mathbb{R}^N$.

Prelude: There are various ways to think about the H^p functions. The standard definition is the uniform estimate on pth-power means. But the definition in terms of harmonic majorants—that a holomorphic function is in H^p if and only if $|f|^p$ has a harmonic majorant—is also of great utility. Walter Rudin called this "Lumer's theory of Hardy spaces." There are also interesting characterizations in terms of maximal functions—see [KRA5] and [STE2].

Proposition 8.3.1 *If $f : B \to \mathbb{R}^+$ is subharmonic, then f has a harmonic majorant if and only if*

$$\sup_{0 < r < 1} \int_{\partial B} f(r\zeta) d\sigma(\zeta) < \infty.$$

Proof: Let u be a harmonic majorant for f. Then

$$\int_{\partial B} f(r\zeta) d\sigma(\zeta) \leq \int_{\partial B} u(r\zeta) d\sigma(\zeta) = \omega_{N-1} \cdot u(0) \equiv C < \infty,$$

as claimed.

Conversely, if f satisfies

$$\sup_r \int_{\partial B} f(r\zeta) d\sigma(\zeta) < \infty,$$

then the functions $f_r : \partial B \to \mathbb{C}$ given by $f_r(\zeta) = f(r\zeta)$ form a bounded subset of $L^1(\partial B) \subseteq \mathcal{M}(\partial B)$. Let $\widetilde{f} \in \mathcal{M}(\partial B)$ be a weak-$*$ accumulation point of the functions f_r. Then $F(r\zeta) \equiv P\widetilde{f}(r\zeta)$ is harmonic on B, and for any $x \in B$ and $1 > r > |x|$ we have

$$0 \leq f(x) \leq \int P(x/r, \zeta) f_r(\zeta) d\sigma(\zeta)$$

$$\to \int P(x, \zeta) d\widetilde{f}(\zeta) = F(x) \qquad \text{as} \quad r \to 1^-. \qquad \square$$

A consequence of Proposition 8.3.1 is that not all subharmonic functions have harmonic majorants. For instance, the function $|e^{1/(1-z)}|$ on the disk has no harmonic majorant. Harmonic majorants play a significant role in the theory of boundary behavior of harmonic and holomorphic functions. Proposition 8.3.1 suggests why growth conditions may, therefore, play a role. The fact that f harmonic implies $|f|^p$ subharmonic only for $p \geq 1$ (exercise—just calculate!) whereas f holomorphic implies $|f|^p$ subharmonic for $p > 0$ (exercise—just calculate!)[1] suggests that we may expect different behavior for harmonic and for holomorphic functions.

---

[1] For simplicity, restrict attention to complex dimension 1 and assume that f is a harmonic function that does not vanish. Then $\triangle |f|^p = p^2 |f|^{p-2} (|\partial f/\partial z|^2 + |\partial \overline{f}/\partial z|^2) + p(p-2)|f|^{p-4} 2\mathrm{Re}\,[(\partial f/\partial z)(\partial f/\partial \overline{z})\overline{f}^2]$. One may verify directly that when $p > 1$, this expression is nonnegative. If now f is holomorphic, then the expression simplifies to $\triangle |f|^p = p^2 |f|^{p-2} |\partial f/\partial z|^2$. This last expression is obviously nonnegative for all $p > 0$. Some of these formulas are attributed to P. Stein.

By way of putting these remarks in perspective and generalizing Proposition 8.3.1, we consider $\mathbf{h}^p(\Omega)$ (resp. $H^p(\Omega)$), with Ω any smoothly bounded domain in $\mathbb{R}^N$ (resp. $\mathbb{C}^n$). First we require some preliminary groundwork.

Let $\Omega \subset\subset \mathbb{R}^N$ be a domain with C^2 boundary. Let $\phi : \mathbb{R} \to [0, 1]$ be a C^∞ function supported in $[-2, 2]$ with $\phi \equiv 1$ on $[-1, 1]$. Then, with $\delta_\Omega(x) \equiv \text{dist}(x, \partial\Omega)$ and $\epsilon_0 > 0$ sufficiently small, we see that

$$\rho(x) = \begin{cases} -\phi(|x|/\epsilon_0)\delta_\Omega(|x|) - (1 - \phi(|x|/\epsilon_0)) & \text{if } x \in \overline{\Omega}, \\ \phi(|x|/\epsilon_0)\delta_\Omega(|x|) + (1 - \phi(|x|/\epsilon_0)) & \text{if } x \notin \overline{\Omega}, \end{cases}$$

is a C^2 defining function for Ω. The implicit function theorem implies that if $0 < \epsilon < \epsilon_0$, then $\partial\Omega_\epsilon \equiv \{x \in \Omega : \rho(x) = -\epsilon\}$ is a C^2 manifold that bounds $\Omega_\epsilon \equiv \{x \in \Omega : \rho_\epsilon(x) \equiv \rho(x) + \epsilon < 0\}$. Now let $d\sigma_\epsilon$ denote area measure on $\partial\Omega_\epsilon$. Then it is natural to let

$$\mathbf{h}^p(\Omega) = \left\{ f \text{ harmonic on } \Omega : \sup_{0 < \epsilon < \epsilon_0} \int_{\partial\Omega_\epsilon} |f(\zeta)|^p d\sigma_\epsilon(\zeta)^{1/p} \right.$$

$$\left. \equiv \|f\|_{\mathbf{h}^p(\Omega)} < \infty \right\}, \qquad 0 < p < \infty,$$

$$\mathbf{h}^\infty(\Omega) = \left\{ f \text{ harmonic on } \Omega : \sup_{x \in \Omega} |f(x)| \equiv \|f\|_{\mathbf{h}^\infty} < \infty \right\}.$$

In case Ω is a subdomain of *complex space*, we define

$$H^p(\Omega) = \mathbf{h}^p(\Omega) \cap \{\text{holomorphic functions}\}, \qquad 0 < p \le \infty.$$

The next lemma serves to free the definitions from their somewhat artificial dependence on δ_Ω and ρ.

Prelude: This simple-minded lemma is important, for it frees the discussion from dependence on a particular defining function. The theory would be unsatisfying without it.

Lemma 8.3.2 (Stein) *Let ρ_1, ρ_2 be two C^2 defining functions for a domain $\Omega \subseteq \mathbb{R}^N$. For $\epsilon > 0$ small and $i = 1, 2$, let*

$$\Omega_\epsilon^i = \{x \in \Omega : \rho_i(x) < -\epsilon\},$$

$$\partial\Omega_\epsilon^i = \{x \in \Omega : \rho_i(x) = -\epsilon\}.$$

Let σ_ϵ^i be area measure on $\partial\Omega_\epsilon^i$. Then for f harmonic on Ω we have

$$\sup_{\epsilon > 0} \int_{\partial\Omega_\epsilon^1} |f(\zeta)|^p d\sigma_\epsilon^1(\zeta) < \infty$$

if and only if

$$\sup_{\epsilon > 0} \int_{\partial \Omega_\epsilon^2} |f(\zeta)|^p d\sigma_\epsilon^2(\zeta) < \infty.$$

(**Note:** *Since f is bounded on compact sets, equivalently the supremum is of interest only as $\epsilon \to 0$, there is no ambiguity in this assertion.*)

Proof: By definition of defining function, grad $\rho_i \neq 0$ on $\partial\Omega$. Since $\partial\Omega$ is compact, we may choose $\epsilon_0 > 0$ so small that there is a constant λ, $0 < \lambda < 1$, with $0 < \lambda \leq |\text{grad } \rho_i(x)| < 1/\lambda$ whenever $x \in \Omega$, $d_\Omega(x) < \epsilon_0$. If $0 < \epsilon < \epsilon_0$, then notice that for $x \in \partial\Omega_\epsilon^2$, we have

$$B(x, \lambda\epsilon/2) \subseteq \Omega$$

and, what is stronger,

$$B(x, \lambda\epsilon/2) \subseteq \left\{ t : -3\epsilon/\lambda^2 < \rho_1(t) < -\lambda^2 \cdot \epsilon/3 \right\} \equiv S(\epsilon). \tag{8.3.2.1}$$

Therefore

$$|f(x)|^p \leq \frac{1}{V(B(x, \lambda\epsilon/2))} \int_{B(x,\lambda\epsilon/2)} |f(t)|^p dV(t).$$

As a result,

$$\int_{\partial\Omega_\epsilon^2} |f(x)|^p d\sigma_\epsilon^2 < C\epsilon^{-N} \int_{\partial\Omega_\epsilon^2} \int_{B(x,\lambda\epsilon/2)} |f(t)|^p dV(t) d\sigma_\epsilon^2(x)$$

$$= C\epsilon^{-N} \int_{\mathbb{R}^N} \int_{\partial\Omega_\epsilon^2} \chi_{B(x,\lambda\epsilon/2)}(t)|f(t)|^p d\sigma_\epsilon^2(x) dV(t)$$

$$\leq C\epsilon^{-N} \int_{S(\epsilon)} |f(t)|^p \int_{\partial\Omega_\epsilon^2 \cap B(t,\lambda\epsilon/2)} d\sigma_\epsilon^2(x) dV(t)$$

$$\leq C'\epsilon^{-N} \epsilon^{N-1} \int_{S(\epsilon)} |f(t)|^p dV(t)$$

$$\leq C'' \sup_\epsilon \int_{\partial\Omega_\epsilon^1} |f(t)|^p d\sigma_\epsilon^1(t).$$

Of course the reverse inequality follows by symmetry. $\square$

One technical difficulty that we face on an arbitrary Ω is that the device (which was so useful on the disk) of considering the dilated functions $f_r(\zeta) = f(r\zeta)$ as harmonic functions on $\overline{\Omega}$ is no longer available. However, this notion is an unnecessary crutch, and it is well to be rid of it. As a substitute, we cover Ω by finitely many domains $\Omega_1, \ldots, \Omega_k$ with the following properties:

(**8.3.3**) $\Omega = \cup_j \Omega_j$.

(8.3.4) For each j, the set $\partial\Omega \cap \partial\Omega_j$ is an $(N-1)$-dimensional manifold with boundary.

(8.3.5) There are an $\epsilon_0 > 0$ and a vector v_j transversal to $\partial\Omega \cap \partial\Omega_j$ and pointing out of Ω such that $\Omega_j - \epsilon v_j \equiv \{x - \epsilon v_j : x \in \Omega_j\} \subset\subset \Omega$, all $0 < \epsilon < \epsilon_0$.

We leave the detailed verification of the existence of the sets Ω_j satisfying **(8.3.3)**–**(8.3.5)** as an exercise. See Figure 8.4 for an illustration of these ideas. For a general C^2 bounded domain, the substitute for dilation will be to fix $j \in \{1, \ldots, k\}$ and consider the translated functions $f_\epsilon(x) = f(x - \epsilon v_j)$, $f_\epsilon : \Omega_j \to \mathbb{C}$, as $\epsilon \to 0^+$.

Prelude: As indicated earlier, there are several different ways to think about H^p and h^p spaces. The next theorem considers several of them. We shall make good use of them all in our ensuing discussions.

Theorem 8.3.6 *Let $\Omega \subseteq \mathbb{R}^N$ be a domain and f harmonic on Ω. Let $1 \le p < \infty$. The following are equivalent:*

(8.3.6.1) $f \in \mathbf{h}^p(\Omega)$.

(8.3.6.2) *If $p > 1$ then there is an $\widetilde{f} \in L^p(\partial\Omega)$ such that*

$$f(x) = \int_{\partial\Omega} P(x, y)\widetilde{f}(y)d\sigma(y)$$

[resp. if $p = 1$ then there is a $\mu \in \mathcal{M}(\partial\Omega)$, the space of regular Borel measures on $\partial\Omega$, such that

$$f(x) = \int_{\partial\Omega} P(x, y)d\mu(y).]$$

Moreover, $\|f\|_{\mathbf{h}^p} \cong \|\widetilde{f}\|_{L^p}$.

(8.3.6.3) $|f|^p$ *has a harmonic majorant on Ω.*

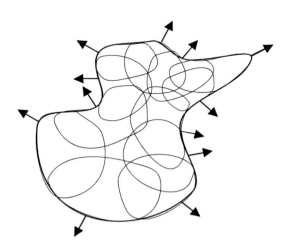

Figure 8.4. Translatable subdomains.

Proof: (2) $\Rightarrow$ (3) If $p > 1$, let

$$h(x) = \int_{\partial\Omega} P(x, y)|\tilde{f}(y)|^p d\sigma(y).$$

Then, treating $P(x, \cdot)d\sigma$ as a positive measure of total mass 1, we have

$$|f(x)|^p = \left|\int_{\partial\Omega} \tilde{f}(y)P(x, y)d\sigma(y)\right|^p$$

$$\overset{\text{(Jensen)}}{\leq} \int_{\partial\Omega} |\tilde{f}(y)|^p P(x, y)d\sigma(y) \equiv h(x).$$

The proof for $p = 1$ is similar.

(3) $\Rightarrow$ (1) If $\epsilon > 0$ is small, $x_0 \in \Omega$ is fixed, and G_Ω is the Green's function for Ω, then $G_\Omega(x_0, \cdot)$ has nonvanishing gradient near $\partial\Omega$ (use Hopf's lemma). Therefore

$$\tilde{\Omega}_\epsilon \equiv \{x \in \Omega : -G_\Omega(x, \cdot) < -\epsilon\}$$

are well-defined domains for ϵ small. Moreover (*check the proof*), the Poisson kernel for $\tilde{\Omega}_\epsilon$ is $P_\epsilon(x, y) = -\nu_y^\epsilon G_\Omega(x, y)$. Here ν_y^ϵ is the normal to $\partial\tilde{\Omega}_\epsilon$ at $y \in \partial\tilde{\Omega}_\epsilon$. Assume that $\epsilon > 0$ is so small that $x_0 \in \tilde{\Omega}_\epsilon$. So if h is the harmonic majorant for $|f|^p$ then

$$h(x_0) = \int_{\partial\tilde{\Omega}_\epsilon} -\nu_y^\epsilon G_\Omega(x_0, y)h(y)d\sigma(y). \tag{8.3.6.4}$$

Let $\pi_\epsilon : \partial\tilde{\Omega}_\epsilon \to \partial\Omega$ be normal projection for ϵ small. Then

$$-\nu_y^\epsilon G_\Omega(x_0, \pi_\epsilon^{-1}(\cdot)) \to -\nu_y G_\Omega(x_0, \cdot)$$

uniformly on $\partial\Omega$ as $\epsilon \to 0^+$. The proof of Proposition 8.2.2 $-\nu_y G_\Omega(x_0, \cdot) \geq c_{x_0} > 0$ for some constant c_{x_0}. Thus $-\nu_y^\epsilon G_\Omega(x_0, \pi_\epsilon^{-1}(\cdot))$ are all bounded below by $c_{x_0}/2$ if ϵ is small enough. As a result, (8.3.6.4) yields

$$\int_{\partial\tilde{\Omega}_\epsilon} h(y)d\sigma(y) \leq 2h(x_0)/c_{x_0}$$

for $\epsilon > 0$ small. In conclusion,

$$\int_{\partial\tilde{\Omega}_\epsilon} |f(y)|^p d\sigma(y) \leq 2h(x_0)/c_{x_0}.$$

(1) $\Rightarrow$ (2) Let Ω_j be as in (8.3.3) through (8.3.5). Fix j. Define on Ω_j the functions $f_\epsilon(x) = f(x - \epsilon\nu_j), 0 < \epsilon < \epsilon_0$. Then the hypothesis and (a small modification of) Lemma 8.3.2 show that $\{f_\epsilon\}$ forms a bounded subset of $L^p(\partial\Omega_j)$. If $p > 1$, let $\tilde{f}_j \in L^p(\partial\Omega_j)$ be a weak-* accumulation point (for the case $p = 1$, replace $\tilde{f}_j$ by a Borel measure $\tilde{\mu}_j$). The crucial observation at this point is that f is the Poisson integral of $\tilde{f}_j$ on Ω_j. Therefore f on Ω_j is completely determined by $\tilde{f}_j$ and conversely (see also the exercises at the end of the section). A moment's reflection now shows that $\widetilde{f}j = \widetilde{f}_k$ almost everywhere $[d\sigma]$ in $\partial\Omega_j \cap \partial\Omega_k \cap \partial\Omega$, so that $\widetilde{f} \equiv \widetilde{f}_j$ on $\partial\Omega_j \cap \partial\Omega$ is well defined. By appealing to a partition of unity on $\partial\Omega$ that is subordinate to the open cover induced by the (relative) interiors of the sets $\partial\Omega_j \cap \partial\Omega$, we see that $f_\epsilon = f \circ \pi_\epsilon^{-1}$ converges weak-* to $\widetilde{f}$ on $\partial\Omega$ when $p > 1$ (resp. $f_\epsilon \to \widetilde{\mu}$ weak-* when $p = 1$).

Referring to the proof of **(3)** $\Rightarrow$ **(1)** we write, for $x_0 \in \Omega$ fixed,

$$f(x_0) = \int_{\partial\Omega_\epsilon} -\nu_y^\epsilon G_\Omega(x_0, y) f(y) d\sigma_\epsilon(y)$$

$$= \int_{\partial\Omega} -\nu_y^\epsilon G_\Omega\left(x_0, \pi_\epsilon^{-1}(y)\right) f\left(\pi_\epsilon^{-1}(y)\right) \mathcal{J}^\epsilon(y) d\sigma(y),$$

where $\mathcal{J}^\epsilon$ is the Jacobian of the mapping $\pi_\epsilon^{-1} : \partial\Omega \to \partial\Omega_\epsilon$. The fact that $\partial\Omega$ is C^2 combined with previous observations implies that the last line tends to

$$\int_{\partial\Omega} -\nu_y G_\Omega(x_0, y) \widetilde{f}(y) d\sigma(y) = \int_{\partial\Omega} P_\Omega(x_0, y) \widetilde{f}(y) d\sigma(y)$$

$$\left(\text{resp.} \int_{\partial\Omega} -\nu_y G_\Omega(x_0, y) d\widetilde{\mu}(y) = \int_{\partial\Omega} P_\Omega(x_0, y) d\widetilde{\mu}(y)\right)$$

as $\epsilon \to 0^+$. $\square$

Exercise for the Reader:

1. Prove the last statement in Theorem 8.3.6.

2. Imitate the proof of Theorem 8.3.6 to show that if u is continuous and *subharmonic* on Ω and if

$$\sup_\epsilon \int_{\partial\Omega_\epsilon} |u(\zeta)|^p d\sigma(\zeta) < \infty, \qquad p \geq 1,$$

then u has a harmonic majorant h. If $p > 1$, then h is the Poisson integral of an L^p function $\widetilde{h}$ on $\partial\Omega$. If $p = 1$, then h is the Poisson integral of a Borel measure $\widetilde{\mu}$ on $\partial\Omega$.

3. Let $\Omega \subseteq \mathbb{R}^N$ be a domain with C^2 boundary and let ρ be a C^2 defining function for Ω. Define $\Omega_\epsilon = \{x \in \Omega : \rho(x) < -\epsilon\}$, $0 < \epsilon < \epsilon_0$. Let $\partial\Omega_\epsilon$ and $d\sigma_\epsilon$ be as usual. Let $\pi_\epsilon : \partial\Omega_\epsilon \to \partial\Omega$ be orthogonal projection. Let $f \in L^p(\partial\Omega)$, $1 \leq p < \infty$. Define

$$F(x) = \int_{\partial\Omega} P_\Omega(x, y) f(y) d\sigma(y).$$

a. Prove that $\int_{\partial\Omega} P_\Omega(x, y) d\sigma(y) = 1$, for any $x \in \Omega$.
b. There is a $C > 0$ such that for any $y \in \partial\Omega$,

$$\int_{\partial\Omega_\epsilon} P_\Omega(x, y) d\sigma_\epsilon(x) \leq C, \qquad \text{for any } 0 < \epsilon < \epsilon_0.$$

c. There is a $C' > 0$ such that

$$\int_{\partial\Omega_\epsilon} |F(x)|^p d\sigma_\epsilon \leq C', \qquad \text{for any } 0 < \epsilon < \epsilon_0.$$

d. If $\phi \in C(\Omega)$ satisfies $\|\phi - f\|_{L^p(\partial\Omega)} < \eta$ and

$$G(x) \equiv \int_{\partial\Omega} P_\Omega(x, y)(\phi(y) - f(y))d\sigma(y),$$

then

$$\int_{\partial\Omega_\epsilon} |G(x)|^p d\sigma_\epsilon(x) \leq C'\eta, \qquad \text{for any } 0 < \epsilon < \epsilon_0.$$

e. With ϕ as in part **d** and $\Phi(x) = \int_{\partial\Omega} P_\Omega(x, y)\phi(y)d\sigma(y)$, then $(\Phi|_{\partial\Omega_\epsilon}) \circ \pi_\epsilon^{-1} \to \phi$ uniformly on $\partial\Omega$.

f. Imitate the proof of Proposition 8.1.2 to see that $F \circ \pi_\epsilon^{-1} \to f$ in the $L^p(\partial\Omega)$ norm.

8.4 More about Pointwise Convergence

Capsule: There are two basic aspects to the boundary behavior of holomorphic functions on a given domain. One is the question of *pointwise* boundary convergence. And the other is the question of *norm* boundary convergence. With Functional Analysis Principles I and II in mind (see Appendix 1), we can imagine that these will depend on different types of estimates. In the present section we shall concentrate on pointwise convergence, and this will in turn rely on a maximal function estimate.

Let $\Omega \subseteq \mathbb{R}^N$ be a domain with C^2 boundary. For $P \in \partial\Omega$, $\alpha > 1$, we define

$$\Gamma_\alpha(P) = \{x \in \Omega : |x - P| < \alpha\delta_\Omega(x)\}.$$

See Figure 8.5. This is the N-dimensional analogue of the Stolz region considered in Section 8.1.

Our theorem is as follows:

Prelude: Classically, Zygmund and others studied H^p and h^p spaces on *particular* concrete domains. It is only with our useful estimates for the Poisson kernel, and the maximal function approach, that we are now able to look at *all* domains in $\mathbb{R}^N$.

Theorem 8.4.1 *Let $\Omega \subset\subset \mathbb{R}^N$ be a domain with C^2 boundary. Let $\alpha > 1$. If $1 < p \leq \infty$ and $f \in h^p$, then*

$$\lim_{\Gamma_\alpha(P)\ni x \to P} f(x) \equiv \tilde{f}(P) \qquad \text{exists for almost every } P \in \partial\Omega.$$

Moreover,

$$\|\tilde{f}\|_{L^p(\partial\Omega)} \cong \|f\|_{h^p(\Omega)}.$$

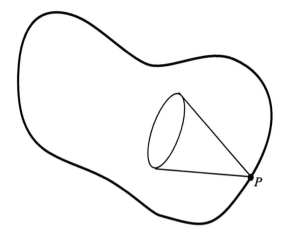

Figure 8.5. A nontangential approach region on a domain in $\mathbb{R}^N$.

Proof: We may as well assume that $p < \infty$. We already know from (8.3.6.2) that there exists an $\widetilde{f} \in L^p(\partial\Omega)$, $\|\widetilde{f}\|_{L^p} \cong \|f\|_{\mathbf{h}^p}$, such that $f = P\widetilde{f}$. It remains to show that $\widetilde{f}$ satisfies the conclusions of the present theorem. This will follow just as in the proof of Theorem 8.1.11 as soon as we prove two things: First,

$$\sup_{x \in \Gamma_\alpha(P)} |P\widetilde{f}(x)| \leq C_\alpha M_1\widetilde{f}(P), \tag{8.4.1.1}$$

where

$$M_1\widetilde{f}(P) \equiv \sup_{R>0} \frac{1}{\sigma(B(P,R) \cap \partial\Omega)} \int_{B(P,R)\cap\partial\Omega} |f(t)|d\sigma(t).$$

Second,

$$\sigma\{y \in \partial\Omega : M_1\widetilde{f}(y) > \lambda\} \leq C\frac{\|\widetilde{f}\|_{L^1(\partial\Omega)}}{\lambda}, \qquad \text{for all } \lambda > 0. \tag{8.4.1.2}$$

Now (8.4.1.1) is proved just as in Proposition 8.1.10. It is necessary to use the estimate given in Proposition 8.2.2. On the other hand, (8.4.1.2) is not so obvious; we supply a proof in the paragraphs that follow.[2]   □

Prelude: One of the important developments of harmonic analysis of the past fifty years is the prominent role of maximal functions in all aspects of the subject. Generally speaking, maximal functions are studied and controlled by way of *covering lemmas*. In this way profound questions of harmonic analysis are reduced to tactile and elementary (but by no means easy) questions of Euclidean geometry. The theory of covering lemmas reaches into other parts of mathematics, including computer graphics and graph theory.

---

[2] It may be mentioned that, once one has (8.4.1.2), then one may couple this with the obvious estimate on L^∞ and the Marcinkiewicz interpolation theorem to obtain an estimate in L^p.

Lemma 8.4.2 (Wiener) *Let $K \subseteq \mathbb{R}^N$ be a compact set that is covered by the open balls $\{B_\alpha\}_{\alpha \in A}$, $B_\alpha = B(c_\alpha, r_\alpha)$. There is a subcover $B_{\alpha_1}, B_{\alpha_2}, \ldots, B_{\alpha_m}$, consisting of pairwise disjoint balls, such that*

$$\bigcup_{j=1}^m B(c_{\alpha_j}, 3r_{\alpha_j}) \supseteq K.$$

Proof: Since K is compact, we may immediately assume that there are only finitely many B_α. Let B_{α_1} be the ball in this collection that has the greatest radius (this ball may not be unique). Let B_{α_2} be the ball that has greatest radius and is also disjoint from B_{α_1}. At the jth step choose the (not necessarily unique) ball of greatest radius that is disjoint from $B_{\alpha_1}, \ldots, B_{\alpha_{j-1}}$. Continue. The process ends in finitely many steps. We claim that the B_{α_j} chosen in this fashion do the job.

It is enough to show that $B_\alpha \subseteq \cup_j B(c_{\alpha_j}, 3r_{\alpha_j})$ for every α. Fix an α. If $\alpha = \alpha_j$ for some j then we are done. If $\alpha \notin \{\alpha_j\}$, let j_0 be the first index with $B_{\alpha_j} \cap B_\alpha \neq \emptyset$ (there must be one; otherwise the process would not have stopped). Then $r_{\alpha_{j_0}} \geq r_\alpha$; otherwise we selected $B_{\alpha_{j_0}}$ incorrectly. But then clearly $B(c_{\alpha_{j_0}}, 3r_{\alpha_{j_0}}) \supseteq B(c_\alpha, r_\alpha)$ as desired. $\square$

Corollary 8.4.3 *Let $K \subseteq \partial\Omega$ be compact, and let $\{B_\alpha \cap \partial\Omega\}_{\alpha \in A}$, $B_\alpha = B(c_\alpha, r_\alpha)$, be an open covering of K by balls with centers in $\partial\Omega$. Then there is a pairwise disjoint subcover $B_{\alpha_1}, B_{\alpha_2}, \ldots, B_{\alpha_m}$ such that $\cup_j \{B(c_\alpha, 3r_\alpha) \cap \partial\Omega\} \supseteq K$.*

Proof: The set K is a compact subset of $\mathbb{R}^N$ that is covered by $\{B_\alpha\}$. Apply the preceding Lemma 8.4.2 and restrict to $\partial\Omega$. $\square$

Lemma 8.4.4 *If $f \in L^1(\partial\Omega)$, then*

$$\sigma\{x \in \partial\Omega : M_1 f(x) > \lambda\} \leq C \frac{\|f\|_{L^1}}{\lambda},$$

for all $\lambda > 0$.

Proof: Let $S_\lambda = \{x \in \partial\Omega : M_1 f(x) > \lambda\}$. Let K be a compact subset of S_λ. It suffices to estimate $\sigma(K)$. Now for each $x \in K$, there is a ball B_x centered at x such that

$$\frac{1}{\sigma(B_x \cap \partial\Omega)} \int_{B_x \cap \partial\Omega} |f(t)| d\sigma(t) > \lambda. \tag{8.4.4.1}$$

The balls $\{\partial\Omega \cap B_x\}_{x \in K}$ cover K. Choose, by Corollary 8.4.3, disjoint balls $B_{x_1}, B_{x_2}, \ldots, B_{x_m}$ such that $\{\partial\Omega \cap 3B_{x_j}\}$ cover K, where $3B_{x_j}$ represents the theefold dilate of B_{x_j} (with the same center). Then

$$\sigma(K) \leq \sum_{j=1}^m \sigma(3B_{x_j} \cap \partial\Omega) \leq C(N, \Omega) \sum_{j=1}^m \sigma(B_{x_j} \cap \partial\Omega),$$

where the constant C will depend on the curvature of $\partial\Omega$. But (8.4.4.1) implies that the last line is majorized by

$$C(N, \partial\Omega) \sum_j \frac{\int_{B_{x_j} \cap \partial\Omega} |f(t)| d\sigma(t)}{\lambda} \le C(N, \partial\Omega) \frac{\|f\|_{L^1}}{\lambda}.$$

This completes the proof of the theorem.   □

8.5 A Preliminary Result in Complex Domains

Capsule: This section presents a "toy" version of our main result. We prove a version of pointwise boundary convergence—not the most obvious or natural one. This is just so that we have a boundary function that we can leverage to get the more sophisticated results that we seek (regarding non-tangential and admissible convergence). It should be stressed that what is new here is the case $p < 1$ for $f \in H^p$. The case $p \ge 1$ has already been covered in the context of harmonic function theory. The ultimate result—our real goal—is about admissible convergence; that will be in the next section.

Everything in Section 8.4 applies a fortiori to domains $\Omega \subseteq \mathbb{C}^n$. However, on the basis of our experience in the classical case, we expect $H^p(\Omega)$ functions to also have pointwise boundary values for $0 < p \le 1$. That this is indeed the case is established in this section by two different arguments.

First, if $\Omega \subset\subset \mathbb{C}^n$ is a C^2 domain and $f \in H^p(\Omega)$, we shall prove through an application of Fubini's theorem (adapted from the paper [LEM]) that f has pointwise boundary limits in a rather special sense at σ-almost every $\zeta \in \partial\Omega$. This argument is self-contained. The logical progression of ideas in this chapter will proceed from the first approach based on [LEM].

Prelude: In the classical theory on the disk, one makes good use of dilations to reduce a Hardy space function to one that is continuous on the closure of the disk. Such a simple device is certainly not available on an arbitrary domain. It is a lovely idea (due to Stein) to instead consider the domains Ω_j—see **(8.3.3)**–**(8.3.5)**. They are a bit harder to grasp, but certainly work just as well to give the results that we need.

Proposition 8.5.1 *Let* $\Omega \subset\subset \mathbb{C}^n$ *have* C^2 *boundary. Let* $0 < p < \infty$ *and* $f \in H^p(\Omega)$. *Write* $\Omega = \cup_{j=1}^k \Omega_j$ *as in* **(8.3.3)** *through* **(8.3.5)**, *and let* $v_1, \ldots, v_k$ *be the associated normal vectors. Then, for each* $j \in \{1, \ldots, k\}$, *it holds that*

$$\lim_{\epsilon \to 0^+} f(\zeta - \epsilon v_j) \equiv \tilde{f}(\zeta)$$

exists for σ*-almost every* $\zeta \in \partial\Omega_j \cap \partial\Omega$.

Proof: We may suppose that $p < \infty$. Fix $1 \le j \le k$. Assume for convenience that $v_j = v_P = (1 + i0, 0, \ldots, 0)$, $P \in \partial\Omega_j$, and that $P = 0$. If $z \in \mathbb{C}^n$, write $z = (z_1, \ldots, z_n) = (z_1, z')$. We may assume that $\Omega_j = \cup_{|z'|<1}\{(z_1, z') : z_1 \in \mathcal{D}_{z'}\}$, where $\mathcal{D}_{z'} \subseteq \mathbb{C}$ is a diffeomorph of $D \subseteq \mathbb{C}$ with C^2 boundary. For each $|z'| < 1$,

$k \in \mathbb{N}, 0 < 1/k < \epsilon_0$, let $b^k_{z'} = (\mathcal{D}_{z'} \times \{z'\}) \cap \{z \in \Omega_j : \text{dist}(z, \partial\Omega_j) = 1/k\}$.
Define $B^k = \cup_{|z'|<1} b^k_{z'}$. Now a simple variant of Lemma 8.3.2 implies that

$$\sup_k \int_{B^k} |f(\zeta)|^p d\sigma_k \leq C_0 < \infty, \tag{8.5.1.1}$$

where σ_k is surface measure on B^k. Formula (8.5.1.1) may be rewritten as

$$\sup_k \int_{|\zeta'|<1} \int_{b^k_{\zeta'}} |f(\zeta_1, \zeta')|^p \, d\tilde{\sigma}_k(\zeta_1) \, dV_{2n-2}(\zeta') \leq C_0, \tag{8.5.1.2}$$

where $\tilde{\sigma}_k$ is surface (= linear) measure on $b^k_{\zeta'}$. If $M > 0$, $k \geq k_0 > 1/\epsilon_0$, we define

$$S^M_k = \left\{ \zeta' : |\zeta'| < 1, \int_{b^k_{\zeta'}} |f(\zeta_1, \zeta')|^p d\tilde{\sigma}_k(\zeta_1) > M \right\}. \tag{8.5.1.3}$$

Then (8.5.1.2), (8.5.1.3), and Chebyshev's inequality together yield

$$V_{2n-1}(S^M_k) \leq \frac{C_0}{M}, \qquad \text{for all } k.$$

Now let

$$S^M = \left\{ \zeta' : |\zeta'| < 1, \int_{b^k_{\zeta'}} |f(\zeta_1, \zeta')|^p d\tilde{\sigma}_k(\zeta_1) \leq M \text{ for only finitely many } k \right\}$$

$$= \bigcup_{\ell=k_0}^{\infty} \bigcap_{k=\ell}^{\infty} S^M_k.$$

Then $V_{2n-1}(S^M) \leq C_0/M$. Since M may be made arbitrarily large, we conclude that for V_{2n-2} almost every $\zeta' \in D^{n-1}(0,1)$, there exist $k_1 < k_2 < \cdots$ such that

$$\int_{b^{k_m}_{\zeta'}} |f(\zeta_1, \zeta')|^p d\tilde{\sigma}_{k_m}(\zeta_1) = O(1) \qquad \text{as } m \to \infty.$$

It follows that the functions $f(\cdot, \zeta') \in H^p(\mathcal{D}_{\zeta'})$ for V_{2n-2} almost every $\zeta' \in D^{n-1}(0,1)$. Now Theorem 8.1.22 yields the desired result. $\square$

8.6 First Concepts of Admissible Convergence

Capsule: In the present section we are finally able to treat the matter of admissible convergence. This involves a dramatically new collection of approach regions. There are some subtleties here. When the domain is the unit ball in $\mathbb{C}^n$, the admissible approach regions may be defined using an explicit formula. For more general domains, there is some delicate geometry involved. The definition is less explicit. There is also an issue of defining corresponding balls in the boundary. Again, when the domain is the unit ball this can be done with traditional formulas. On more general domains the definition is less explicit. In the end the decisive meshing of the balls in the boundary and the admissible approach regions in the interior yields the boundary behavior result for holomorphic functions that we seek (this point of view is developed in detail in [KRA7]). The result certainly requires, as we might expect, a maximal function estimate.

Let $B \subseteq \mathbb{C}^n$ be the unit ball. The Poisson kernel for the ball has the form

$$P(z, \zeta) = c_n \frac{1 - |z|^2}{|z - \zeta|^{2n}},$$

whereas the Poisson–Szegő kernel has the form

$$\mathcal{P}(z, \zeta) = c_n \frac{(1 - |z|^2)^n}{|1 - z \cdot \overline{\zeta}|^{2n}}.$$

As we know, an analysis of the convergence properties of these kernels entails dominating them by appropriate maximal functions. The maximal function involves the use of certain balls, and the shape of the ball should be compatible with the singularity of the kernel. That is why, when we study the *real analysis* of the Poisson kernel, we consider balls of the form

$$\beta_1(\zeta, r) = \{\xi \in \partial B : |\xi - \zeta| < r\}, \quad \zeta \in \partial B, \quad r > 0.$$

[Here the singularity of the kernel has the form $|\xi - \zeta|$—so it fits the balls.]
    In studying the *complex analysis* of the Poisson–Szegő kernel (equivalently, the Szegő kernel), it is appropriate to use the balls

$$\beta_2(\zeta, r) = \{\xi \in \partial B : |1 - \xi \cdot \overline{\zeta}| < r\}, \quad \zeta \in \partial B, \quad r > 0.$$

[Here the singularity of the kernel has the form $|1 - \xi \cdot \overline{\zeta}|$—so it fits the balls.] These new nonisotropic balls are fundamentally different from the classical (or *isotropic*) balls β_1, as we shall now see. Assume without loss of generality that $\zeta = \mathbf{1} = (1, 0, \ldots, 0)$. Write $z' = (z_2, \ldots, z_n)$. Then

$$\beta_2(\mathbf{1}, r) = \{\xi \in \partial B : |1 - \xi_1| < r\}.$$

Notice that for $\xi \in \partial B$,

$$|\xi'|^2 = 1 - |\xi_1|^2$$
$$= (1 - |\xi_1|)(1 + |\xi_1|)$$
$$\leq 2|1 - \xi_1|;$$

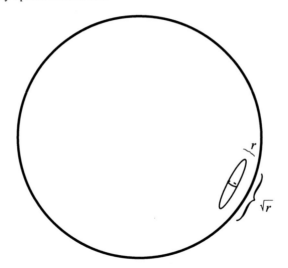

Figure 8.6. The nonisotropic balls.

hence
$$\beta_2(\mathbf{1}, r) \subseteq \{\xi \in \partial B : |1 - \xi_1| < r, |\xi'| < \sqrt{2r}\}.$$

A similar computation shows that

$$\beta_2(\mathbf{1}, r) = \{\xi \in \partial B : |1 - \xi_1| < r, |\xi'| = \sqrt{1 - |\xi_1|^2}\}$$

$$\supseteq \partial B \cap \{\xi : |\operatorname{Im} \xi_1| < r/2, \ 1 - r/2 < \operatorname{Re} \xi_1 < 1, \ |\xi'| < \sqrt{r}\}.$$

In short, the balls we now are considering have dimension $\approx r$ in the complex space containing ν_1 and dimension $\approx \sqrt{r}$ in the orthogonal complement (see Figure 8.6). The word "nonisotropic" means that we have different geometric behavior in different directions.

In the classical setup, on the domain the unit ball B, we considered *cones* modeled on the balls β_1:

$$\Gamma_\alpha(P) = \{z \in B : |z - P| < \alpha(1 - |z|)\}, \quad P \in \partial B, \quad \alpha > 1.$$

In the new situation we consider *admissible regions* modeled on the balls β_2:

$$\mathcal{A}_\alpha(P) = \{z \in B : |1 - z \cdot \overline{P}| < \alpha(1 - |z|)\}.$$

Our new theorem about boundary limits of H^p functions is as follows:

Prelude: This next is a version of Adam Koranyi's famous theorem that changed the nature of Fatou theorems for domains in $\mathbb{C}^n$ forever. It is still a matter of considerable study to determine the sharp Fatou theorem on any domain in $\mathbb{C}^n$ (see, for example, [DIK]).

Theorem 8.6.1 *Let $f \in H^p(B)$, $0 < p \le \infty$. Let $\alpha > 1$. Then the limit*

$$\lim_{\mathcal{A}_\alpha(P) \ni x \to P} f(P) \equiv \tilde{f}(P)$$

exists for σ-almost every $P \in \partial B$.

Since the Poisson–Szegő kernel is known explicitly on the ball, then for $p \ge 1$ the proof is deceptively straightforward: One defines, for $P \in \partial B$ and $f \in L^1(\partial B)$,

$$M_2 f(P) = \sup_{r>0} \frac{1}{\sigma(\beta_2(P, r))} \int_{\beta_2(P,r)} |f(\zeta)| d\sigma(\zeta).$$

Also set $f(z) = \int_{\partial B} \mathcal{P}(z, \zeta) f(\zeta) d\sigma(\zeta)$ for $z \in B$. Then, by explicit computation similar to the proof of Proposition 8.4.1,

$$f_2^{*,\alpha}(P) \equiv \sup_{x \in \mathcal{A}_\alpha(P)} |f(z)| \le C_\alpha M_2 f(P), \qquad \text{for all} \quad f \in L^1(\partial B).$$

This crucial fact, together with appropriate estimates on the operator M_2, enables one to complete the proof along classical lines for $p \ge 1$. For $p < 1$, matters are more subtle.

We forgo the details of the preceding argument on B and instead develop the machinery for proving an analogue of Theorem 8.6.1 on an arbitrary C^2 bounded domain in $\mathbb{C}^n$. In this generality, there is no hope of obtaining an explicit formula for the Poisson–Szegő kernel; indeed, there are no known techniques for obtaining estimates for this kernel on arbitrary domains (however, see [FEF] and [NRSW] for estimates on strongly pseudoconvex domains and on domains of finite type in $\mathbb{C}^2$). Therefore we must develop more geometric methods that *do not* rely on information about kernels. The results that we present were proved on the ball and on bounded symmetric domains by A. Koranyi [KOR1], [KOR2]. Many of these ideas were also developed independently in Gong Sheng [GOS1], [GOS2]. The paper [HOR1] of Hörmander was remarkably prescient for many of these ideas. All of the principal ideas for arbitrary Ω are due to E.M. Stein [STE4].

Our tasks, then, are as follows: **(1)** to define the balls β_2 on the boundary of an arbitrary smoothly bounded Ω; **(2)** to define admissible convergence regions $\mathcal{A}_\alpha$; **(3)** to obtain appropriate estimates for the corresponding maximal function; and **(4)** to couple the maximal estimates, together with the fact that "radial" boundary values are already known to exist (see Theorem 8.4.1) to obtain the admissible convergence result.

If z, w are vectors in $\mathbb{C}^n$, we continue to write $z \cdot \overline{w}$ to denote $\sum_j z_j \overline{w}_j$. (*Warning:* It is also common in the literature to use the notation $z \cdot w = \sum_j z_j \overline{w}_j$ or $\langle z, w \rangle = \sum_j z_j \overline{w}_j$.) Also, for $\Omega \subseteq \mathbb{C}^n$ a domain with C^2 boundary, $P \in \partial\Omega$, we let ν_P be the unit outward normal at P. Let $\mathbb{C}\nu_P$ denote the complex line generated by ν_P: $\mathbb{C}\nu_P = \{\zeta\nu_P : \zeta \in \mathbb{C}\}$.

By dimensional considerations, if $T_P(\partial\Omega)$ is the $(2n-1)$-dimensional real tangent space to $\partial\Omega$ at P, then $\ell = \mathbb{C}\nu_P \cap T_P(\partial\Omega)$ is a (one-dimensional) real line. Let

$$\mathcal{T}_P(\partial\Omega) = \{w \in \mathbb{C}^n : w \cdot \overline{\nu}_P = 0\}$$

$$= \{w \in \mathbb{C}^n : w \cdot \overline{\xi} = 0 \ \forall \xi \in \mathbb{C}\nu_P\}.$$

A fortiori, $\mathcal{T}_P(\partial\Omega) \subseteq T_P(\partial\Omega)$ since $\mathcal{T}_P$ is the orthogonal complement of ℓ in T_P. If $w \in \mathcal{T}_P(\partial\Omega)$, then $iw \in \mathcal{T}_P(\partial\Omega)$. Therefore $\mathcal{T}_P(\partial\Omega)$ may be thought of as an $(n-1)$-dimensional complex subspace of $T_P(\partial\Omega)$. Clearly, $\mathcal{T}_P(\partial\Omega)$ is the complex subspace of $T_P(\partial\Omega)$ of maximal dimension. It contains all complex subspaces of $T_P(\partial\Omega)$. [The reader should check that $\mathcal{T}_P(\partial\Omega)$ is the same complex tangent space that was introduced when we first studied the Levi form.]

Now let us examine the matter from another point of view. The complex structure is nothing other than a linear operator J on $\mathbb{R}^{2n}$ that assigns to $(w_1, w_2, \dots, w_{2n-1}, w_{2n})$ the vector $(-w_2, w_1, -w_4, w_3, \dots, -w_{2n}, w_{2n-1})$ (think of multiplication by i). With this in mind, we have that $J : \mathcal{T}_P(\partial\Omega) \to \mathcal{T}_P(\partial\Omega)$ both injectively and surjectively. So J preserves the complex tangent space. On the other hand, notice that $J\nu_P \in T_P(\partial\Omega)$ while $J(J\nu_P) = -\nu_P \notin T_P(\partial\Omega)$. Thus J does not preserve the real tangent space.

We call $\mathbb{C}\nu_P$ the *complex normal* space to $\partial\Omega$ at P and $\mathcal{T}_P(\partial\Omega)$ the *complex tangent space* to $\partial\Omega$ at P. Let $\mathcal{N}_P = \mathbb{C}\nu_P$. Then we have $\mathcal{N}_P \perp \mathcal{T}_P$ and

$$\mathbb{C}^n = \mathcal{N}_P \oplus_{\mathbb{C}} \mathcal{T}_P,$$

$$T_P = \mathbb{R}J\nu_P \oplus_{\mathbb{R}} \mathcal{T}_P.$$

Example 8.6.2 Let $\Omega = B \subseteq \mathbb{C}^n$ be the unit ball and $P = \mathbf{1} = (1, 0, \dots, 0) \in \partial\Omega$. Then $\mathbb{C}\nu_P = \{(z_1, 0, \dots, 0) : z_1 \in \mathbb{C}\}$ and $\mathcal{T}_P = \{(0, z') : z' \in \mathbb{C}^{n-1}\}$.

Exercise for the Reader:

1. Let $\Omega \subseteq \mathbb{C}^n$ be a domain with C^2 boundary. Let J be the real linear operator on $\mathbb{R}^{2n}$ that gives the complex structure. Let $P \in \partial\Omega$. Let $z = (z_1, \dots, z_n) = (x_1 + iy_1, \dots, x_n + iy_n) \approx (x_1, y_1, \dots, x_n, y_n)$ be an element of $\mathbb{C}^n \cong \mathbb{R}^{2n}$. The following are equivalent:

  (i) $w \in \mathcal{T}_P(\Omega)$;
  (ii) $Jw \in \mathcal{T}_P(\Omega)$;
  (iii) $Jw \perp \nu_P$ and $w \perp \nu_P$.

2. With notation as in the previous exercise, let $A = \sum_j a_j(z)\partial/\partial z_j$, $B = \sum_j b_j(z)\partial/\partial z_j$ satisfy $A\rho|_{\partial\Omega} = 0$, $B\rho|_{\partial\Omega} = 0$, where ρ is any defining function for Ω. Then the vector field $[A, B]$ has the same property. (However, note that $[A, \overline{B}]$ does *not* annihilate ρ on $\partial\Omega$ if Ω is the ball, for instance.) Therefore the holomorphic part of $\mathcal{T}_P$ is integrable (see [FOK]).

3. If $\Omega = B \subseteq \mathbb{C}^2$, $P = (x_1 + iy_1, x_2 + iy_2) \approx (x_1, y_1, x_2, y_2) \in \partial B$, then $\nu_P = (x_1, y_1, x_2, y_2)$ and $J\nu_P = (-y_1, x_1, -y_2, x_2)$. Also $\mathcal{T}_P$ is spanned over $\mathbb{R}$ by $(y_2, x_2, -y_1, -x_1)$ and $(-x_2, y_2, x_1, -y_1)$.

The next definition is best understood in light of the foregoing discussion and the definition of $\beta_2(P, r)$ in the boundary of the unit ball B. Let $\Omega \subset\subset \mathbb{C}^n$ have C^2 boundary. For $P \in \partial\Omega$, let $\pi_P : \mathbb{C}^n \to \mathcal{N}_P$ be (real or complex) orthogonal projection.

Definition 8.6.3 If $P \in \partial\Omega$ let

$$\beta_1(P, r) = \{\zeta \in \partial\Omega : |\zeta - P| < r\};$$

$$\beta_2(P, r) = \{\zeta \in \partial\Omega : |\pi_P(\zeta - P)| < r, |\zeta - P| < r^{1/2}\}.$$

Exercises for the Reader: The ball $\beta_1(P, r)$ has diameter $\approx r$ in all $(2n - 1)$ directions in the boundary. Therefore $\sigma(\beta_1(P, r)) \approx r^{2n-1}$.

The ball $\beta_2(P, r)$ has diameter $\approx \sqrt{r}$ in the $(2n - 2)$ complex tangential directions and diameter $\approx r$ in the one (complex normal) direction. Therefore $\sigma(\beta_2(P, r)) \approx (\sqrt{r})^{2n-2} \cdot r \approx Cr^n$.

If $z \in \Omega, P \in \partial\Omega$, we let

$$d_P(z) = \min\{\text{dist}(z, \partial\Omega), \text{dist}(z, T_P(\Omega))\}.$$

Notice that if Ω is convex, then $d_P(z) = \delta_\Omega(z)$, where $\delta_\Omega(z)$ (as before) denotes the ordinary Euclidean distance of z to $\partial\Omega$.

Definition 8.6.4 If $P \in \partial\Omega, \alpha > 1$, let

$$\mathcal{A}_\alpha = \{z \in \Omega : |(z - P) \cdot \overline{v}_P| < \alpha d_P(z), |z - P|^2 < \alpha d_P(z)\}.$$

Observe that d_P is used because near nonconvex boundary points, we still want $\mathcal{A}_\alpha$ to have the fundamental geometric shape of (paraboloid $\times$ cone) as shown in Figure 8.7. We call $\mathcal{A}_\alpha(P)$ an *admissible approach region* at the point P. It is strictly larger than a nontangential approach region. Any theorem about the boundary behavior of holomorphic functions that is expressed in the language of $\mathcal{A}_\alpha$ will be a stronger result than one expressed in the language of Γ_α.

Definition 8.6.5 If $f \in L^1(\partial\Omega)$ and $P \in \partial\Omega$ then we define

$$M_j f(P) = \sup_{r>0} \sigma(\beta_j(P, r))^{-1} \int_{\beta_j(P,r)} |f(\zeta)| d\sigma(\zeta), \qquad j = 1, 2.$$

Definition 8.6.6 If $f \in C(\Omega), P \in \partial\Omega$, then we define

$$f_2^{*,\alpha}(P) = \sup_{z \in \mathcal{A}_\alpha(P)} |f(z)|.$$

The first step of our program is to prove an estimate for M_2. This will require a covering lemma (indeed, it is known that weak-type estimates for operators like M_j are logically equivalent to covering lemmas—see [CORF1], [CORF2]). We exploit a rather general paradigm due to K.T. Smith [SMI] (see also [HOR1]):

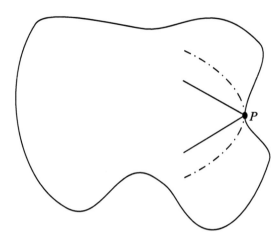

Figure 8.7. Shape of the admissible approach region.

Definition 8.6.7 Let X be a topological space equipped with a positive Borel measure m, and suppose that for each $x \in X$ and each $r > 0$, there is a "ball" $B(x, r)$. The "K.T. Smith axioms" for this setting are:[3]

(8.6.7.1) Each $B(x, r)$ is an open set of finite, positive measure that contains x.
(8.6.7.2) If $r_1 \leq r_2$ then $B(x, r_1) \subseteq B(x, r_2)$.
(8.6.7.3) There is a constant $c_0 > 0$ such that if $B(x, r) \cap B(y, s) \neq \emptyset$ and $r \geq s$ then $B(x, c_0 r) \supseteq B(y, s)$.
(8.6.7.4) There is a constant K such that $m(B(x_0, c_0 r)) \leq K m(B(x_0, r))$ for all r.

Now we have the following result:

Theorem 8.6.8 *Let the topological space X, measure m, and balls $B(x, r)$ be as in Definition 8.6.7. Let K be a compact subset of x and $\{B(x_a, r_a)\}_{a \in A}$ a covering of K by balls.*

Then there is a finite pairwise disjoint subcollection $B(x_{a_1}, r_{a_1}), \ldots,$ $B(x_{a_m}, r_{a_m})$ such that $K \subseteq \cup_{j=1}^{k} B(x_{a_j}, c_0 r_{a_j})$.

It follows that if we define

$$M f = \sup_{r > 0} (B(x, r))^{-1} \int_{B(x, r)} |f(t)| \, dm(t), \qquad f \in L^1(X, dm),$$

then

$$m\{x : M f(x) > \lambda\} \leq C \frac{\|f\|_{L^1}}{\lambda}.$$

Proof: Exercise for the reader: Imitate the proofs of Lemmas 8.4.2 and 8.4.4. □

---

[3] In Chapter 9 we shall revisit these ideas in the guise of "space of homogeneous type".

Thus we need to see that the $\beta_2(P, r)$ on $X = \partial\Omega$ with $m = \sigma$ satisfy **(8.6.7.1)**–**(8.6.7.4)**. Now **(8.6.7.1)** and **(8.6.7.2)** are trivial. Also **(8.6.7.4)** is easy if one uses the fact that $\partial\Omega$ is C^2 and compact (use the exercise for the reader following Definition 8.6.3). Thus it remains to check **(8.6.7.3)** (in many applications, this is the most difficult property to check).

Suppose that $\beta_2(z, r) \cap \beta_2(z', s) \neq \emptyset$. Thus there is a point $a \in \beta_2(z, r) \cap \beta_2(z', s)$. We may assume that $r = s$ by **(8.6.7.2)**. We thus have $|z - a| \leq r^{1/2}$, $|z' - a| \leq r^{1/2}$, hence $|z - z'| \leq 2r^{1/2}$. Let the constant $M \geq 2$ be chosen such that $\|\pi_z - \pi_{z'}\| \leq M|z - z'|$. (We must use here the fact that the boundary is C^2.) We claim that $\beta_2(z, (3 + 4M)r) \supseteq \beta_2(z', r)$. To see this, let $v \in \beta_2(z', r)$.

The easy half of the estimate is

$$|z - v| \leq |z - z'| + |z' - v| \leq 2r^{1/2} + r^{1/2} = 3r^{1/2}.$$

Also

$$\pi_z(z - v) = \pi_z(z - a) + \pi_{z'}(a - v) + \{\pi_z - \pi_{z'}\}(a - v).$$

Therefore

$$|\pi_z(z - v)| \leq r + 2r + \|\pi_z - \pi_{z'}\||a - v|$$

$$\leq 3r + M|z - z'| \cdot |a - v|$$

$$\leq 3r + M2r^{1/2}(|a - z'| + |z' - v|)$$

$$\leq (3 + 4M)r.$$

This proves **(8.6.7.3)**. Thus we have the following:

Corollary 8.6.9 *If $f \in L^1(\partial\Omega)$, then*

$$\sigma\{\zeta \in \partial\Omega : M_2 f(\zeta) > \lambda\} \leq C\frac{\|f\|_{L^1(\partial\Omega)}}{\lambda}, \qquad \text{for all} \quad \lambda > 0.$$

Proof: Apply the theorem.   □

Corollary 8.6.10 *The operator M_2 maps $L^2(\partial\Omega)$ to $L^2(\partial\Omega)$ boundedly.*

Proof: Exercise. The maximal operator is trivially bounded on L^∞, so apply the Marcinkiewicz interpolation theorem.   □

The next lemma is the heart of the matter: it is the technical device that allows us to estimate the behavior of a holomorphic function in the interior (in particular, on an admissible approach region) in terms of a maximal function on the boundary. The argument comes from [STE4] and [BAR].

Prelude: One of the triumphs of Stein's approach to Fatou-type theorems is that he reduced the entire question of boundary behavior of holomorphic functions to a result on plurisubharmonic functions like the one that follows.

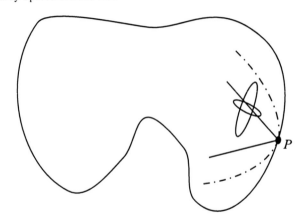

Figure 8.8. A canonical polydisk.

Lemma 8.6.11 *Let $u \in C(\overline{\Omega})$ be nonnegative and plurisubharmonic on Ω. Define $f = u|_{\partial\Omega}$. Then*
$$u_2^{*,\alpha}(P) \leq C_\alpha M_2(M_1 f)(P)$$
for all $P \in \partial\Omega$ and any $\alpha > 1$.

Proof: After rotating and translating coordinates, we may suppose that $P = 0$ and $\nu_P = (1 + i0, 0, \ldots, 0)$. Let $\alpha' > \alpha$. Then there is a small positive constant k such that if $z = (x_1 + iy_1, z_2, \ldots, z_n) \in \mathcal{A}_\alpha(P)$ then $\mathcal{D}(z) = D(z_1, -kx_1) \times D^{n-1}((z_2, \ldots, z_n), \sqrt{-kx_1}) \subseteq \mathcal{A}_{\alpha'}(P)$ (see Figure 8.8).

We restrict attention to $z \in \Omega$ so close to $P = 0$ that the projection along ν_P given by
$$z = (x_1 + iy_1, \ldots, x_n + iy_n) \rightarrow (\tilde{x}_1 + iy_1, x_2 + iy_2, \ldots, x_n + iy_n) \equiv \tilde{z} \in \partial\Omega$$

makes sense. [Observe that points z that are far from $P = 0$ are trivial to control using our estimates on the Poisson kernel.] The projection of $\mathcal{D}(z)$ along ν_P into the boundary lies in a ball of the form $\beta_2(\tilde{z}, K|x_1|)$—*this observation is crucial.*

Notice that the subharmonicity of u implies that $u(z) \leq Pf(z)$. Also there is a $\beta > 1$ such that $z \in \mathcal{A}_{\alpha'}(0) \Rightarrow z \in \Gamma_\beta(\tilde{z})$. Therefore the standard argument leading up to (8.4.1.1) yields that

$$|u(z)| \leq |Pf(z)| \leq C_\alpha M_1 f(\tilde{z}). \tag{8.6.11.1}$$

Now we bring the complex analysis into play. For we may exploit the plurisubharmonicity of $|u|$ on $\mathcal{D}(z)$ by invoking the subaveraging property in each dimension in succession. Thus

$$|u(z)| \leq \left(\pi|kx_1|^2\right)^{-1} \cdot \left(\pi(\sqrt{-kx_1})^2\right)^{-(n-1)} \int_{\mathcal{D}(z)} |u(\zeta)| dV(\zeta)$$

$$= Cx_1^{-n-1} \int_{\mathcal{D}(z)} |u(\zeta)| dV(\zeta).$$

Notice that if $z \in \mathcal{A}_\alpha(P)$ then each ζ in the last integrand is in $\mathcal{A}_{\alpha'}(P)$. Thus the last line is

$$\leq C' x_1^{-n-1} \int_{\mathcal{D}(z)} M_1 f(\widetilde{\zeta}) dV(\zeta)$$

$$\leq C'' x_1^{-n-1} \cdot x_1 \int_{\beta_2(\widetilde{z}, K x_1)} M_1 f(t) d\sigma(t)$$

$$\leq C''' x_1^{-n} \int_{\beta_2(0, K' x_1)} M_1 f(t) d\sigma(t)$$

$$\leq C'''' \left(\sigma \left(\beta_2(0, K' x_1) \right) \right)^{-1} \int_{\beta_2(0, K' x_1)} M_1 f(t) d\sigma(t)$$

$$\leq C''''' M_2(M_1 f)(0). \qquad \square$$

Prelude: Next is a version of Stein's main theorem. The result is not sharp, because the formulation of the optimal result will depend on the Levi geometry of the domain in question (see [DIK]). Even so, this theorem was revolutionary. The idea of proving a result of this power on an arbitrary domain was virtually unheard of at the time.

Now we may prove our main result:

Theorem 8.6.12 Let $0 < p \leq \infty$. Let $\alpha > 1$. If $\Omega \subset\subset \mathbb{C}^n$ has C^2 boundary and $f \in H^p(\Omega)$, then for σ-almost every $P \in \partial\Omega$ we have that

$$\lim_{\mathcal{A}_\alpha(P) \ni z \to P} f(z)$$

exists. In fact, the limit equals the quantity $\widetilde{f}(P)$ that we constructed in Theorems 8.4.1, 8.5.2, 8.5.3.

Proof: We already know that the limit exists almost everywhere *in the special sense of* Proposition 8.5.1. Call the limit function $\widetilde{f}$. Our job now is to show that the limit exists *in the admissible sense*.

We need only consider the case $p < \infty$. Let $\Omega = \cup_{j=1}^k \Omega_j$ as usual. It suffices to concentrate on Ω_1. Let $\nu = \nu_1$ be the outward normal given by **(8.3.5)**. Then by Proposition 8.5.1, the Lebesgue dominated convergence theorem implies that for $\partial\widetilde{\Omega}_1 \equiv \partial\Omega \cap \partial\Omega_1$,

$$\lim_{\epsilon \to 0} \int_{\partial\widetilde{\Omega}_1} |f(\zeta - \epsilon\nu) - \widetilde{f}(\zeta)|^p d\sigma(\zeta) = 0. \qquad (8.6.12.1)$$

For each $j, k \in \mathbb{N}$, consider the function $f_{j,k} : \Omega_1 \to \mathbb{C}$ given by

$$f_{j,k}(z) = |f(z - \nu/j) - f(z - \nu/k)|^{p/2}.$$

Then $f_{j,k} \in C(\overline{\Omega}_1)$ and is plurisubharmonic on Ω_1. Therefore a trivial variant of Lemma 8.6.11 yields

$$\int_{\partial\tilde\Omega_1} |(f_{j,k})_2^{*,a}(\zeta)|^2 d\sigma(\zeta) \le C_a \int_{\partial\tilde\Omega_1} |M_2(M_1 f_{j,k}(\zeta))|^2 d\sigma(\zeta)$$

$$\le C_a' \int_{\partial\tilde\Omega_1} |M_1 f_{j,k}(\zeta)|^2 d\sigma(\zeta)$$

$$\le C_a'' \int_{\partial\tilde\Omega_1} |f_{j,k}(\zeta)|^2 d\sigma(\zeta),$$

where we have used Corollary 8.6.9, Lemma 8.3.5, and the *proof* of Lemma 8.3.2. Now let $j \to \infty$ and apply (8.6.11.1) to obtain

$$\int_{\zeta \in \partial\tilde\Omega_1} \sup_{z \in \mathcal{A}_a(\zeta)} |f(z) - f(z - v/k)|^p \, d\sigma(\zeta)$$

$$\le C_a'' \int_{\zeta \in \partial\tilde\Omega_1} |\tilde{f}(\zeta) - f(\zeta - v/k)|^p \, d\sigma(\zeta). \tag{8.6.12.2}$$

Let $\epsilon > 0$. Then

$$\sigma\{\zeta \in \partial\tilde\Omega_1 : \limsup_{\mathcal{A}_a(\zeta)\ni z \to \zeta} |f(z) - \tilde{f}(\zeta)| > \epsilon\}$$

$$\le \sigma\{\zeta \in \partial\tilde\Omega_1 : \limsup_{\mathcal{A}_a(\zeta)\ni z \to \zeta} |f(z) - f(z - v/k)| > \epsilon/3\}$$

$$+ \sigma\{\zeta \in \partial\tilde\Omega_1 : \limsup_{\mathcal{A}_a(\zeta)\ni z \to \zeta} |f(z - v/k) - f(\zeta - v/k)| > \epsilon/3\}$$

$$+ \sigma\{\zeta \in \partial\tilde\Omega_1 : \limsup_{\mathcal{A}_a(\zeta)\ni z \to \zeta} |f(\zeta - v/k) - \tilde{f}(\zeta)| > \epsilon/3\}$$

$$\le C \int_{\partial\tilde\Omega_1} \sup_{z \in \mathcal{A}_a(\zeta)} |f(z) - f(z - v/k)|^p d\sigma(\zeta)/\epsilon^p + 0$$

$$+ C \int_{\partial\tilde\Omega_1} |\tilde{f}(\zeta) - f(\zeta - v/k)|^p d\sigma(\zeta)/\epsilon^p,$$

where we have used (the proof of) Chebyshev's inequality. By (8.6.12.2), the last line does not exceed

$$C' \int_{\partial\tilde\Omega_1} |\tilde{f}(\zeta) - f(\zeta - v/k)|^p d\sigma(\zeta)/\epsilon^p.$$

Now (8.6.11.1) implies that as $k \to \infty$, this last quantity tends to 0. Since $\epsilon > 0$ was arbitrary, we conclude that

$$\limsup_{\mathcal{A}_a(\zeta)\ni z \to \zeta} |f(z) - \tilde{f}(\zeta)| = 0$$

almost everywhere. $\qquad\qquad\qquad\qquad\qquad\qquad\qquad\qquad\qquad\qquad\qquad\square$

The theorem says that f has "admissible limits" at almost every boundary point of Ω. It is not difficult to see (indeed, by an inspection of the arguments in the present section) that Theorem 8.6.12 is best possible only for strongly pseudoconvex domains. At the boundary point $(1, 0)$ of the domain $\{(z_1, z_2) : |z_1|^2 + |z_2|^{2m} < 1\}$, the natural interior polydisks to study are of the form

$$\{(1 - \delta + \xi_1, \xi_2) : |\xi_1| < c \cdot \delta, |\xi_2| < c \cdot \delta^{1/2m}\}.$$

This observation, together with an examination of the proof of Corollary 8.6.10, suggests that the aperture in complex tangential directions of the approach regions should vary from boundary point to boundary point—and this aperture should depend on the Levi geometry of the point. A theory of boundary behavior for H^p functions taking these observations into account, for a special class of domains in $\mathbb{C}^2$, is enunciated in [NSW]. A more general paradigm for theories of boundary behavior of holomorphic functions is developed in S.G. Krantz [KRA7]. Related ideas also appear in [KRA8] and [KRA9]. The key tool in the last two references is the Kobayashi metric.

8.7 Real-Variable Methods

Capsule: In this section, looking ahead to the next, we provide a bridge between the (classical) holomorphic-functions viewpoint of Hardy space theory and the more modern real-variable viewpoint. The two approaches stand on their own, but the richest theory comes from the interaction of the two, because this gives a marriage of two sets of techniques and yields powerful new kinds of results. The atomic theory for Hardy spaces, just as an instance, arises from the real-variable point of view.

This chapter is *not* part of the main stream of the book. It is provided in order to give the reader some perspective on modern developments in the subject. Thus far in the book we have measured the utility and the effectiveness of an integral operator by means of the action on L^p spaces. This point of view is limited in a number of respects; a much broader perspective is gained when one expands the horizon to include real-variable Hardy spaces and functions of bounded mean oscillation (and, for that matter, Lipschitz spaces). We stress that we *shall not actually prove anything about real-variable Hardy spaces or BMO in this book*. We could do so, but it would take us far afield. Our purpose here is to provide background and context for the ideas that we *do* present in detail.

Hardy spaces are a venerable part of modern analysis. Originating in work of M. Riesz, O. Toeplitz, and G.H. Hardy in the early twentieth century, these spaces proved to be a fruitful venue both for function theory and for operator-theoretic questions. In more recent times, thanks to work of Fefferman, Stein, and Weiss (see [STG2], [FES], [STE2]), we see the Hardy spaces as artifacts of the real-variable theory. In this guise, they serve as substitutes for the L^p spaces when $0 < p \leq 1$.

Their functional-analytic properties have proved to be of seminal importance in modern harmonic analysis.

In the present chapter we have first reviewed the classical point of view concerning the Hardy spaces. Then we segue into the more modern, real-variable theory. It is the latter that will be the foundation for our studies later in the book. The reader should make frequent reference to Section 8.1 and the subsequent material to find context for the ideas that follow.

8.8 Hardy Spaces from the Real-Variable Point of View

Capsule: In this section we give a very brief introduction to the real-variable theory of Hardy spaces. Inspired by the theorem of Burkholder–Gundy–Silverstein about a maximal function characterization of the real parts of classical Hardy space functions on the unit disk in the plane, this is a far-reaching theory that can define Hardy spaces even on an arbitrary manifold or Lie group (see, for instance, [COIW1], [COIW2]). Today it is safe to say that the real-variable Hardy spaces are more important than the classical holomorphic Hardy spaces. The main reason is that we understand clearly how the important classes of integral operators act on the real-variable Hardy spaces; this makes them part of our toolkit.

We saw in Section 8.1 that a function f in the Hardy class $H^p(D)$ on the disk may be identified in a natural way with its boundary function, which we continue to call $\widetilde{f}$. Fix attention for the moment on $p = 1$.

If $\phi \in L^1(\partial D)$ and is real-valued, then we may define a harmonic function u on the disk by

$$u(re^{i\theta}) = \frac{1}{2\pi} \int_0^{2\pi} \phi(e^{i\psi}) \frac{1-r^2}{1-2r\cos(\theta-\psi)+r^2} \, d\psi.$$

Of course this is just the usual Poisson integral of ϕ. As was proved in Section 8.1, the function ϕ is the "boundary function" of u in a natural manner. Let v be the harmonic conjugate of u on the disk (we may make the choice of v unique by demanding that $v(0) = 0$). Thus $h \equiv u + iv$ is holomorphic. We may then ask whether the function v has a boundary limit function $\widetilde{\phi}$.

To see that $\widetilde{\phi}$ exists, we reason as follows: Suppose that the original function ϕ is nonnegative (any real-valued ϕ is the difference of two such functions, so there is no loss in making this additional hypothesis). Since the Poisson kernel is positive, it follows that $u > 0$. Now consider the holomorphic function

$$F = e^{-u-iv}.$$

The positivity of u implies that F is bounded. Thus $F \in H^\infty$. By Theorem 8.1.14, we may conclude that F has radial boundary limits at almost every point of ∂D. Unraveling our notation (and thinking a moment about the ambiguity caused by multiples of 2π), we find that v itself has radial boundary limits almost everywhere. We define thereby the function $\tilde{\phi}$.

Of course the function $h = u + iv$ can be expressed (up to an additive factor of $1/2$ and a multiplicative factor of $1/2$—see the calculations that led up to (2.1.3)) as the Cauchy integral of ϕ. The real part of the Cauchy kernel is the Poisson kernel (again up to a multiplicative and additive factor of $1/2$), so it makes sense that the real part of F on D converges back to ϕ. By the same token, the imaginary part of F is the integral of ϕ against the imaginary part of the Cauchy kernel, and it will converge to $\tilde{\phi}$. It behooves us to calculate the imaginary part of the Cauchy kernel.

Of course we know from our studies in Chapter 2 that this leads to a conjugate kernel and thence to the kernel of the Hilbert transform. That, in turn, is the primordial example of a singular integral kernel.

And our calculations will now give us a new way to think about the Hardy space $H^1(D)$. For if ϕ and $\tilde{\phi}$ are, respectively, the boundary functions of $\operatorname{Re} f$ and $\operatorname{Im} f$ for an $f \in H^1$, then $\phi, \tilde{\phi} \in L^1$, and our preceding discussion shows that (up to our usual correction factors) $\tilde{\phi} = H\phi$. *But this relationship is worth special note:* We have already proved that the Hilbert transform is not bounded on L^1, yet we see that the functions ϕ that arise as boundary functions of the real parts of functions in H^1 have the property that $\phi \in L^1$ and (notably) $H\phi \in L^1$. These considerations motivate the following *real-variable definition of the Hardy space H^1*:

Definition 8.8.2 A function $f \in L^1$ (on the circle, or on the real line) is said to be in the *real-variable Hardy space H^1_{Re}* if the Hilbert transform of f is also in L^1. The H^1_{Re} norm of f is given by

$$\|f\|_{H^1_{\operatorname{Re}}} \equiv \|f\|_{L^1} + \|Hf\|_{L^1}.$$

In higher dimensions the role of the Hilbert transform is played by a family of N singular integral operators. On $\mathbb{R}^N$, let

$$K_j(x) = \frac{x_j/|x|}{|x|^N}, \quad j = 1, \dots, N.$$

Notice that each K_j possesses the three defining properties of a Calderón–Zygmund kernel: it is smooth away from the origin, homogeneous of degree $-N$, and satisfies the mean-value condition because $\Omega_j(z) \equiv x_j/|x|$ is odd. We set $R_j f$ equal to the singular integral operator with kernel K_j applied to f, and call this operator the jth *Riesz transform*. Since the kernel of the jth Riesz transform is homogeneous of degree $-N$, it follows that the Fourier multiplier for this operator is homogeneous of degree zero (see Proposition 2.2.8). It is possible to calculate (though we shall not do it) that this multiplier has the form $c \cdot \xi_j/|\xi|$ (see [STE1] for the details). Observe that on $\mathbb{R}^1$ there is just one Riesz transform, and it is the Hilbert transform.

It turns out that the N-tuple $(K_1(z), \dots K_N(z))$ behaves naturally with respect to rotations, translations, and dilations in just the same way that the Hilbert kernel $1/t$ does in $\mathbb{R}^1$ (see [STE1] for the particulars). These considerations, and the ideas leading up to the preceding definition, give us the following:

Definition 8.8.3 (Stein–Weiss [STG2]) Let $f \in L^1(\mathbb{R}^N)$. We say that f is in the real-variable Hardy space of order 1, and write $f \in H^1_{\text{Re}}$, if $R_j f \in L^1$, $j = 1, \ldots, N$. The norm on this new space is

$$\|f\|_{H^1_{\text{Re}}} \equiv \|f\|_{L^1} + \|R_1 f\|_{L^1} + \cdots + \|R_N f\|_{L^1}.$$

We have provided a motivation for this last definition by way of the theory of singular integrals. It is also possible to provide a motivation via the Cauchy–Riemann equations. We now explain:

Recall that in the classical complex plane, the Cauchy–Riemann equations for a C^1, complex-valued function $f = u + iv$ with complex variable $z = x + iy$ are

$$\frac{\partial v}{\partial y} = \frac{\partial u}{\partial x},$$

$$\frac{\partial v}{\partial x} = -\frac{\partial u}{\partial y}.$$

The function f will satisfy this system of two linear first-order equations if and only if f is holomorphic (see [GRK12] for details).

On the other hand, if (v, u) is the gradient of a real-valued harmonic function F then

$$\frac{\partial u}{\partial x} = \frac{\partial^2 F}{\partial x \partial y} = \frac{\partial^2 F}{\partial y \partial x} = \frac{\partial v}{\partial y}$$

and

$$\frac{\partial u}{\partial y} = \frac{\partial^2 F}{\partial y^2} = -\frac{\partial^2 F}{\partial x^2} = -\frac{\partial v}{\partial x}.$$

[Note how we have used the fact that $\triangle F = 0$.] These are the Cauchy–Riemann equations. One may also use elementary ideas from multivariable calculus (see [GRK12]) to see that a pair (v, u) that satisifies the Cauchy–Riemann equations must be the gradient of a harmonic function.

Passing to N variables, let us now consider a real-valued function $f \in H^1_{\text{Re}}(\mathbb{R}^N)$. Set $f_0 = f$ and $f_j = R_j f$, $j = 1, \ldots, N$. By definition, $f_j \in L^1(\mathbb{R}^N)$, $j = 1, \ldots, N$. Thus it makes sense to consider

$$u_j(x, y) = P_y * f_j(x), \quad j = 0, 1, \ldots, N,$$

where

$$P_y(x) \equiv c_N \frac{y}{(|x|^2 + y^2)^{(N+1)/2}}$$

is the standard Poisson kernel for the upper half-space

$$\mathbb{R}^N_+ \equiv \{(x, y) : x \in \mathbb{R}^N, y > 0\}.$$

See also [KRA10].

A formal calculation (see [STG1]) shows that

$$\frac{\partial u_j}{\partial x_k} = \frac{\partial u_k}{\partial x_j} \tag{8.8.4}$$

for $j, k = 0, 1, \ldots, N$ and

$$\sum_{j=1}^{N} \frac{\partial u_j}{\partial x_j} = 0. \tag{8.8.5}$$

These are the *generalized Cauchy–Riemann equations*. The two conditions (8.8.4) and (8.8.5) taken together are equivalent to the hypothesis that the $(N + 1)$-tuple $(u_0, u_1, \ldots, u_N)$ is the gradient of a harmonic function F on $\mathbb{R}_+^{N+1}$. See [STG1] for details.

Both the singular integrals point of view and the Cauchy–Riemann equations point of view can be used to define H_{Re}^p for $0 < p < 1$. These definitions, however, involve algebraic complications that are best avoided in the present book. [Details may be found in [FES] and [STG2].] In the next section we present another point of view for Hardy spaces that treats all values of $p, 0 < p \leq 1$, simultaneously.

8.9 Maximal-Function Characterizations of Hardy Spaces

Capsule: This section is an outgrowth of the last two. It concentrates on the maximal-function approach to Hardy spaces; this is of course an aspect of the real-variable theory. We shall see in the next section how this in turn leads to the atomic theory of Hardy spaces. One advantage of the new approach presented here is that it frees the ideas up from the convolution (or translation-invariant) structure of Euclidean space.

Recall (see Section 2.5) that the classical Hardy–Littlewood maximal function

$$Mf(x) \equiv \sup_{r>0} \frac{1}{m[B(x,r)]} \int_{B(x,r)} |f(t)| dt$$

is not bounded on L^1. Part of the reason for this failure is that L^1 is not a propitious space for harmonic analysis, and another part of the reason is that the characteristic function of a ball is not smooth. To understand this last remark, we set $\phi = [1/m(B(0, 1))] \cdot \chi_{B(0,1)}$, the normalized characteristic function of the unit ball, and note that the classical Hardy–Littlewood maximal operator

$$Mf(z) = \sup_{R>0} \left| (\alpha^R \phi) * f(z) \right|$$

(where α^R is the dilation operator defined in Section 3.2).[4] It is natural to ask what would happen if we were to replace the nonsmooth kernel expression

---

[4] Observe that there is no loss of generality, and no essential change, in omitting the absolute values around f that were originally present in the definition of M. For if M is restricted to positive f, then the usual maximal operator results.

$\phi = [1/m(B(0, 1))] \cdot \chi_{B(0,1)}$ in the definition of Mf with a smooth testing function ϕ. This we now do.

We fix a function $\phi_0 \in C_c^\infty(\mathbb{R}^N)$ and, for technical reasons, we assume that $\int \phi_0 \, dx = 1$. We define

$$f^*(z) = \sup_{R>0} \left| (\alpha^R \phi_0) * f(z) \right|$$

for $f \in L^1(\mathbb{R}^N)$. We say that $f \in H^1_{\max}(\mathbb{R}^N)$ if $f^* \in L^1$. The following theorem, whose complicated proof we omit (but see [FES] or [STE2]), justifies this new definition:

Theorem 8.9.1 Let $f \in L^1(\mathbb{R}^N)$. Then $f \in H^1_{\mathrm{Re}}(\mathbb{R}^N)$ if and only if $f \in H^1_{\max}(\mathbb{R}^N)$.

It is notable that the equivalence enunciated in this last theorem is valid *no matter what the choice of ϕ_0 is*.

It is often said that in mathematics, a good theorem will spawn an important new definition. That is what will happen for us right now:

Definition 8.9.2 Let $f \in L^1_{\mathrm{loc}}(\mathbb{R}^N)$ and $0 < p \leq 1$. We say that $f \in H^p_{\max}(\mathbb{R}^N)$ if $f^* \in L^p$.

It turns out that this definition of H^p is equivalent to the definitions using singular integrals or Cauchy–Riemann equations that we alluded to, but did not enunciate, at the end of the last section. For convenience, we take Definition 8.9.2 to be our definition of real-variable H^p when $p < 1$. In the next section we shall begin to explore what has come to be considered the most flexible approach to Hardy spaces. It has the advantage that it requires a minimum of machinery, and can be adapted to a variety of situations—boundaries of domains, manifolds, Lie groups, and other settings as well. This is the so-called *atomic theory* of Hardy spaces.

8.10 The Atomic Theory of Hardy Spaces

Capsule: An atom is in some sense the most basic type of Hardy space function. Based on an idea of C. Fefferman, these elements have been developed today into a sophisticated theory (see [COIW1], [COIW2]). Any element of H^p_{Re} may be written as a sum—in a suitable sense—of atoms in H^p_{Re}. Thus any question, for instance, about the action of an integral operator can be reduced to that same question applied only to atoms. Atoms are also very useful in understanding the duals of the Hardy spaces. Today much of this theory may be subsumed into the modern wavelet theory.

We first formulate the basic ideas concerning atoms for $p = 1$. Then we shall indicate the generalization to $p < 1$. The complete story of the atomic theory of Hardy spaces may be found in [STE2]. Foundational papers in the subject are [COIW1], [COIW2], [COI], and [LATT].

Although the atomic theory fits very naturally into the context of spaces of homogeneous type (see Chapter 9), we shall content ourselves for now with a development in $\mathbb{R}^N$. Let $a \in L^1(\mathbb{R}^N)$. We impose three conditions on the function a:

(8.10.1) The support of a lies in some ball $B(x, r)$.
(8.10.2) We have the estimate

$$|a(t)| \leq \frac{1}{m(B(x, r))}$$

for every t.
(8.10.3) We have the mean-value condition

$$\int a(t) dt = 0.$$

A function a that enjoys these three properties is called a 1-*atom*.

Notice the mean-value-zero property in Axiom **(8.10.3)**. Assuming that atoms are somehow basic or typical H^1 functions (and this point we shall treat momentarily), we might have anticipated this vanishing-moment condition as follows. Let $f \in H^1(\mathbb{R}^N)$ according to the classical definition using Riesz transforms. Then $f \in L^1$ and $R_j f \in L^1$ for each j. Taking Fourier transforms, we see that

$$\widehat{f} \in C_0 \quad \text{and} \quad \widehat{R_j f}(\xi) = c \frac{\xi_j}{|\xi|} \widehat{f}(\xi) \in C_0, \quad j = 1, \ldots, N.$$

The only way that the last N conditions could hold—in particular that $[\xi_j/|\xi|] \cdot \widehat{f}(\xi)$ could be continuous at the origin—is for $\widehat{f}(0)$ to be 0. But this says that $\int f(t) dt = 0$. That is the mean-value-zero condition that we are now mandating for an atom.

Now let us discuss p-atoms for $0 < p < 1$. It turns out that we must stratify this range of p's into infinitely many layers, and treat each layer separately. Fix a value of p, $0 < p \leq 1$. Let a be a measurable function. We impose three conditions for a to be a p-atom:

(8.10.4$_p$) The support of a lies in some ball $B(x, r)$.
(8.10.5$_p$) We have the estimate

$$|a(t)| \leq \frac{1}{m(B(x, r))^{1/p}} \quad \text{for all } t.$$

(8.10.6$_p$) We have the mean-value condition

$$\int a(t) \cdot t^\beta \, dt = 0 \quad \text{for all multi-indices } \beta \text{ with } |\beta| \leq N \cdot (p^{-1} - 1).$$

The aforementioned stratification of values of p now becomes clear: if k is a nonnegative integer, then when

$$\frac{N}{N + k + 1} < p \leq \frac{N}{N + k},$$

we demand that a p-atom a have vanishing moments up to and including order k. This means that the integral of a against any monomial of degree less than or equal to k must be zero.

The basic fact about the atomic theory is that a p-atom is a "typical" H^p_{Re} function. More formally:

Prelude: The atomic theory of Hardy spaces is an idea that C. Fefferman offered as a comment in a conversation. It has turned out to be an enormously powerful and influential approach in the subject. In particular, it has made it possible to define Hardy spaces on the boundaries of domains and, more generally, on manifolds.

Theorem 8.10.7 *Let* $0 < p \le 1$. *For each* $f \in H^p_{\text{Re}}(\mathbb{R}^N)$ *there exist* p-*atoms* a_j *and complex numbers* β_j *such that*

$$f = \sum_{j=1}^{\infty} \beta_j a_j \tag{8.10.7.1}$$

and the sequence of numbers $\{\beta_j\}$ *satisfies* $\sum_j |\beta_j|^p < \infty$. *The sense in which the series representation (8.10.7.1) for* f *converges is a bit subtle (that is, it involves distribution theory) when* $p < 1$, *and we shall not discuss it here. When* $p = 1$ *the convergence is in* L^1.

The converse to the decomposition (8.10.7.1) holds as well: any sum as in (8.10.7.1) represents an $H^p_{\text{Re}}(\mathbb{R}^N)$ *function (where we may take this space to be defined by any of the preceding definitions).*

If one wants to study the action of a singular integral operator, or a fractional integral operator, on H^p, then by linearity it suffices to check the action of that operator on an atom.

One drawback of the atomic theory is that a singular integral operator will not generally send atoms to atoms. Thus the program described in the last paragraph is not quite as simple as it sounds. To address this problem, a theory of "molecules" has been invented. Just as the name suggests, a p-molecule is an agglomeration of atoms—subject to certain rules. And it is a theorem that a singular integral operator will map molecules to molecules. See [TAW] for further details.

In Section 9.8, when we further consider spaces of homogeneous type, we shall continue our development of the atomic theory in a more general setting.

As an exercise, the reader may wish to consider what space of functions is obtained when the mean-value condition (Axiom (**8.10.3**)) is omitted from the definition of 1-atom. Of course the resulting space is L^1, and this will continue to hold in Chapter 9 when we are in the more general setting of spaces of homogeneous type. Matters are more complicated for either $p < 1$ or $p > 1$.

8.11 The Role of *BMO*

Capsule: One of the results that opened up the modern real-variable theory of Hardy spaces was C. Fefferman's theorem that the dual of H^1_{Re} is the space *BMO* of John and Nirenberg. This is a truly profound and original theorem, and the techniques for proving it are important for the subject. The paper [FES] contains this result, and it truly launched a whole new era of work on real-variable Hardy spaces.

The space of functions of bounded mean oscillation was first treated by F. John and L. Nirenberg (see [JON]) in their study of certain nonlinear partial differential equations that arise in the study of minimal surfaces. Their ideas were, in turn, inspired by deep ideas of J. Moser [MOS1], [MOS2].

A function $f \in L^1_{\text{loc}}(\mathbb{R}^N)$ is said to be in *BMO* (the functions of *bounded mean oscillation*) if

$$\|f\|_* \equiv \sup_Q \frac{1}{|Q|} \int_Q |f(z) - f_Q| \, dx < \infty. \tag{8.11.1}$$

Here Q ranges over all cubes in $\mathbb{R}^N$ with sides parallel to the axes and $f_Q = [1/|Q|] \int_Q f(t) dt$ denotes the average of f over the cube Q; we use the expression $|Q|$ to denote the Lebesgue measure, or volume, of Q. There are a number of equivalent definitions of *BMO*; we mention two of them:

$$\inf_{c \in \mathbb{C}} \sup_Q \frac{1}{|Q|} \int_Q |f(z) - c| \, dx < \infty \tag{8.11.2}$$

and

$$\sup_Q \left[\frac{1}{|Q|} \int_Q |f(z) - f_Q|^q \, dx \right]^{1/q} < \infty, \quad \text{for some } 1 \le q < \infty. \tag{8.11.3}$$

The latter definition, when $q = 2$, is particularly useful in martingale and probability theory.

It is easy to see that definition (8.11.2) for *BMO* implies the original definition (8.11.1) of *BMO*. For

$$\frac{1}{|Q|} \int_Q |f(x) - f_Q| \, dx$$

$$\le \frac{1}{|Q|} \int_Q |f(x) - c| \, dx + \frac{1}{|Q|} \int_Q |c - f_Q| \, dx$$

$$= \frac{1}{|Q|} \int_Q |f(x) - c| \, dx + \frac{1}{|Q|} \int_Q \left| c - \frac{1}{|Q|} \int_Q f(t) dt \right| dx$$

$$= \frac{1}{|Q|} \int_Q |f(x) - c| \, dx + \frac{1}{|Q|} \int_Q \left| \frac{1}{|Q|} \int_Q [c - f(t)] dt \right| dx$$

$$\le \frac{1}{|Q|} \int_Q |f(x) - c| \, dx + \frac{1}{|Q|} \int_Q \frac{1}{|Q|} \int_Q |c - f(t)| dt \, dx$$

$$\le \frac{1}{|Q|} \int_Q |f(x) - c| \, dx + \frac{1}{|Q|} \int_Q |c - f(t)| dt.$$

The converse implication is immediate. Also definition (8.11.3) of *BMO* implies the original definition (8.11.1) by an application of Hölder's inequality. The converse implication requires the John–Nirenberg inequality (see the discussion below, as well as [JON]); it is difficult, and we omit it.

It is obvious that $L^\infty \subseteq BMO$. A nontrivial calculation (see [JON]) shows that $\ln|x| \in BMO(\mathbb{R})$. It is noteworthy that $(\ln|x|) \cdot \operatorname{sgn} x \notin BMO(\mathbb{R})$. We invite the reader to do some calculations to verify these assertions. In particular, accepting these facts, we see that the *BMO* "norm" is a measure both of size and of smoothness.[5]

Observe that the $\|\ \|_*$ norm is oblivious to (additive) constant functions. So the *BMO* functions are really defined modulo additive constants. Equipped with a suitable quotient norm, *BMO* is a Banach space. Its first importance for harmonic analysis arose in the following result of Stein: If T is a Calderón–Zygmund operator, then T maps L^∞ to *BMO* (see [FES] for the history of this result). If we take Stein's result for granted, then we can begin to explore how *BMO* fits into the infrastructure of harmonic analysis.

To do so, we think of H^1_{Re} in the following way:

$$H^1_{\mathrm{Re}} \ni f \longleftrightarrow (f, R_1 f, R_2 f, \ldots, R_N f) \in (L^1)^{N+1}. \tag{8.11.4}$$

Suppose that we are interested in calculating the dual of the Banach space H^1_{Re}. Let $\beta \in (H^1_{\mathrm{Re}})^*$. Then, using (8.11.4) and the Hahn–Banach theorem, there is a continuous extension of β to an element of $[(L^1)^{N+1}]^*$. But we know the dual of L^1, so we know that the extension (which we continue to denote by β) can be represented by integration against an element of $(L^\infty)^{N+1}$. Say that $(g_0, g_1, \ldots, g_N) \in (L^\infty)^{N+1}$ is the representative for β. Then we may calculate, for $f \in H^1_{\mathrm{Re}}$, that

$$\beta(f) = \int_{\mathbb{R}^N} (f, R_1 f, \ldots, R_N f) \cdot (g_0, g_1, \ldots, g_N)\, dx$$

$$= \int_{\mathbb{R}^N} f \cdot g_0\, dx + \sum_{j=1}^N \int_{\mathbb{R}^N} (R_j f) g_j\, dx$$

$$= \int_{\mathbb{R}^N} f \cdot g_0\, dx - \sum_{j=1}^N f(R_j g_j)\, dx.$$

Here we have used the elementary observation that the adjoint of a convolution operator with kernel K is the convolution operator with kernel $\widetilde{K}(z) \equiv K(-z)$. We finally rewrite the last line as

$$\beta(f) = \int_{\mathbb{R}^N} f\left[g_0 - \sum_{j=1}^N R_j g_j \right] dx.$$

---

[5] Technically speaking, the *BMO* norm is not a norm because it assigns size 0 to all constant functions. So, technically, it is a norm on a quotient space. It is nonetheless common for people to abuse language and refer to $\|\ \|_*$ as a norm; we carry on the tradition here.

But by the remarks two paragraphs ago, $R_j g_j \in BMO$ for each j. And we have already noted that $L^\infty \subseteq BMO$. As a result, the function

$$g_0 - \sum_{j=1}^{N} R_j g_j \in BMO.$$

We conclude that the dual space of H^1_{Re} has a natural embedding into *BMO*. Thus we might wonder whether $[H^1_{\text{Re}}]^* = BMO$. The answer to this question is affirmative, and is a deep result of C. Fefferman (see [FES]). We shall not prove it here. However, we shall use our understanding of atoms to have a look at the result.

The atomic theory has taught us that a typical H^1_{Re} function is an atom a. Let us verify that any such 1-atom a pairs with a *BMO* function ϕ. We suppose for simplicity that a is supported in the ball $B = B(0, r)$.

To achieve our goal, we examine

$$\int_{\mathbb{R}^N} a(x)\phi(x)\, dx = \int_B a(x)\phi(x)\, dx$$

for ϕ a testing function. We use the fact that a has mean value zero to write this last as

$$\int_B a(x)[\phi(x) - \phi_B]\, dx.$$

Here ϕ_B is the average of ϕ over the ball B. Then

$$\left| \int_{\mathbb{R}^N} a(x)\phi(x)\, dx \right| \leq \int_B |a(x)||\phi(x) - \phi_B|\, dx$$

$$\leq \frac{1}{m(B)} \int_B |\phi(x) - \phi_B|\, dx$$

$$\leq \|\phi\|_*.$$

This calculation shows that ϕ pairs with any atom, and the bound on the pairing is independent of the particular atom (indeed it depends only on the $\| \ \|_*$ norm of ϕ). It follows by linearity that any *BMO* function will pair with any H^1_{Re} function.

A fundamental fact about *BMO* functions is the John–Nirenberg inequality (see [JON]). It says, in effect, that a *BMO* function ϕ has distribution function μ that is comparable to the distribution function of an exponentially integrable function (i.e., a function f such that $e^{c|f|}$ is integrable for some small positive constant c). Here is a more precise statement:

Prelude: The space of functions of bounded mean oscillation (*BMO*) were invented in the paper [JON]. They arose originally in the context of the study (in the context of minimal surfaces) of partial differential equations of gradient type with L^∞ coefficients. Today, in harmonic analysis, the space is important because (Fefferman's theorem—see [KRA5]) it is the dual of the real-variable Hardy space H^1_{Re}. The John–Nirenberg inequality was the first hard analytic fact about the space that demonstrated its centrality and importance.

Theorem 8.11.5 (The John–Nirenberg inequality) *Let f be a function that lies in* $BMO(Q_0)$, *where Q_0 is a fixed cube lying in* $\mathbb{R}^N$ *(here we are mandating that f satisfy the BMO condition for subcubes of Q_0 only). Then, for appropriate constants* $c_1, c_2 > 0$,

$$m\{x \in Q_0 : |f(x) - f_{Q_0}| > \lambda\} \leq c_1 e^{-c_2 \lambda} |Q_0|. \tag{8.11.5.1}$$

Lurking in the background here is the concept of distribution function. This will be of interest for us later, so we say a little about it now. Let f be a measurable, nonnegative, real-valued function on $\mathbb{R}^N$. For $t > 0$ we define

$$\lambda_f(t) = \lambda(t) = m\{x \in \mathbb{R}^N : f(x) > t\}.$$

It is a fact, which the reader may verify with an elementary argument (or see [SAD]) that

$$\int f(x)^p\, dx = \int_0^\infty p t^{p-1} \cdot \lambda(t) dt = -\int_0^\infty t^p\, d\lambda(t).$$

The John–Nirenberg theorem is a statement about the distribution function of a *BMO* function.

It follows from (8.11.5.1) that f is in every L^p class (at least locally) for $p < \infty$ (exercise). But, as noted at the beginning of this section, *BMO* functions are not necessarily L^∞.

We have noted elsewhere that the function $f(x) = \log|x|$ is in *BMO*. It is a remarkable result of Garnett and Jones [GAJ] that *any BMO* function is a superposition of logarithmic functions. Their proof uses the John–Nirenberg theorem in a key manner. Theirs is a useful structure theorem for this important space.

For many purposes, *BMO* functions are the correct Ersatz for L^∞ in the context of harmonic analysis. For instance, it can be shown that any Calderón–Zygmund operator maps *BMO* to *BMO*. By duality (since the adjoint of a Calderón–Zygmund operator is also a Calderón–Zygmund operator) it follows that any Calderón–Zygmund operator maps H^1_{Re} to H^1_{Re}. Thus we see that the space H^1_{Re} is a natural substitute for L^1 in the context of harmonic analysis. The last assertion can be seen very naturally using atoms.

The study of real-variable Hardy spaces, and corresponding constructs such as the space *BMO* of John and Nirenberg, has changed the face of harmonic analysis in the past twenty-five years. We continue to make new discoveries about these spaces (see, for example, [CHKS1], [CHKS2]).

9

Introduction to the Heisenberg Group

Prologue: This chapter and the next constitute the climax of the present book. We have tried to lay the groundwork so that the reader may see how it is natural to identify the boundary of the unit ball in $\mathbb{C}^n$ with the Heisenberg group and then to do harmonic analysis on that group.

Analysis on the Heisenberg group is fascinating because it is topologically Euclidean but analytically non-Euclidean. Many of the most basic ideas of analysis must be developed again from scratch. Even the venerable triangle inequality, the concept of dilation, and the method of polar coordinates, must be rethought. One of the main points of our work will be to define, and then to prove estimates for, singular integrals on the Heisenberg group.

One of the really big ideas here is that the critical singularity—the singularity for a singular integral kernel—will not be the same as the topological dimension of the space (recall that on $\mathbb{R}^N$, the critical index is N). Thus we must develop the concept of "homogeneous dimension." It is also the case that the Fourier transform—while it certainly exists on the Heisenberg group—is not nearly as useful a tool as it was in classical Euclidean analysis. The papers [GEL1]–[GEL3] and [GELS] provide some basis for analysis using the Fourier transform on the Heisenberg group. While this theory is rich and promising, it has not borne the sort of fruit that classical Euclidean Fourier theory has.

The reader will see in this chapter all the foundational ideas that we have laid in the preceding nine chapters—brought into sharp focus by their application in a new context. It is hoped that the resulting tapestry will prove to be both enlightening and rewarding.

9.1 The Classical Upper Half-Plane

Capsule: It almost seems like a step backward—after all the machinery we have built up—to now revert to the classical upper half-plane. But we shall

S.G. Krantz, *Explorations in Harmonic Analysis*, Applied and Numerical
Harmonic Analysis, DOI 10.1007/978-0-8176-4669-1_9,
© Birkhäuser Boston, a part of Springer Science+Business Media, LLC 2009

first understand the Heisenberg group as a Lie group that acts on the boundary of the Siegel upper half-space. And that is a direct generalization of the classical upper half-plane. So our first task is to understand that half-plane in a new light, and with somewhat new language. We shall in particular analyze the group of holomorphic self-maps of the upper half-plane, and perform the Iwasawa decomposition for that group (thought of as a Lie group). This will yield a new way to think about translations, dilations, and Möbius transformations.

As usual, we let $U = \{\zeta \in \mathbb{C} : \text{Im}\,\zeta > 0\}$ be the upper half-plane. Of course the unit disk D is conformally equivalent to U by way of the map

$$c : D \longrightarrow U,$$

$$\zeta \longmapsto i \cdot \frac{1 - \zeta}{1 + \zeta}.$$

If Ω is any planar domain then we let $\text{Aut}\,(\Omega)$ denote the group of conformal self-maps of Ω, with composition as the binary operation. We call this the *automorphism group* of Ω. We equip the automorphism group with a topology by using uniform convergence on compact sets (equivalently, the compact-open topology). We sometimes refer to convergence in the automorphism group as "normal convergence." Then the automorphism groups of D and U are canonically isomorphic (both algebraically and topologically) by way of

$$\text{Aut}\,(D) \ni \varphi \longmapsto c \circ \varphi \circ c^{-1} \in \text{Aut}\,(U).$$

We now wish to understand $\text{Aut}\,(D)$ by way of the so-called *Iwasawa decomposition* of the group. This is a decomposition of the form

$$\text{Aut}\,(D) = K \cdot A \cdot N,$$

where K is compact, A is abelian, and N is nilpotent. We shall use the Iwasawa decomposition as a guide to our thoughts, but we shall not prove it here (see [HEL] for the chapter and verse on this topic). We also shall not worry for the moment what "nilpotent" means. The concept will be explained in the next section. In the present context, the "nilpotent" piece is actually abelian, and that is a considerably stronger condition.

Now let K be those automorphisms of D that fix the origin. By Schwarz's lemma, these are simply the rotations of the disk. And that is certainly a compact group, for it may be canonically identified (both algebraically and topologically) with the unit circle. Alternatively, it is easy to see (using Montel's theorem) that any sequence of rotations contains a subsequence that converges to another rotation.

To understand the abelian part A of the group, it is best to work with the unbounded realization U (that is, the upper half-plane). Then consider the group of dilations

$$\widetilde{\alpha}_\delta(\xi) = \delta\xi$$

for $\delta > 0$. This is clearly a subgroup of the full automorphism group of U, and it is certainly abelian. Let us examine the group that it corresponds to in the automorphism group action on D. Obviously we wish to consider

$$\alpha_\delta(\zeta) = c^{-1} \circ \tilde{\alpha}_\delta \circ c \in \text{Aut}(D). \qquad (9.1.1)$$

We know that $c^{-1}(\xi) = [i - \xi]/[i + \xi]$. So we may calculate the quantity in (9.1.1) to find that

$$\alpha_\delta(\zeta) = \frac{(1 - \delta) + \zeta(1 + \delta)}{(1 + \delta) + \zeta(1 - \delta)}. \qquad (9.1.2)$$

This is the "dilation group action" on the disk. Clearly the dilations are much easier to understand, and the abelian nature of the subgroup more transparent, when we examine the action on the upper half-plane U.

Next we look at the nilpotent piece, which in the present instance is in fact abelian. We again find it most convenient to examine the group action on the upper half-plane U. This subgroup is the translations:

$$\tilde{\tau}_a(\xi) = \xi + a \,,$$

where $a \in \mathbb{R}$. Then the corresponding automorphism on the disk is

$$\tau_a(\zeta) = c^{-1} \circ \tilde{\tau}_a \circ c \in \text{Aut}(D).$$

With some tedious calculation we find that

$$\tau_a(\zeta) = \frac{-a + \zeta(2i - a)}{(2i + a) + a\zeta} = \frac{2i - a}{2i + a} \cdot \frac{\zeta + a/(a - 2i)}{1 + [a/(a + 2i)]\zeta}.$$

Again, the "translation" nature of the automorphism group elements is much clearer in the group action on the unbounded realization U, and the abelian property of the group is also much clearer.

Notice that the group of translations acts simply transitively on the boundary of U. For if $\tilde{\tau}_a$ is a translation then $\tilde{\tau}_a(0) = a + i0 \in \partial U$. Conversely, if $z = x + i0 \in \partial U$ then $\tau_z(0) = x + i0$. So we may identify the translation group with the boundary, and vice versa. This simple observation is the key to Fourier analysis in this setting.

We will use these elementary calculations of the Iwasawa decomposition in one complex dimension as an inspiration for our more sophisticated calculations on the ball in $\mathbb{C}^n$ that we carry out in the next section.

9.2 Background in Quantum Mechanics

The Heisenberg group derives its name from the fact that the commutator relations in its Lie algebra mimic those satisfied by the classical operators of quantum theory. The details of these assertions are too technical for the present context, and we refer the reader to [STE2, pp. 547–553] for the details.

9.3 The Role of the Heisenberg Group in Complex Analysis

Capsule: In this section we begin to familiarize ourselves with the unit ball in $\mathbb{C}^n$. We examine the automorphism group of the ball, and we detail its Iwasawa decomposition. The subgroup of the automorphism group that plays the role of "translations" in the classical setting turns out to be the Heisenberg group. There are certainly other ways to discover the Heisenberg group, but this one turns out to be most natural for us. It is important to understand that the Heisenberg group acts simply transitively on the boundary; this makes possible the identification of the group with the boundary. One may also do harmonic analysis on the boundary by exploiting the unitary group action. We shall not explore that approach here, but see [FOL].

Complex analysis and Fourier analysis on the unit disk $D = \{\zeta \in \mathbb{C} : |\zeta| < 1\}$ work well together because there is a group—namely the group of *rotations*—that acts naturally on ∂D. Complex analysis and Fourier analysis on the upper half-plane $U = \{\zeta \in \mathbb{C} : \operatorname{Im} \zeta > 0\}$ are symbiotic because there is a group—namely the group of *translations*—that acts naturally on ∂U.[1] We also might note that the group of dilations $\zeta \mapsto \delta\zeta$ acts naturally on U for $\delta > 0$. One of the main points here is that the disk D and the upper half-plane U are conformally equivalent. The *Cayley transform*

$$c : D \to U$$

is given explicitly by

$$c(\zeta) = i \cdot \frac{1 - \zeta}{1 + \zeta}.$$

Notice that c is both one-to-one and onto. Its inverse is given by

$$c^{-1}(\mu) = \frac{i - \mu}{i + \mu}.$$

We would like a similar situation to obtain for the domain the unit ball $B = \{(z_1, \ldots, z_n) \in \mathbb{C}^n : |z_1|^2 + \cdots + |z_n|^2 < 1\}$. It turns out that in this situation, the unbounded realization[2] of the domain B is given by

$$\mathcal{U} = \left\{ (w_1, \ldots, w_n) \in \mathbb{C}^n : \operatorname{Im} w_1 > \sum_{j=2}^{n} |w_j|^2 \right\}.$$

---

[1] It must be noted, however, that the rotations on the disk and the translations on the upper half-plane do not "correspond" in any natural way; certainly the Cayley transform does not map the one group to the other. This anomaly is explored in the fine text [HOF].

[2] We use here the classical terminology of Siegel upper half-spaces. Such an upper half-space is defined with an inequality using a quadratic form. The resulting space is unbounded. However, when the quadratic form is positive definite then the domain has a bounded realization—that is to say, it is biholomorphically equivalent to a bounded domain. See [KAN] for details of this theory.

It is convenient to write $w' \equiv (w_2, \ldots, w_n)$. We refer to our domain $\mathcal{U}$ as the *Siegel upper half-space*, and we write its defining equation as $\text{Im } w_1 > |w'|^2$.

Now the mapping that shows B and $\mathcal{U}$ to be biholomorphically equivalent is given by

$$\Phi : B \longrightarrow \mathcal{U},$$

$$(z_1, \ldots, z_n) \longmapsto \left(i \cdot \frac{1 - z_1}{1 + z_1}, \frac{z_2}{1 + z_1}, \ldots, \frac{z_n}{1 + z_1} \right).$$

We leave it to the reader to perform the calculations to verify that Φ maps B to $\mathcal{U}$ in a holomorphic, one-to-one, and onto fashion. The inverse of the mapping Φ may also be calculated explicitly.

Just as in one dimension, if $\Omega \subseteq \mathbb{C}^n$ is any domain, we let $\text{Aut}(\Omega)$ denote the collection of biholomorphic self-mappings of Ω. This set forms a group when equipped with the binary operation of composition of mappings. In fact, it is a topological group with the topology of uniform convergence on compact sets (which is the same as the compact-open topology). Further, it can be shown that, at least when Ω is a bounded domain, $\text{Aut}(\Omega)$ is a real Lie group (never a complex Lie group—see [KOB]). We shall not make much use of this last fact, but it a helpful touchstone in our discussions. There is a natural isomorphism between $\text{Aut}(B)$ and $\text{Aut}(\mathcal{U})$ given by

$$\text{Aut}(B) \ni \varphi \longmapsto \Phi \circ \varphi \circ \Phi^{-1} \in \text{Aut}(\mathcal{U}). \tag{9.3.1}$$

It turns out that we can understand the automorphism group of B more completely by passing to the automorphism group of $\mathcal{U}$. We used this technique earlier in the present chapter to understand the automorphism group of the disk. We shall again indulge in that conceit right now. We shall use, as we did in the more elementary setting of the disk, the idea of the Iwasawa decomposition $G = KAN$.

The compact part of $\text{Aut}(B)$ is the collection of all automorphisms that fix the origin. It is easy to prove, using a version of the Schwarz lemma (see [RUD]), that any such automorphism is a unitary rotation. This is an $n \times n$ complex matrix whose rows (or columns) form a Hermitian orthonormal basis of $\mathbb{C}^n$ and that has determinant 1. Let us denote this subgroup by K. We see that the group is compact just using a normal families argument: if $\{\varphi_j\}$ is a sequence in K then Montel's theorem guarantees that there will be a subsequence converging uniformly on compact sets. Of course the limit function will be a biholomorphic mapping that fixes 0 (i.e., a unitary transformation).

Implicit in our discussion here is a fundamental idea of H. Cartan: If Ω is any bounded domain and $\{\varphi_j\}$ a sequence of automorphisms of Ω, and if the φ_j converge uniformly on compact subsets of Ω, then the limit mapping φ_0 is either itself an automorphism, or else it maps the entire domain Ω into the boundary. The proof of this result (which we omit, but see [NAR]) is a delicate combination of Hurwitz's principle and the open mapping theorem. In any event, since the mappings in the last paragraph all fix the origin, it is clear that the limit mapping cannot map the entire domain into the boundary. Hence, by Cartan, the limit mapping must itself be an automorphism.

One may utilize the isomorphism (9.3.1)—this is an explicit and elementary calculation—to see that the subgroup of Aut $(\mathcal{U})$ that corresponds to K is $\widetilde{K}$, which is the subgroup of automorphisms of $\mathcal{U}$ that fix the point $(i, 0, \ldots, 0)$. Although $\widetilde{K}$ is a priori a compact Lie group, one may also verify this property by a direct argument as in the last two paragraphs.

Thus we have disposed of the compact piece of the automorphism group of the unit ball. Now let us look at the abelian piece. For this part, it is most convenient to begin our analysis on $\mathcal{U}$. Let us consider the group of dilations, which consists of the nonisotropic mappings

$$\widetilde{\alpha}_\delta : \mathcal{U} \longrightarrow \mathcal{U}$$

given by

$$\widetilde{\alpha}_\delta(w_1, \ldots, w_n) = (\delta^2 w_1, \delta w_2, \delta w_3, \ldots, \delta w_n)$$

for any $\delta > 0$. Check for yourself that $\widetilde{\alpha}_\delta$ maps $\mathcal{U}$ to $\mathcal{U}$. We call these mappings non-isotropic (meaning "acts differently in different directions") because they treat the w_1 variable differently from the $w_2, \ldots, w_n$ variables. The group is clearly abelian. It corresponds, under the mapping Φ, to the group of mappings on B given by

$$\alpha_\delta(z_1, \ldots, z_n) = \Phi^{-1} \circ \widetilde{\alpha}_\delta \circ \Phi(z). \tag{9.3.2}$$

Now it is immediate to calculate that

$$\Phi^{-1}(w) = \left(\frac{i - w_1}{i + w_1}, \frac{2i w_2}{i + w_1}, \ldots, \frac{2i w_n}{i + w_n} \right).$$

Of course it is just a tedious algebra exercise to determine α_δ. The answer is

$$\alpha_\delta(z) = \left(\frac{(1 - \delta^2) + z_1(1 + \delta^2)}{(1 + \delta^2) + z_1(1 - \delta^2)}, \frac{2\delta z_2}{(1 + \delta^2) + z_1(1 - \delta^2)}, \ldots, \right.$$

$$\left. \frac{2\delta z_n}{(1 + \delta^2) + z_1(1 - \delta^2)} \right).$$

One may verify directly that $z \in B$ if and only if $\alpha_\delta(z) \in B$.

It is plain that the dilations are much easier to understand on the unbounded realization $\mathcal{U}$. And the group structure, in particular the abelian nature of the group, is also much more transparent in that context.

We shall find that the "nilpotent" piece of the automorphism group is also much easier to apprehend in the context of the unbounded realization. We shall explore that subgroup in the next section.

9.4 The Heisenberg Group and Its Action on $\mathcal{U}$

Capsule: Here we study the action of the Heisenberg group on the Siegel upper half-space and on its boundary. We find that the upper half-space decomposes into level sets that are "parallel" to the boundary (similar to the horizontal level lines in the classical upper half-plane), and that the Heisenberg group acts on each of these. We set up the basics for the convolution structure on the Heisenberg group.

If G is a group and $g, h \in G$, then we define a *first commutator* of g and h to be the expression $\lambda(g, h) \equiv ghg^{-1}h^{-1}$. [Clearly if the group is abelian then this expression will always equal the identity; otherwise not.] If $g, h, k \in G$ then a *second commutator* is an expression of the form $\lambda(\lambda(g, h), k)$. Of course higher-order commutators are defined inductively.[3]

Let m be a nonnegative integer. We say that the group G is *nilpotent of order m* (or step m) if all commutators of order $m + 1$ in G are equal to the identity, and if m is the least such integer. Clearly an abelian group is nilpotent of order 0. It turns out that the collection of "translations" on $\partial \mathcal{U}$ is a nilpotent group of order 1. In fact, that group can be *identified* in a natural way with $\partial \mathcal{U}$ (in much the same way that the ordinary left-right translations of the boundary of the classical upper half-space U can be identified with ∂U). We now present the details of this idea.

The *Heisenberg group* of order $n - 1$, denoted by $\mathbb{H}_{n-1}$, is an algebraic structure that we impose on $\mathbb{C}^{n-1} \times \mathbb{R}$. Let (ζ, t) and (ξ, s) be elements of $\mathbb{C}^{n-1} \times \mathbb{R}$. Then the binary Heisenberg group operation is given by

$$(\zeta, t) \cdot (\xi, s) = (\zeta + \xi, t + s + 2\mathrm{Im}\,(\zeta \cdot \bar{\xi})).$$

It is clear, because of the Hermitian inner product $\zeta \cdot \bar{\xi} = \zeta_1 \bar{\xi}_1 + \cdots + \zeta_{n-1} \bar{\xi}_{n-1}$, that this group operation is nonabelian (although in a fairly subtle fashion). Later on we shall have a convenient means to verify the nilpotence, so we defer that question for now.

Now an element of $\partial \mathcal{U}$ has the form $(\mathrm{Re}\,w_1 + i|(w_2, \ldots, w_n)|^2, w_2, \ldots, w_n) = (\mathrm{Re}\,w_1 + i|w'|^2, w')$, where $w' = (w_2, \ldots, w_n)$. We identify this boundary element with the Heisenberg group element $(w', \mathrm{Re}\,w_1)$, and we call the corresponding mapping Ψ. Now we can specify how the Heisenberg group acts on $\partial \mathcal{U}$. If $w = (w_1, w') \in \partial \mathcal{U}$ and $g = (z', t) \in \mathbb{H}_{n-1}$ then we have the action

$$g[w] = \Psi^{-1}[g \cdot \Psi(w)] = \Psi^{-1}[g \cdot (w', \mathrm{Re}\,w_1)] = \Psi^{-1}[(z', t) \cdot (w', \mathrm{Re}\,w_1)].$$

More generally, if $w \in \mathcal{U}$ is *any element* then we write

$$w = (w_1, w_2, \ldots, w_n) = (w_1, w')$$

$$= ((\mathrm{Re}\,w_1 + i|w'|^2) + i(\mathrm{Im}\,w_1 - |w'|^2), w_2, \ldots, w_n)$$

$$= (\mathrm{Re}\,w_1 + i|w'|^2, w') + (i(\mathrm{Im}\,w_1 - |w'|^2), 0, \ldots, 0).$$

---

[3] In some sense it is more natural to consider commutators in the Lie algebra of the group. By way of the exponential map and the Campbell–Baker–Hausdorff formula (see [SER]), the two different points of view are equivalent. We shall describe some of the Lie algebra approach in the material below. For now, the definition of commutators in the context of the group is a quick-and-dirty way to get at the idea we need to develop right now.

The first expression in parentheses is an element of $\partial \mathcal{U}$. It is convenient to let $\rho(w) = \text{Im } w_1 - |w'|^2$. We think of ρ as a "height function." In short, we are expressing an arbitrary element $w \in \mathcal{U}$ as an element in the boundary plus a translation "up" to a certain height in the i direction of the first variable.

Now we let g act on w by

$$g[w] = g[(\text{Re } w_1 + i|w'|^2, w') + (i(\text{Im } w_1 - |w'|^2), 0, \dots, 0)]$$

$$\equiv g[(\text{Re } w_1 + i|w'|^2, w')] + (i(\text{Im } w_1 - |w'|^2), 0, \dots, 0). \qquad (9.4.1)$$

In other words, we let g act on level sets of the height function.

It is our job now to calculate this last line and to see that it is a holomorphic action on $\mathcal{U}$. We have, for $g = (z', t)$,

$g[w]$

$$= g[(\text{Re } w_1 + i|w'|^2, w')] + (i(\text{Im } w_1 - |w'|^2), 0, \dots, 0)$$

$$= \Psi^{-1} \left[g \cdot (w', \text{Re } w_1) \right] + (i(\text{Im } w_1 - |w'|^2), 0, \dots, 0)$$

$$= \Psi^{-1}[(z' + w', t + \text{Re } w_1 + 2\text{Im } (z' \cdot \overline{w}'))] + (i(\text{Im } w_1 - |w'|^2), 0, \dots, 0)$$

$$\equiv (t + \text{Re } w_1 + 2\text{Im } (z' \cdot \overline{w}') + i|z' + w'|^2, z' + w')$$

$$\quad + (i(\text{Im } w_1 - |w'|^2), 0, \dots, 0)$$

$$= (t + \text{Re } w_1 + (-i)[z' \cdot \overline{w}' - \overline{x}' \cdot w'] + i|z'|^2$$

$$\quad + i|w'|^2 + 2i\text{Re } z' \cdot \overline{w}' + i\text{Im } w_1 - i|w'|^2, z' + w')$$

$$= (t + w_1 + i|z'|^2 + i[2\text{Re } \overline{x}' \cdot w' + 2i\text{Im } \overline{x}' \cdot w'], z' + w')$$

$$= (t + i|z'|^2 + w_1 + i2\overline{x}' \cdot w', z' + w').$$

This mapping is plainly holomorphic in w (but *not* in z!). Thus we see explicitly that the action of the Heisenberg group on $\mathcal{U}$ is a (bi)holomorphic mapping.

As we have mentioned previously, the Heisenberg group acts simply transitively on the boundary of $\mathcal{U}$. Thus the group may be identified with the boundary in a natural way. Let us now make this identification explicit. First observe that $\mathbf{0} \equiv (0, \dots, 0) \in \partial \mathcal{U}$. If $g = (z', t) \in \mathbb{H}_{n-1}$ then

$$g[\mathbf{0}] = \Psi^{-1}[(z', t) \cdot (0', 0)] = \Psi^{-1}[(z', t)] = (t + i|z'|^2, z') \in \partial \mathcal{U}.$$

Conversely, if $(\text{Re } w_1 + i|w'|^2, w') \in \partial \mathcal{U}$ then let $g = (w', \text{Re } w_1)$. Hence

$$g[\mathbf{0}] = \Psi^{-1}[(w', \text{Re } w_1)] = (\text{Re } w_1 + i|w'|^2, w') \in \partial \mathcal{U}.$$

Compare this result with the similar but much simpler result for the classical upper half-plane U that we discussed in the last section.

The upshot of the calculations in this section is that analysis on the boundary of the ball B may be reduced to analysis on the boundary of the Siegel upper half-space $\mathcal{U}$. And that in turn is equivalent to analysis on the Heisenberg group $\mathbb{H}_{n-1}$. The Heisenberg group is a step-one nilpotent Lie group. In fact, all the essential tools of analysis may be developed on this group, just as they were in the classical Euclidean setting. That is our goal in the next several sections.

9.5 The Geometry of $\partial\mathcal{U}$

Capsule: The boundary of the classical upper half-plane is *flat*. It is geometrically flat and it is complex analytically flat. In fact, it is a line. Not so with the Siegel upper half-space. It is strongly pseudoconvex (in point of fact, the upper half-space is biholomorphic to the unit ball). This boundary is naturally "curved" in a complex analytic sense, and *it cannot be flattened*. These are ineluctable facts about the Siegel upper half-space that strongly influence the analysis of this space that we are about to learn.

The boundary of the Siegel upper half-space $\mathcal{U}$ is strongly pseudoconvex. This fact may be verified directly—by writing out the Levi form and calculating its eigenvalues—or it may be determined by invoking an important theorem of S. Bell [BEL], just because we already know that the ball is strongly pseudoconvex.

As such, we see that the boundary of $\mathcal{U}$ cannot be "flattened." That is to say, it would be convenient if there were a biholomorphic mapping of $\mathcal{U}$ to a Euclidean half-space, but in fact this is impossible because the boundary of a Euclidean half-space is Levi flat. And Bell's paper says in effect that a strongly pseudoconvex domain can be biholomorphic only to another strongly pseudoconvex domain.

There are other ways to understand the geometry of $\partial\mathcal{U}$. In Section 9.7 we discuss the commutators of vector fields—in the context of the Heisenberg group. The main point of that discussion is that the Heisenberg group is a *step-one nilpotent Lie group*. This means that certain first-order commutators in the Heisenberg group are nonzero, but all other commutators are zero. This idea also has a complex-analytic formulation which we now treat briefly.

For simplicity let us restrict attention to $\mathbb{C}^2$. And let $\Omega = \{z \in \mathbb{C}^2 : \rho(z) < 0\} \subseteq \mathbb{C}^2$ be a smoothly bounded domain. If $P \in \partial\Omega$ satisfies $\partial\rho/\partial z_1(P) \neq 0$ then the vector field

$$L = \frac{\partial\rho}{\partial z_1}(P)\frac{\partial}{\partial z_2} - \frac{\partial\rho}{\partial z_2}(P)\frac{\partial}{\partial z_1}$$

is *tangent* to $\partial\Omega$ at P just because $L\rho(P) = 0$. Likewise $\overline{L}$ is also a tangent vector field.

If P is a strongly pseudoconvex point then it may be calculated that the commutator $[L, \overline{L}]\rho(P)$ is *not* zero—that is to say, $[L, \overline{L}]$ has a nonzero component in the normal direction. This is analogous to the Lie algebra structure of the Heisenberg group. And in fact, Folland and Stein [FOST1] have shown that the analysis of a strongly pseudoconvex point may be accurately modeled by the analysis of the

Heisenberg group. It is safe to say that much of what we present in the last two chapters of the present book is inspired by [FOST1].

The brief remarks made here will be put into a more general context, and illustrated with examples, in Section 9.6.

9.6 The Lie Group Structure of the Heisenberg Group $\mathbb{H}^n$

> **Capsule:** The Heisenberg group is a step-one nilpotent Lie group. This is a very strong statement about the complexity of the Lie algebra of the group. In particular, it says something about the Lie brackets of invariant vector fields on the group. These will in turn shape the analysis that we do on the group. It will lead to the notion of homogeneous dimension.

Let us denote the elements of $\mathbb{C}^{n-1} \times \mathbb{R}$ by $[\zeta, t], \zeta \in \mathbb{C}^{n-1}$ and $t \in \mathbb{R}$. Then the space $\mathbb{C}^{n-1} \times \mathbb{R}$ (which is now the Heisenberg group $\mathbb{H}^{n-1}$) has the group structure defined as follows:

$$[\zeta, t] \cdot [\zeta^*, t^*] = [\zeta + \zeta^*, t + t^* + 2\mathrm{Im}\zeta \cdot \overline{\zeta^*}],$$

where $\zeta \cdot \overline{\zeta^*} = \zeta_1 \overline{\zeta_1^*} + \cdots + \zeta_n \overline{\zeta_n^*}$. The identity element is $[0, 0]$ and $[\zeta, t]^{-1} = [-\zeta, -t]$. Check the associativity:

$$[z, t] \cdot ([w, s] \cdot [\eta, u]) = [z, t] \cdot [w + \eta, s + u + 2\mathrm{Im}w \cdot \overline{\eta}]$$

$$= [z + w + \eta, t + s + u + 2\mathrm{Im}w \cdot \overline{\eta} + 2\mathrm{Im}z \cdot \overline{(w + \eta)}]$$

and

$$([z, t] \cdot [w, s]) \cdot [\eta, u] = [z + w, t + s + 2\mathrm{Im}z \cdot \overline{w}] \cdot [\eta, u]$$

$$= [z + w + \eta, t + s + u + 2\mathrm{Im}z \cdot \overline{w} + 2\mathrm{Im}(z + w) \cdot \overline{\eta}].$$

9.6.1 Distinguished 1-Parameter Subgroups of the Heisenberg Group

The Heisenberg group $\mathbb{H}^{n-1}$ has $2n - 1$ real dimensions and we can define the differentiation of a function in each direction consistent with the group structure by considering 1-parameter subgroups in each direction.

Let $g = [z, t] \in \mathbb{H}^{n-1}$, where $z = (z_1, \ldots, z_{n-1}) = (x_1 + iy_1, \ldots, x_{n-1} + iy_{n-1})$ and $t \in \mathbb{R}$. If we let

$$\gamma_{2j-1}(s) = [(0, \ldots, s + i0, \ldots, 0), 0],$$

$$\gamma_{2j}(s) = [(0, \ldots, 0 + is, \ldots, 0), 0],$$

for $1 \le j \le n - 1$ and the s term in the jth slot, and if we let

$$\gamma_{2n-1}(s) = \gamma_t(s) = [0, s]$$

[with $(n - 1)$ zeros and one s], then each forms a one-parameter subgroup of $\mathbb{H}^n$. Just as an example,

$$[(0, \ldots, s+i0, \ldots, 0), 0] \cdot [(0, \ldots, s'+i0, \ldots, 0), 0] = [(0, \ldots, s+s'+i0, \ldots, 0), 0].$$

We define the differentiation of f at $g = [z, t]$ in each one-parameter group direction as follows:

$$X_j f(g) \equiv \frac{d}{ds} f(g \cdot \gamma_{2j-1}(s))|_{s=0}$$

$$= \frac{d}{ds} f([(x_1 + iy_1, \ldots, x_j + s + iy_j, \ldots, x_{n-1} + iy_{n-1}), t + 2y_j s])|_{s=0}$$

$$= \left(\frac{\partial f}{\partial x_j} + 2y_j \frac{\partial f}{\partial t} \right) [z, t], \quad 1 \le j \le n - 1,$$

$$Y_j f(g) \equiv \frac{d}{ds} f(g \cdot \gamma_{2j}(s))|_{s=0}$$

$$= \frac{d}{ds} f([(x_1 + iy_1, \ldots, x_j + i(y_j + s), \ldots, x_{n-1} + iy_{n-1}), t - 2x_j s])|_{s=0}$$

$$= \left(\frac{\partial f}{\partial y_j} - 2x_j \frac{\partial f}{\partial t} \right) [z, t], \quad 1 \le j \le n - 1,$$

$$T f(g) \equiv \frac{d}{ds} f(g \cdot \gamma_t(s))|_{s=0}$$

$$= \frac{d}{ds} f([x, t + s])|_{s=0}$$

$$= \frac{\partial f}{\partial t} [z, t].$$

We think of X_j, Y_j, and T as vector fields on the Heisenberg group. These three objects embody the structure of the group in an analytic manner (as we shall see below).

9.6.2 Commutators of Vector Fields

Central to geometric analysis and symplectic geometry is the concept of the commutator of vector fields. We review the idea here in the context of $\mathbb{R}^N$.

A *vector field* on a domain $U \subseteq \mathbb{R}^N$ is a function

$$\lambda : U \to \mathbb{R}^N$$

with $\lambda(x) = \sum_{j=1}^{N} a_j(x)\partial/\partial x_j$. We think of $\partial/\partial x_1, \dots, \partial/\partial x_N$ as a basis for the range space $\mathbb{R}^N$. If λ_1, λ_2 are two such vector fields then we define their *commutator* to be

$$[\lambda_1, \lambda_1] = \lambda_1 \lambda_2 - \lambda_2 \lambda_1. \tag{9.6.2.1}$$

Of course a vector field is a linear partial differential operator. It acts on the space of testing functions. So, if $\varphi \in C_c^\infty$ then it is useful to write (9.6.2.1) as

$$[\lambda_1, \lambda_2]\varphi = \lambda_1(\lambda_2\varphi) - \lambda_2(\lambda_1\varphi).$$

Let us write this out in coordinates. We set

$$\lambda_1 = \sum_{j=1}^{N} a_j^1(x)\frac{\partial}{\partial x_j}$$

and

$$\lambda_2 = \sum_{j=1}^{N} a_j^2(x)\frac{\partial}{\partial x_j}.$$

Then

$$[\lambda_1, \lambda_2]\varphi = \lambda_1(\lambda_2\varphi) - \lambda_2(\lambda_1\varphi)$$

$$= \sum_{j=1}^{N} a_j^1(x)\frac{\partial}{\partial x_j}\left(\sum_{j=1}^{N} a_j^2(x)\frac{\partial}{\partial x_j}\varphi\right)\psi$$

$$- \sum_{j=1}^{N} a_j^2(x)\frac{\partial}{\partial x_j}\left(\sum_{j=1}^{N} a_j^1(x)\frac{\partial}{\partial x_j}\varphi\right)\varphi$$

$$= \left[\sum_{j,\ell=1}^{N} a_j^1(x)\left(\frac{\partial}{\partial x_j}a_\ell^2(x)\right)\frac{\partial}{\partial x_\ell} + \sum_{j,\ell=1}^{N} a_j^1(x)a_\ell^2(x)\frac{\partial^2}{\partial x_j \partial x_\ell}\right]\varphi$$

$$- \left[\sum_{j,\ell=1}^{N} a_j^2(x)\left(\frac{\partial}{\partial x_j}a_\ell^1(x)\right)\frac{\partial}{\partial x_\ell} + \sum_{j,\ell=1}^{N} a_j^2(x)a_\ell^1(x)\frac{\partial^2}{\partial x_j \partial x_\ell}\right]\varphi$$

$$= \left(\sum_{j,\ell=1}^{N} a_j^1(x)\left(\frac{\partial}{\partial x_j}a_\ell^2(x)\right)\frac{\partial}{\partial x_\ell} - \sum_{j,\ell=1}^{N} a_j^2(x)\left(\frac{\partial}{\partial x_j}a_\ell^1(x)\right)\frac{\partial}{\partial x_\ell}\right)\varphi.$$

The main thing to notice is that $[\lambda_1, \lambda_2]$ is ostensibly—by its very definition—a second-order linear partial differential operator. But in fact, the top-order terms cancel out, so that in the end, $[\lambda_1, \lambda_2]$ is a *first-order* linear partial differential operator. In other words—and this point is absolutely essential—*the commutator of two vector fields is another vector field*. This is what will be important for us in our study of the Heisenberg group.

9.6.3 Commutators in the Heisenberg Group

Let

$$X_j = \frac{\partial}{\partial x_j} + 2y_j \frac{\partial}{\partial t}, \quad 1 \le j \le n-1,$$

$$Y_j = \frac{\partial}{\partial y_j} - 2x_j \frac{\partial}{\partial t}, \quad 1 \le j \le n-1,$$

$$T = \frac{\partial}{\partial t}.$$

See Section 9.6.1. Note that $[X_j, X_k] = [Y_j, Y_k] = [X_j, T] = [Y_j, T] = 0$ for all $1 \le j, k \le n$ and $[X_j, Y_k] = 0$ if $j \ne k$. The only nonzero commutator in the Heisenberg group is $[X_j, Y_j]$, and we calculate that right now:

$$[X_j, Y_j] = \left(\frac{\partial}{\partial x_j} + 2y_j \frac{\partial}{\partial t} \right) \left(\frac{\partial}{\partial y_j} - 2x_j \frac{\partial}{\partial t} \right)$$

$$- \left(\frac{\partial}{\partial y_j} - 2x_j \frac{\partial}{\partial t} \right) \left(\frac{\partial}{\partial x_j} + 2y_j \frac{\partial}{\partial t} \right)$$

$$= -2 \left(\frac{\partial}{\partial x_j} x_j \right) \frac{\partial}{\partial t} - 2 \left(\frac{\partial}{\partial y_j} y_j \right) \frac{\partial}{\partial t}$$

$$= -4 \frac{\partial}{\partial t}$$

$$= -4T.$$

Thus we see that

$$[X_j, Y_j] = -4T.$$

To summarize: all commutators $[X_j, X_k]$ for $j \ne k$ and $[X_j, T]$ equal 0. The only nonzero commutator is $[X_j, Y_j] = -4T$. One upshot of these simple facts is that any *second-order commutator* $[[A, B], C]$ will be zero—just because $[A, B]$ will be either 0 or $-4T$. Thus the vector fields on the Heisenberg group form a nilpotent Lie algebra of step one.

9.6.4 Additional Information about the Heisenberg Group Action

We have discussed in Section 9.4 how the Heisenberg group acts holomorphically on the Siegel upper half-space. Here we collect some facts about the invariant measure for this action.

Definition 9.6.1 Let G be a topological group that is locally compact and Hausdorff. A Haar measure on G is a Radon measure that is invariant under the group operation. Among other things, this means that if K is a compact set and $g \in G$ then the measure of K and the measure of $g \cdot K \equiv \{g \cdot k : k \in K\}$ are equal.

Exercise for the Reader: In $\mathbb{H}^n$, Haar measure coincides with the Lebesgue measure. [This is an easy calculation using elementary changes of variable.]

Let $g = [z, t] \in \mathbb{H}^{n-1}$. The dilation on $\mathbb{H}^{n-1}$ is defined to be

$$\alpha_\delta g = [\delta z, \delta^2 t].$$

We can easily check that α_δ is a group homomorphism:

$$\alpha_\delta\left([z, t] \cdot [z^*, t^*]\right) = \alpha_\delta[z, t] \cdot \alpha_\delta[z^*, t^*].$$

A ball with center $[z, t]$ and radius r is defined as

$$B([z, t], r) = \{[\zeta, s] : |\zeta - z|^4 + |s - t|^2 < r^2\}.$$

[Later on, in Section 9.9, we shall examine this idea in the language of the Heisenberg group norm.] For $f, g \in L^1(\mathbb{H}^n)$, we can define the convolution of f and g:

$$f * g(x) = \int f(y^{-1} \cdot x) g(y) \, dy.$$

9.7 A Fresh Look at Classical Analysis

Capsule: In this section we begin to reexamine the most elementary artifacts of analysis for our new context. Dilations, translations, the triangle inequality, polar coordinates, differentiation, and integration are just some of the tools that we must reconfigure for our new mission. The path is both entertaining and enlightening, for it will cause us to see the Euclidean tools that we already know in a new light. The result is a deeper understanding of the analytic world.

In preparation for our detailed hard analysis of the Heisenberg group in Section 9.9, we use this section to review a number of ideas from classical real analysis. This will include the concept of space of homogeneous type, and various ideas about fractional integration and singular integrals.

9.7.1 Spaces of Homogeneous Type

These are fundamental ideas of K.T. Smith [SMI] and L. Hörmander [HOR4] that were later developed by R.R. Coifman and Guido Weiss [COIW1], [COIW2].

Definition 9.7.1 We call a set X a *space of homogeneous type* if it is equipped with a collection of open balls $B(x, r)$ and a Borel regular measure μ, together with positive constants C_1, C_2, such that

(9.7.1.1) The Positivity Property: $0 < \mu(B(x, r)) < \infty$ for $x \in X$ and $r > 0$;

(9.7.1.2) The Doubling Property: $\mu(B(x, 2r)) \leq C_1\mu(B(x, r))$ for $x \in X$ and $r > 0$;

(9.7.1.3) The Enveloping Property: If $B(x, r) \cap B(y, s) \neq \emptyset$ and $r \geq s$, then $B(x, C_2 r) \supseteq B(y, s)$.

We frequently use the notation (X, μ) to denote a space of homogeneous type. In some contexts a space of homogeneous type is equipped with a metric as well, but we opt for greater generality here.

Example 9.7.2 The Euclidean space $\mathbb{R}^N$ is a space of homogeneous type when equipped with the usual isotropic Euclidean balls and μ Lebesgue measure.

Example 9.7.3 Let $\Omega \subseteq \mathbb{R}^N$ be a smoothly bounded domain and X its boundary. Let the balls $B(x, r)$ be the intersection of ordinary Euclidean balls from Euclidean space with X. Let $d\mu$ be $(2N - 1)$-dimensional Hausdorff measure on X (see Section 9.9.3). Then X, so equipped, is a space of homogeneous type.

Example 9.7.4 Let X be a compact Riemannian manifold. Let $B(x, r)$ be the balls that come from the Riemannian metric. Let $d\mu$ be Hausdorff measure on X. Let K be a compact subset of X. Then one may use the exponential map to verify the axioms of a space of homogeneous type for (X, μ). See [COIW1] for the details.

Theorem 9.7.5 (Wiener's covering lemma) *Let (X, μ) be a space of homogeneous type. Let K be a compact subset of X. Let $\{B_\alpha\}_{\alpha \in A}$ be a collection of balls, $B_\alpha = B(x_\alpha, r_\alpha)$ such that $\cup_{\alpha \in A} B_\alpha \supseteq K$. Then $\exists B_{\alpha_1}, \ldots, B_{\alpha_m}$ pairwise disjoint such that the C_2-fold dilation (note that $B(x_\alpha, C_2 r_\alpha)$ is the C_2-fold dilation of $B(x_\alpha, r_\alpha)$) of the selected balls covers K.*

Proof: We proved a version of this result in Section 8.6. $\square$

Exercise for the Reader [Besicovitch]: In $\mathbb{R}^N$, there is a universal constant $M = M(N)$ that satisfies the following: Suppose that $\mathcal{B} = \{B_1, \ldots, B_k\}$ is a collection of Euclidean balls in $\mathbb{R}^N$. Assume that no ball contains the center of any other. Then we may write $\mathcal{B} = \mathcal{B}_1 \cup \mathcal{B}_2 \cup \cdots \cup \mathcal{B}_M$, where each $\mathcal{B}_j$ consists of pairwise disjoint balls. [*Hint:* Use the same proof strategy as for the Wiener covering lemma.]

We can now define the Hardy–Littlewood maximal function on $L^1(X, \mu)$. If $f \in L^1(X, \mu)$, then define

$$Mf(x) \equiv \sup_{R > 0} \frac{1}{\mu(B(x, R))} \int_{B(x, r)} |f(t)| d\mu(t).$$

Proposition 9.7.6 *M is weak-type $(1, 1)$.*

Proof: We proved a version of this result in Section 8.6. $\square$

Since M is obviously strong-type (∞, ∞) and weak-type $(1, 1)$, we may apply the Marcinkiewicz interpolation theorem to see that M is strong-type (p, p), $1 < p < \infty$.

Remark: The Marcinkiewicz interpolation theorem works for sublinear operators T (i.e., $T(f+g) \leq Tf + Tg$), whereas the Riesz–Thorin interpolation theorem works only for linear operators. See [STG1] for more on these matters.

9.7.2 The Folland–Stein Theorem

Let (X, μ) be a measure space and $f : X \to \mathbb{C}$ a measurable function. We say f is *weak-type* r, $0 < r < \infty$, if there exists some constant C such that

$$\mu\{x : |f(x)| > \lambda\} \leq \frac{C}{\lambda^r}, \quad \text{for any } \lambda > 0.$$

Remark: If $f \in L^r$, then f is weak-type r, but not vice versa. For suppose that $f \in L^r$; then

$$C \geq \int |f|^r d\mu \geq \int_{\{|f|>\lambda\}} |f|^r d\mu \geq \lambda^r \cdot \mu\{|f| > \lambda\};$$

hence f is of weak-type r. For the other assertion suppose that $X = \mathbb{R}^+$. Then $f(x) = \frac{1}{x^{1/r}}$ is weak-type r but not rth-power integrable.

Prelude: In the classical theory of fractional integration—due to Riesz and others—the L^p mapping properties of the fractional integral operators were established using particular Euclidean properties of the kernels $|x|^{-N+\alpha}$. It was a remarkable insight of Folland and Stein that all that mattered was the distribution of values of the kernel. This fact is captured in the next theorem.

Theorem 9.7.7 (Folland, Stein (CPAM, 1974)) *Let $(X, \mu), (Y, \nu)$ be measurable spaces. Let*

$$k : X \times Y \to \mathbb{C}$$

satisfy

$$\mu\{x : |k(x, y)| > \lambda\} \leq \frac{C}{\lambda^r} \quad \text{(for fixed } y\text{)},$$

$$\nu\{y : |k(x, y)| > \lambda\} \leq \frac{C'}{\lambda^r} \quad \text{(for fixed } x\text{)},$$

where C and C' are independent of y and x respectively and $r > 1$. Then

$$f \longmapsto \int_Y f(y)k(x, y)d\nu(y)$$

maps $L^p(X)$ to $L^q(X)$, where $\frac{1}{q} = \frac{1}{p} + \frac{1}{r} - 1$, for $1 < p < \frac{r}{r-1}$.

Remark: Certainly you should compare and contrast this result with the classical Riesz fractional integration result that we treated in Chapter 5. The Folland–Stein result is a far-reaching generalization that frees the result from the structure of Euclidean space and shows quite plainly how the key idea is measure-theoretic.

It is worth noting here that the norms of the operators *must* blow up as $p \to 1$ or $p \to r/(r - 1)$. This is so because if not, then one could use the semicontinuity of the integral to derive boundedness at the endpoints.

Prelude: Schur's lemma is probably *the* most basic fact about integral kernels. Many of the more sophisticated ideas—including the Folland–Stein theorem—are based on Schur. We include below the most fundamental formulation of Schur.

A key tool in our proof of the Folland–Stein result is the following idea of Isaiah Schur (which is in fact a rather basic version of the Folland–Stein theorem):

Lemma 9.7.8 (Schur) *Let* $1 \le r \le \infty$. *Let* (X, μ), (Y, ν) *be measurable spaces and let* $k : X \times Y \to \mathbb{C}$ *satisfy*

$$\left(\int |k(x, y)|^r d\mu(x) \right)^{1/r} \le C,$$

$$\left(\int |k(x, y)|^r d\nu(x) \right)^{1/r} \le C',$$

where C *and* C' *are independent of* y *and* x *respectively. Then*

$$f \longmapsto \int_Y k(x, y) f(y) dy$$

maps $L^p(X)$ *to* $L^q(X)$, *where* $\frac{1}{q} = \frac{1}{p} + \frac{1}{r} - 1$, *for* $1 \le p \le \frac{r}{r-1}$.

Schur's lemma is a standard result, with an easy proof, and the details may be found in [FOL3] or in our Lemma A1.5.5. Note that for Schur's lemma, which of course has a stronger hypothesis than Folland–Stein, the boundedness in the conclusions is also true at the endpoints.

In order to prove the Folland–Stein theorem, we shall use the idea of distribution function that was introduced in Sections 8.5 and 8.6.

Proof of Theorem 9.7.7: By the Marcinkiewicz interpolation theorem, it is enough to show that $f \longmapsto Tf$ is weak-type (p, q) for each p and corresponding q. Fix $s > 0$. Let $\kappa > 0$ be a constant to be specified later. Let us define

$$k_1(x, y) = \begin{cases} k(x, y) & \text{if} \quad |k(x, y)| \ge \kappa, \\ 0 & \text{otherwise,} \end{cases}$$

$$k_2(x, y) = \begin{cases} k(x, y) & \text{if} \quad |k(x, y)| < \kappa \\ 0 & \text{otherwise,} \end{cases}$$

i.e., $k(x, y) = k_1(x, y) + k_2(x, y)$ and k_2 is bounded. Let

$$T_1 f(x) = \int k_1(x, y) f(y) d\nu(y),$$

$$T_2 f(x) = \int k_2(x, y) f(y) dv(y).$$

Then $Tf = T_1 f + T_2 f$. Therefore

$$\begin{aligned}
\alpha_{Tf}(2s) &= \mu\{|Tf| > 2s\} \\
&= \mu\{|T_1 f + T_2 f| > 2s\} \\
&\leq \mu\{|T_1 f| + |T_2 f| > 2s\} \\
&\leq \mu\{|T_1 f| > s\} + \mu\{|T_2 f| > s\} \\
&= \alpha_{T_1 f}(s) + \alpha_{T_2 f}(s).
\end{aligned} \tag{9.7.7.1}$$

Let $f \in L^p(X)$ and assume $\|f\|_{L^p} = 1$. Choose p' such that $\frac{1}{p} + \frac{1}{p'} = 1$. Then we get

$$\begin{aligned}
|T_2 f(x)| &= \left| \int k_2(x, y) f(y) dv(y) \right| \\
&\leq \left(\int |k_2(x, y)|^{p'} dv(y) \right)^{\frac{1}{p'}} \left(\int |f(y)|^p dv(y) \right)^{\frac{1}{p}}
\end{aligned}$$

and from Section 8.5,

$$\begin{aligned}
\int |k_2(x, y)|^{p'} dv(y) &= \int_0^\kappa p' s^{p'-1} \alpha_{k_2(x,\cdot)}(s) ds \\
&\leq \int_0^\kappa p' s^{p'-1} \frac{C}{s^r} ds = Cp' \int_0^\kappa s^{p'-1-r} ds = C' \kappa^{p'-r}.
\end{aligned}$$

The last equality holds since

$$p' - 1 - r = \frac{1}{1 - \frac{1}{p}} - 1 - r = \frac{p}{p-1} - 1 - r > -1.$$

Thus we get

$$|T_2 f(x)| \leq (C' \kappa^{p'-r})^{\frac{1}{p'}} \|f\|_{L^p} = C'' \kappa^{1 - \frac{r}{p'}}.$$

Let $\kappa = \left(D \frac{s}{C''} \right)^{\frac{q}{r}}$. Then

$$|T_2 f(x)| \leq C'' \left(\frac{s}{C''} \right)^{\frac{q}{r} \left(1 - \frac{r}{p'} \right)} = s.$$

Therefore we get $\alpha_{T_2 f}(s) = 0$. Hence, from (9.8.7.1), we get

$$\alpha_{Tf}(2s) \le \alpha_{T_1 f}(s).$$

Since $|k_1(x, y)| \ge \kappa$, we have $\alpha_{k_1(x,\cdot)}(s) = \alpha_{k_1(x,\cdot)}(\kappa)$ if $s \le \kappa$. Thus

$$\int_Y |k_1(x, y)| dv(y) = \int_0^\infty \alpha_{k_1(x,\cdot)}(s) ds$$

$$= \int_0^\kappa \alpha_{k_1(x,\cdot)}(s) ds + \int_\kappa^\infty \alpha_{k_1(x,\cdot)}(s) ds$$

$$\le \kappa \alpha_{k_1(x,\cdot)}(\kappa) + \int_\kappa^\infty \frac{C}{s^r} ds$$

$$\le \kappa \frac{C}{\kappa^r} + \frac{C}{1-r} \kappa^{1-r}$$

$$= C' \kappa^{1-r}. \tag{9.7.7.2}$$

Similarly, we get

$$\int_x |k_1(x, y)| d\mu(x) \le C \kappa^{1-r}.$$

Recall that if $m(x, y)$ is a kernel and

$$\int |m(x, y)| d\mu(x) \le C,$$

$$\int |m(x, y)| dv(y) \le C,$$

then, by Schur's lemma, $f \longmapsto \int f(y) m(x, y) dy$ is bounded on $L^p(X)$, $1 \le p \le \infty$. Thus, T_1 is bounded on L^p:

$$\|T_1 f\|_{L^p(X)} \le C \kappa^{1-r} \|f\|_{L^p} = C \kappa^{1-r}.$$

By Chebyshev's inequality,

$$\mu\{x : |f(x)| > \lambda\} \le \frac{\|f\|_{L^p}^p}{\lambda^p}.$$

Therefore

$$\alpha_{T_1 f}(s) \le \frac{\|T_1 f\|_{L^p}^p}{s^p} \le \frac{(C \kappa^{1-r})^p}{s^p} = C' \frac{\left(\frac{s}{C''}\right)^{\frac{q}{r} \cdot (1-r)p}}{s^p} = C'' s^{\frac{qp(1-r)}{r} - p} = C''' \frac{1}{s^q}.$$

Therefore

$$\alpha_{Tf}(2s) \le \frac{C'''}{s^q}.$$

This is the weak-type estimate that we seek. □

Now we have a rather universal fractional integral result at our disposal, and it certainly applies to fractional integration on the Heisenberg group $\mathbb{H}^n$. For example, set $|\cdot|_h$ on $\mathbb{H}^n$ to be equal to

$$|g|_h = (|z|^4 + t^2)^{1/4}$$

when $g = (z, t) \in \mathbb{H}^n$ (we discuss this norm in greater detail below). If

$$k_\beta(x) = |x|_h^{-2n-2+\beta}, \quad 0 < \beta < 2n + 2,$$

is a kernel and we wish to consider the operator

$$\mathcal{I}^\beta : f \mapsto f * k_\beta$$

on $\mathbb{H}^n$, then the natural way to proceed now is to calculate the weak-type of k_β. Then the Folland–Stein theorem will instantly tell us the mapping properties of $\mathcal{I}^\beta$. Now

$$m\{x : |k_\beta(x)| > \lambda\} = m\left\{x : |x|_h \leq \left(\frac{1}{\lambda}\right)^{1/[2n+2-\beta]}\right\}$$

$$\leq c \cdot \left(\frac{1}{\lambda}\right)^{[2n+2]/[2n+2-\beta]}.$$

We see immediately that k_β is of weak-type $[2n+2]/[2n+2-\beta]$. Thus the hypotheses of the Folland–Stein theorem are satisfied with $r = [2n+2]/[2n+2-\beta]$. We conclude that $\mathcal{I}^\beta$ maps L^p to L^q with

$$\frac{1}{q} = \frac{1}{p} - \frac{\beta}{2n+2}, \quad 1 < p < \frac{2n+2}{\beta}.$$

9.7.3 Classical Calderón–Zygmund Theory

We now wish to turn our attention to singular integral operators. One of the key tools for the classical approach to this subject is the Whitney decomposition of an open set. That is an important tool in geometric analysis that first arose in the context of the Whitney extension theorem. We begin with a review of that idea.

Prelude: Extending a smooth function from a smooth submanifold of $\mathbb{R}^N$ to all of $\mathbb{R}^N$ is an intuitively obvious and appealing process. For the implicit function theorem tells us that we may as well assume that the submanifold is a linear subspace, and then the extension process is just the trivial extension in the orthogonal variables. Extending from a more arbitrary subset is a very meaningful idea, but much more subtle. It was Hassler Whitney who cracked the problem and wrote the fundamental papers [WHI1], [WHI2].

Theorem 9.7.9 (Whitney extension theorem) *Suppose $E \subseteq \mathbb{R}^N$ is any closed set. Let $f \in C^k(E)$.[4] Then f can be extended to a C^k function on all $\mathbb{R}^N$.*

---

[4] This idea requires some explanation. If E is a closed set without interior, or more generally just any old set, then what does $C^k(E)$ mean? The answer is rather technical, but quite natural in view of the theory of Taylor series. The tract [FED] contains all the details.

The idea of the proof is that one decomposes ^cE into carefully chosen boxes, and then extends f to each box in the appropriate manner. We shall not provide the details here, but see [FED, p. 225]. As already noted, one of the main tools in the proof of Whitney's result is this decomposition theorem:

Prelude: The Whitney decomposition theorem arose originally as a fundamental tool in the proof of the Whitney extension theorem. The extension of the function was achieved box by box. Today Whitney's decomposition is a basic tool in harmonic analysis, used to prove the Calderón–Zygmund theorem and other martingale-type results.

Theorem 9.7.10 (Whitney decomposition theorem) *Let $F \subseteq \mathbb{R}^N$ be a closed set and $\Omega = \mathbb{R}^N \setminus F$. Then there exists a collection of closed cubes $\mathcal{F} = \{Q_j\}_{j=1}^{\infty}$ such that*

1. *$\cup_j Q_j = \Omega$.*
2. *Q_j's have pairwise disjoint interiors.*
3. *There exist constants $C_1 < C_2$ such that $C_1 \cdot \operatorname{diam}(Q_j) \leq \operatorname{dist}(Q_j, F) \leq C_2 \cdot \operatorname{diam}(Q_j)$.*

Remark: It may be noted that in one real dimension, things are deceptively simple. For if F is a closed subset of $\mathbb{R}^1$ then the complement is open and may be written as the disjoint union of open intervals. Nothing like this is true in several real variables. The Whitney decomposition is a substitute for that simple and elegant decomposition.

Proof: Examine Figure 9.1 as you read the proof. Consider the collection of cubes $\mathcal{C}_0$ in $\mathbb{R}^N$ with vertices having integer coordinates and side length 1. Let $\mathcal{C}_1$ be the set of cubes obtained by slicing the cubes in $\mathcal{C}_0$ in half in each coordinate direction. The set $\mathcal{C}_2$ is gotten from $\mathcal{C}_1$ by slicing in half in each coordinate direction. Continue the process.

Also, we may get a collection of cubes $\mathcal{C}_{-1}$ such that slicing the cubes in $\mathcal{C}_{-1}$ in half in each direction produces $\mathcal{C}_0$. We obtain $\mathcal{C}_{-2}, \mathcal{C}_{-3}$, and so on in a similar fashion.

Ultimately, we have collections of cubes $\mathcal{C}_j$, $j \in \mathbb{Z}$, where the cubes in $\mathcal{C}_j$ have side length 2^{-j} and diameter $\sqrt{N} 2^{-j}$.

Define

$$\Omega_j = \{x \in \mathbb{R}^N : C \cdot 2^{-j} < \operatorname{dist}(x, F) \leq C \cdot 2^{-j+1}\},$$

where the constant C will be specified later. Then obviously $\Omega = \cup_{j=1}^{\infty} \Omega_j$. Now we select a cube in $\mathcal{C}_j$ if it has nonempty intersection with Ω_j. Then the collection of such cubes, say $\{Q_\alpha\}$, covers Ω:

$$\Omega \subset \cup_\alpha Q_\alpha.$$

If we let $C = 2\sqrt{N}$, then each selected cube Q is disjoint from F. Suppose that $Q \in \mathcal{C}_j$. Then $\exists x \in Q \cap \Omega_j$. By the definition of Ω_j, we get

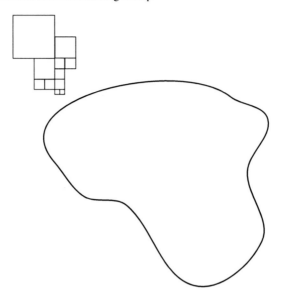

Figure 9.1. The Whitney decomposition.

$$2 \operatorname{diam} Q = 2\sqrt{N}2^{-j} \leq \operatorname{dist}(x, F) \leq 2\sqrt{N}2^{-j+1} = 4 \operatorname{diam} Q. \qquad (9.7.10.1)$$

Therefore

$$\operatorname{dist}(Q, F) \geq \operatorname{dist}(x, F) - \operatorname{diam} Q \geq 2 \operatorname{diam} Q - \operatorname{diam} Q = \operatorname{diam} Q > 0.$$

Hence, for each selected Q_α, $Q_\alpha \cap F = \emptyset$ and thus $\cup_\alpha Q_\alpha \subset \Omega$. Therefore we have

$$\Omega = \cup_\alpha Q_\alpha.$$

The key fact about our cubes is that if two cubes have nontrivially intersecting interiors then one is contained in the other. Thus we can find a disjoint collection of cubes Q'_α such that $\Omega = \cup_\alpha Q'_\alpha$. □

Prelude: This is the result, previously mentioned, that depends on the Whitney decomposition. When the Calderón–Zygmund decomposition was first proved it was a revelation: a profound geometric way to think about singular integrals. It continues today to be influential and significant. Certainly it has affected the way that the subject has developed. The atomic theory, pseudodifferential operators, the David–Journé theorem, and many other essential parts of our subject have been shaped by the Calderón–Zygmund theorem.

Theorem 9.7.11 (Calderón–Zygmund decomposition) *Let f be a nonnegative, integrable function in $\mathbb{R}^N$. Then, for $\alpha > 0$ fixed, there is a decomposition of $\mathbb{R}^N$ such that*

1. $\mathbb{R}^N = F \cup \Omega$, $F \cap \Omega = \emptyset$, F is closed.

2. $f(x) \le \alpha$ for almost every $x \in F$.

3. $\Omega = \cup_j Q_j$, where the Q_j's are closed cubes with disjoint interiors and f satisfies

$$\alpha < \frac{1}{m(Q_j)} \int_{Q_j} f(x)\, dx \le 2^N \alpha.$$

($m(Q_j)$ denotes the measure of the cube Q_j.)

Proof: Decompose $\mathbb{R}^N$ into an equal mesh of diadic cubes with size chosen such that

$$\frac{1}{m(Q)} \int_Q f\, dx \le \alpha$$

for every cube Q in the mesh. Since f is integrable, it is possible to choose such cubes if we let $m(Q)$ be large enough.

Bisect each cube Q a total of N times to create 2^N congruent pieces with side length half of the original cubes. Say that Q' is such a subcube. Then we have two cases:

$$\frac{1}{m(Q')} \int_{Q'} f\, dx > \alpha \quad \text{and} \quad \frac{1}{m(Q')} \int_{Q'} f\, dx \le \alpha.$$

If the first case holds, then we keep the cube. If it is the second case, then we subdivide Q' again. For each subdivided cube we have two cases as above and we repeat the same procedure.

Consider one of the selected cubes Q. We have

$$\alpha < \frac{1}{m(Q)} \int_Q f\, dx$$

and if $\tilde{Q}$ is the father cube[5] of Q, we have

$$\alpha \ge \frac{1}{m(\tilde{Q})} \int_{\tilde{Q}} f\, dx \ge \frac{1}{2^N m(Q)} \int_Q f\, dx.$$

Hence

$$\alpha < \frac{1}{m(Q)} \int_Q f(x)\, dx \le 2^N \alpha.$$

We let Ω be the union of selected cubes and $F = \mathbb{R}^N \setminus \Omega$.

If $x \in F$, then x is contained in a decreasing sequence of cubes Q_j, $Q_1 \supset Q_2 \supset \cdots$, on which

$$\frac{1}{m(Q_j)} \int_{Q_j} f(x)\, dx \le \alpha.$$

Therefore, by the Lebesgue differentiation theorem, $f(x) \le \alpha$ for a.e. $x \in F$. □

---

[5] A cube $\tilde{Q}$ is a "father cube" of Q if $\tilde{Q}$ has twice the side length of Q, $\tilde{Q}$ comes from the previous generation in our construction, and $\tilde{Q} \supseteq Q$.

Theorem 9.7.12 (Lebesgue differentiation theorem) *If $f \in L^1_{loc}(\mathbb{R}^N)$, then*

$$\lim_{\delta \to 0} \frac{1}{Q_\delta(x)} \int_{Q_\delta(x)} f(t)dt = f(x), \quad \text{for a.e. } x.$$

Here $Q_\delta(x)$ is a cube with side length δ, sides parallel to the axes, and center x.

Proof: Let $\mathcal{Q} \equiv \{q_j\}$ be a sequence of positive real numbers that tends to 0. Suppose $f \in L^1(\mathbb{R}^N)$. We let

$$T_j f(x) = \frac{1}{Q_j(x)} \int_{Q_j(x)} f(t)dt,$$

where $Q_j(x)$ is a cube with side length q_j and center x. Apply Functional Analysis Principle II (Appendix 1) with dense set $C_c(\mathbb{R}^N) \subset L^1(\mathbb{R}^N)$. Then, for $f \in C_c(\mathbb{R}^N)$, we have $T_j f(x) \to f(x)$ for a.e. x. Also the maximal function of $T_j f$,

$$T^* f(x) = \sup_j |T_j f(x)| = \sup_j \frac{1}{Q_j(x)} \int_{Q_j(x)} f(t)dt,$$

coincides with the Hardy–Littlewood maximal operator (at least for nonnegative f), which is weak-type $(1, 1)$. Since the choice of $\mathcal{Q}$ was arbitrary, we have pointwise convergence for a.e. x. $\qquad\square$

Next we present an alternative formulation of the Calderón–Zygmund decomposition:

Theorem 9.7.13 (Another useful decomposition) *Suppose f is a nonnegative integrable function on $\mathbb{R}^N$ and $\alpha > 0$ is a fixed constant. Then there exists a decomposition of $\mathbb{R}^N$ such that*

1. *$\mathbb{R}^N = F \cup \Omega$, where F is closed, $F \cap \Omega = \emptyset$ and there exists a constant A such that*

$$m(\Omega) \le \frac{A}{\alpha} \|f\|_1.$$

2. *$\Omega = \cup_j Q_j$, where the Q_j's are closed cubes with disjoint interiors and there exists a constant B such that*

$$\frac{1}{m(Q_j)} \int_{Q_j} f\, dx \le B\alpha \quad \text{for each } j.$$

Also diam $Q_j \approx \text{dist}(Q_j, F)$.

Proof: Define $F = \{Mf(x) \ge \alpha\}$, where $Mf(x)$ is the Hardy–Littlewood maximal function:

$$Mf(x) = \sup_{r>0} \frac{1}{m(B(x,r))} \int_{B(x,r)} |f(t)|dt.$$

Let $\Omega = \mathbb{R}^N \setminus F = \{Mf(x) > \alpha\}$. Then, since $f \longmapsto Mf$ is weak-type $(1, 1)$, we have

$$m(\Omega) \leq A \frac{\|f\|_1}{\alpha}$$

for some constant A.

Since Ω is open, we can use the Whitney decomposition so that $\Omega = \cup_j Q_j$, where the Q_j's are closed cubes with disjoint interiors and from (9.7.9.1),

$$\text{diam } Q_j \approx \text{dist}\,(x, F), \quad \forall x \in Q_j.$$

Suppose Q is one of the cubes. Since F is closed, there exists $p \in F$ such that $\text{dist}\,(Q, F) = \text{dist}\,(Q, p)$. Then we have $Mf(p) \leq \alpha$. Let $r = \text{dist}\,(p, Q) + \text{diam }Q$. Then $Q \subset B(p, r)$ and $r \approx \text{diam }Q$. Therefore

$$\alpha \geq Mf(p) \geq \frac{1}{m(B(p, r))} \int_{B(p,r)} f \, dx \geq \frac{1}{m(B(p, r))} \int_Q f \, dx \overset{\gtrsim}{\sim} \frac{1}{m(Q)} \int_Q f \, dx.$$

$\square$

Prelude: From Stein's point of view, in his original book [STE1] on singular integrals, the Marcinkiewicz integral is the key to the L^p boundedness of the operators. Today there are many other approaches to singular integral theory (which do *not* use Marcinkiewicz's idea), but the Marcinkiewicz integral remains an important tool.

Definition 9.7.14 (The Marcinkiewicz integral) Fix a closed set $F \subseteq \mathbb{R}^N$ and let $\delta(x) = \text{dist}\,(x, F)$. We define the Marcinkiewicz integral as follows:

$$I_*(x) = \int_{\mathbb{R}^N} \frac{\delta(x + y)}{|y|^{N+1}} \, dy.$$

Remark: We consider only the case $x \in F$, because if $x \in {}^cF$, then $\delta(x) > 0$. Hence the integral is singular at 0.

Proposition 9.7.15 *We have*

$$\int_F I_*(x) \, dx \leq Cm({}^cF).$$

Proof: Now

$$\int_F I_*(x) \, dx = \int_F \int_{\mathbb{R}^N} \frac{\delta(x + t)}{|t|^{N+1}} \, dt \, dx$$

$$= \int_F \int_{\mathbb{R}^N} \frac{\delta(t)}{|t - x|^{N+1}} \, dt \, dx$$

$$= \int_F \int_{{}^cF} \frac{\delta(t)}{|t - x|^{N+1}} \, dt \, dx$$

$$= \int_{{}^cF} \left(\int_F \frac{1}{|t - x|^{N+1}} \, dx \right) \delta(t) dt. \qquad (9.7.15.1)$$

For $t \in {}^c F$, we have $|t - x| \geq \delta(t)$, $\forall x \in F$. Therefore,

$$\int_F \frac{1}{|t - x|^{N+1}} \, dx \leq \int_S \int_{\delta(t)}^\infty \frac{1}{r^{N+1}} r^{N-1} \, dr \, d\sigma = \int_S \int_{\delta(t)}^\infty \frac{1}{r^2} dr \, d\sigma = C \frac{1}{\delta(t)}.$$
$$(9.7.15.2)$$

Hence, from (9.7.15.1) and (9.7.15.2), we have

$$\int_F I_*(x) \, dx \leq \int_{{}^c F} C \frac{1}{\delta(t)} \delta(t) dt = Cm({}^c F). \qquad \square$$

Definition 9.7.16 (Calderón–Zygmund kernel) A Calderón–Zygmund kernel $K(x)$ in $\mathbb{R}^N$ is one having the form

$$K = \frac{\Omega(x)}{|x|^N},$$

where

1. $\Omega(x)$ is homogeneous of degree 0.
2. $\Omega(x) \in C^1(\mathbb{R}^N \setminus \{0\})$.
3. $\int_\Sigma \Omega(x) d\sigma(x) = 0$ (Σ is the unit sphere in $\mathbb{R}^N$).

Example 9.7.17 The Riesz kernels

$$\frac{x_1}{|x|^{N+1}}, \ldots, \frac{x_N}{|x|^{N+1}}$$

are certainly examples of Calderón–Zygmund kernels. For we may write

$$\frac{x_j}{|x|^{N+1}} = \frac{x_j/|x|}{|x|^N}$$

and notice that the numerator is homogeneous of degree zero and (by parity) has the mean-value-zero property.

Integration against K of a Calderón–Zygmund kernel induces a distribution in a natural way. Let $\phi \in C_c^\infty(\mathbb{R}^N)$. Then we recall that

$$\lim_{\epsilon \to 0} \int_{|x| > \epsilon} \frac{\Omega(x)}{|x|^N} \phi(x) \, dx = \lim_{\epsilon \to 0} \int_{|x| > \epsilon} \frac{\Omega(x)}{|x|^N} [\phi(x) - \phi(0)] dx$$

$$= \lim_{\epsilon \to 0} \int_{|x| > \epsilon} \frac{\Omega(x)}{|x|^N} O(|x|) dx$$

and for $0 < \epsilon_1 < \epsilon_2 \ll 1$, we have

$$\int_{|x| > \epsilon_1} \frac{\Omega(x)}{|x|^N} O(|x|) dx - \int_{|x| > \epsilon_2} \frac{\Omega(x)}{|x|^N} O(|x|) dx = \int_{\epsilon_1 < |x| < \epsilon_2} \frac{\Omega(x)}{|x|^N} O(|x|) dx.$$

Therefore

$$\left\| \int_{\epsilon_1 < |x| < \epsilon_2} \frac{\Omega(x)}{|x|^N} O(|x|) dx \right\| \leq \int_{\epsilon_1 < |x| < \epsilon_2} C \frac{1}{|x|^{N-1}} dx$$

$$= \int_\Sigma \int_{\epsilon_1}^{\epsilon_2} \frac{C}{r^{N-1}} r^{N-1} \, dr \, d\sigma$$

$$= C'(\epsilon_2 - \epsilon_1) \to 0.$$

If we let $K(x) = \frac{\Omega(x)}{|x|^N}$, then, for $\alpha > 0$,

$$\widehat{K}(\alpha x) = \int_{\mathbb{R}^N} e^{-i\alpha x \cdot t} \frac{\Omega(t)}{|t|^N} \, dt = \int_{\mathbb{R}^N} e^{-ix \cdot t} \frac{\Omega(\frac{t}{\alpha})}{|\frac{t}{\alpha}|^N} \frac{1}{\alpha^N} \, dt$$

$$= \int_{\mathbb{R}^N} e^{-ix \cdot t} \frac{\Omega(t)}{|t|^N} \, dt = \widehat{K}(x).$$

Therefore, $\widehat{K}(x)$ is homogeneous of degree 0. Hence $\|\widehat{K}(x)\|_\infty \le \sup_{|x|=1} |\widehat{K}(x)| < \infty$.

Remark: It requires a small additional argument to see that $\widehat{K}$ is not simply a distribution that is homogeneous of degree 0 but is in fact a *function* that is homogeneous of degree 0. One sees this by writing K itself as a limit of L^1 functions. Each of those L^1 functions has Fourier transform that is bounded and continuous, and $\widehat{K}$ is the limit of those bounded functions in a suitable topology.

Lemma 9.7.18

$$T : f \longmapsto K * f$$

is bounded on $L^2(\mathbb{R}^N)$.

Proof: For $\phi \in C_c^\infty(\mathbb{R}^N)$, consider

$$T : \phi \longmapsto K * \phi.$$

Then $(\widehat{T\phi}) = \widehat{K * \phi} = \widehat{K}\widehat{\phi}$. Therefore, by Plancherel's theorem,

$$\|T\phi\|_2 = \|\widehat{T\phi}\|_2 = \|\widehat{K}\widehat{\phi}\|_2 \le C\|\widehat{\phi}\|_2 = C\|\phi\|_2.$$

Since $C_c^\infty(\mathbb{R}^N) \subset L^2(\mathbb{R}^N)$ is a dense subset, then by Functional Analysis Principle I (Appendix 1), T is bounded on $L^2(\mathbb{R}^N)$. $\qquad\square$

The next result is the key to our study of singular integral kernels and operators. At first, for technical convenience, we do not formulate the theorem in the classical language of Calderón and Zygmund (that is to say, there is no mention of "homogeneity of degree zero" nor of "mean value zero"). That will come later.

Prelude: The Calderón–Zygmund theorem is one of *the* seminal results of twentieth-century mathematics. It was profound because it produced the right generalization of the Hilbert transform to higher dimensions, and it was important because of the new proof techniques that it introduced. It is safe to say that it shaped an entire subject for many years.

Theorem 9.7.19 (Calderón–Zygmund) *Let $K \in L^2(\mathbb{R}^N)$. Assume that*

1. $|\widehat{K}| \leq B$.
2. $K \in C^1(\mathbb{R}^N \setminus \{0\})$ and $|\nabla K(x)| \leq C|x|^{-N-1}$.

For $1 < p < \infty$, and $f \in L^1 \cap L^p(\mathbb{R}^N)$, set

$$Tf(x) = K * f(x) = \int_{\mathbb{R}^N} K(x - t) f(t) dt.$$

Then there exists a constant A_p such that

$$\|Tf\|_p \leq A_p \|f\|_p.$$

Proof: We know that T is bounded on L^2. So if we can prove that T is weak type $(1, 1)$ then, by the Marcinkiewicz interpolation theorem, T is bounded on L^p, for $1 < p < 2$. Then, by duality, T is bounded on L^p, for $p > 1$.

Now we prove that T is weak-type $(1, 1)$.

Let $f \in L^1(\mathbb{R}^N)$ and fix $\alpha > 0$. Apply Theorem 9.7.11 to $|f|$ and α. Then we get $\mathbb{R}^N = F \cup \Omega$, $F \cap \Omega = \emptyset$, F closed. We have $Mf(x) \leq \alpha$ on F. Thus $f(x) \leq \alpha$ a.e. $x \in F$. We write $\Omega = \cup_j Q_j$, where the Q_j's are closed cubes with pairwise disjoint interiors and with diameter comparable to the distance from F. By Theorem 9.7.11 we have

$$\frac{1}{m(Q_j)} \int_{Q_j} |f| dx < C\alpha \tag{9.7.19.1}$$

and

$$m(\Omega) \leq C' \frac{\|f\|_1}{\alpha}. \tag{9.7.19.2}$$

Let

$$g(x) = \begin{cases} f(x) & \text{if} \quad x \in F, \\ \frac{1}{m(Q_j)} \int_{Q_j} f(t) dt & \text{if} \quad x \in Q_j^\circ, \end{cases}$$

and $b(x) = f(x) - g(x)$,

$$b(x) = \begin{cases} 0 & \text{if} \quad x \in F, \\ f(x) - \frac{1}{m(Q_j)} \int_{Q_j} f(t) dt & \text{if} \quad x \in Q_j^\circ. \end{cases}$$

Then we have

$$m\{|Tf| > \alpha\} = m\{|Tg + Tb| > \alpha\}$$

$$\leq m\{|Tg| + |Tb| > \alpha\}$$

$$\leq m\left\{|Tg| > \frac{\alpha}{2}\right\} + m\left\{|Tb| > \frac{\alpha}{2}\right\}$$

and, from (9.7.19.1) and (9.7.19.2),

$$\|g\|_2^2 = \int_{\mathbb{R}^N} |g(x)|^2 dx$$

$$= \int_F |g(x)|^2 dx + \int_\Omega |g(x)|^2 dx$$

$$\leq \int_F \alpha |g(x)| dx + \int_\Omega C^2 \alpha^2 dx$$

$$\leq \alpha \|f\|_1 + C^2 \alpha^2 m(\Omega)$$

$$\leq \alpha \|f\|_1 + C^2 \alpha^2 \frac{1}{\alpha} \|f\|_1$$

$$\leq C'\alpha \|f\|_1 < \infty. \tag{9.7.19.3}$$

Thus $g \in L^2(\mathbb{R}^N)$ and, by Lemma 9.7.18, $Tg \in L^2(\mathbb{R}^N)$. Hence, by Chebyshev's inequality and (9.7.19.3), we get

$$m\left\{|Tg| > \frac{\alpha}{2}\right\} \leq \frac{\|Tg\|_2^2}{(\alpha/2)^2} \leq C\frac{\|g\|_2^2}{\alpha^2} \leq C\frac{\alpha \|f\|_1}{\alpha^2} = \frac{C\|f\|_1}{\alpha}.$$

From the above calculations, we need only show that

$$m\left\{|Tb| > \frac{\alpha}{2}\right\} \leq C\frac{\|f\|_1}{\alpha}.$$

Furthermore,

$$m\left\{x \in \mathbb{R}^N : |Tb(x)| > \frac{\alpha}{2}\right\} \leq m\left\{x \in F : |Tb(x)| > \frac{\alpha}{2}\right\} + m(\Omega)$$

$$= m\left\{x \in F : |Tb(x)| > \frac{\alpha}{2}\right\} + \frac{\|f\|_1}{\alpha}.$$

Therefore it is enough to consider $m\{x \in F : |Tb(x)| > \frac{\alpha}{2}\}$.

Let $b_j(x) = b(x)\chi_{Q_j}(x)$. Then

$$b_j(x) = \begin{cases} b(x) & \text{if } x \in Q_j, \\ 0 & \text{otherwise,} \end{cases}$$

and $b(x) = \sum_j b_j(x)$. Thus $Tb(x) = \sum_j Tb_j(x)$. We have

$$Tb_j(x) = \int_{\mathbb{R}^N} K(x-t)b_j(t)dt = \int_{Q_j} K(x-t)b_j(t)dt.$$

Since

$$\int_{Q_j} b_j(t)dt = \int_{Q_j} f(t) - \left[\frac{1}{m(Q_j)} \int_{Q_j} f(t)dt\right]dt = 0,$$

we can rewrite the preceding expression as

$$Tb_j(x) = \int_{Q_j} \big(K(x-t) - K(x-t_j)\big) b_j(t)dt,$$

where t_j is the center of Q_j. By the mean value theorem and the hypothesis of the theorem, we get

$$|K(x-t) - K(x-t_j)| \leq |\nabla K(x - \tilde{t})| \cdot |t - t_j| \leq C \frac{\text{diam } Q_j}{|x - \tilde{t}|^{N+1}},$$

where $\tilde{t}$ lies on the segment connecting t_j and t. Hence we have

$$|Tb_j(x)| \leq C \int_{Q_j} \frac{\text{diam } Q_j}{|x - \tilde{t}|^{N+1}} |b_j(t)| dt.$$

We also have

$$\int_{Q_j} |b_j(t)| dt \leq \int_{Q_j} |f| dt + \int_{Q_j} C\alpha \, dt \leq C\alpha \cdot m(Q_j) + C\alpha \cdot m(Q_j) \leq C\alpha \cdot m(Q_j).$$

Since diam $Q_j \approx \text{dist}(Q_j, F) \leq \delta(t)$, we get

$$\int_{Q_j} \text{diam } Q_j |b_j(t)| dt \leq C(\text{diam } Q_j)\alpha \cdot m(Q_j) \leq C\alpha \int \delta(t) dt.$$

Therefore we find that

$$|Tb_j(x)| \overset{<}{\sim} \alpha \int_{Q_j} \frac{\delta(t)}{|x - t|^{N+1}} dt.$$

Hence

$$|Tb(x)| \leq \sum_j |Tb_j(x)| \overset{<}{\sim} \alpha \sum_j \int_{Q_j} \frac{\delta(t)}{|x - t|^{N+1}} dt$$

$$= \alpha \int_{\Omega} \frac{\delta(t)}{|x - t|^{N+1}} dt$$

$$\leq \alpha \int_{\mathbb{R}^N} \frac{\delta(t)}{|x - t|^{N+1}}$$

$$= I_*(x),$$

where $I_*(x)$ is the Marcinkiewicz integral that satisfies

$$\int_F I_*(x) dx \overset{<}{\sim} m(\Omega) \overset{<}{\sim} \frac{\|f\|_1}{\alpha}.$$

Hence we get

$$\int_F |Tb(x)| dx \overset{<}{\sim} \|f\|_1.$$

Therefore

$$\frac{a}{2} m\left\{x \in F : |Tb(x)| > \frac{a}{2}\right\} \leq \int_F |Tb(x)|\,dx \overset{<}{\sim} \|f\|_1.$$

Hence

$$m\left\{x \in F : |Tb(x)| > \frac{a}{2}\right\} \overset{<}{\sim} \frac{\|f\|_1}{a}.$$

The theorem is proved. $\qquad\qquad\qquad\qquad\qquad\qquad\qquad\qquad\qquad$ □

Theorem 9.7.20 Let $K(x)$ be a Calderón–Zygmund kernel. Then $f \longmapsto K * f$ is bounded on L^p, $1 < p < \infty$.

Proof: Let $\epsilon_1 > \epsilon_2 > \cdots \to 0$ and $K_j(x) = K(x)\chi_{\{|t|>\epsilon_j\}}$. Define $T_j f = f * K_j$. Then K_j satisfies the hypotheses of Theorem 9.7.19, which we proved above. Thus T_j is bounded on L^p for $1 < p < \infty$. If we let $\|T_j\| = A_p$, then $A_p \sim \frac{C}{p-1}$ as $p \to 1^+$ and $A_p \sim p$ as $p \to \infty$.

Applying Functional Analysis Principle I with a dense subset $C_c^\infty \subset L^p$, we get that $K * f = \lim_{j \to \infty} T_j f$ is bounded on L^p, for $1 < p < \infty$. $\qquad$ □

9.8 Analysis on $\mathbb{H}^n$

Capsule: Here we jump in and begin to do basic analysis on the Heisenberg group. Along the way, we treat Hausdorff measure, and other basic tools of wide utility are examined and developed in our new context.

When we were thinking of the Heisenberg group as the boundary of a domain in $\mathbb{C}^n$, then the appropriate Heisenberg group to consider was $\mathbb{H}^{n-1}$, since that Lie group has dimension $2n - 1$ (the correct dimension for a boundary). Now we are about to study the Heisenberg intrinsically, in its own right, so it is appropriate (and it simplifies the notation a bit) to focus our attention on $\mathbb{H}^n$.

In $\mathbb{H}^n = \mathbb{C}^n \times \mathbb{R}$, the group operation is defined as

$$(z, t) \cdot (z', t') = (z + z', t + t' + 2\mathrm{Im}\, z \cdot \overline{z'}), \quad z, z' \in \mathbb{C}^n, \; t, t' \in \mathbb{R}.$$

Let $g = (z, t) = (z_1, z_2, \ldots, z_n, t) = (x_1 + iy_1, x_2 + iy_2, \ldots, x_n + iy_n, t) \in \mathbb{H}^n$. The dilation $\delta(g) = \delta(z, t) = (\delta z, \delta^2 t)$ is a group isomorphism. We write

$$dV(g) = d\mathrm{Vol}(g) = dx_1 dy_1 \cdots dx_n dy_n dt,$$

so that

$$dV(\delta g) = d(\delta x_1)\,d(\delta y_1)\cdots d(\delta x_n)\,d(\delta y_n)\,d\delta^2 t = \delta^{2n+2}\,dV(g).$$

We call $2n + 2$ the homogeneous dimension of $\mathbb{H}^n$. [Note that the topological dimension of $\mathbb{H}^n$ is $2n + 1$.] The critical index N for a singular integral is such that

$$\int_{B(0,1)} \frac{1}{|z|^{\alpha}} \, d\mathrm{Vol}(z) = \begin{cases} \infty & \text{if} \quad \alpha \geq N, \\ < \infty & \text{if} \quad 0 < \alpha < N, \end{cases}$$

and the critical index coincides with the homogeneous dimension. Thus the critical index for a singular integral in $\mathbb{H}^n$ is $2n + 2$, which is different from the topological dimension.

Note by contrast that the critical index for a singular integral in $\mathbb{C}^n \times \mathbb{R} = \mathbb{R}^{2n+1}$, equipped with usual commutative, isotropic Euclidean structure, is $2n + 1$; this is the same as the topological dimension.

9.8.1 The Norm on $\mathbb{H}^n$

In earlier parts of the book we have alluded to the Heisenberg group norm. Now we define it carefully.

We define the norm $| \cdot |_h$ on $\mathbb{H}^n$ to be

$$|g|_h = (|z|^4 + t^2)^{1/4}$$

when $g = (z, t) \in \mathbb{H}^n$. Then $| \cdot |_h$ satisfies the following desiderata:

1. $|g|_h \geq 0$ and $|g|_h = 0$ if and only if $g = 0$;
2. $g \longmapsto |g|_h$ is a continuous function from $\mathbb{H}^n$ to $\mathbb{R}$ and is C^{∞} on $\mathbb{H}^n \setminus \{0\}$;
3. $|\delta(g)|_h - \delta|g|_h$.

Note that $\delta g = \delta(g)$ denotes the dilation (in the Heisenberg structure) of g by a positive factor of δ. The preceding three properties do not uniquely determine the norm. If ϕ is positive, smooth away from 0, and homogeneous of degree 0 in the Heisenberg group dilation structure, then $|g|_h^* \equiv \phi(g)|g|_h$ is another norm.

The Heisenberg group $\mathbb{H}^n = \mathbb{C}^n \times \mathbb{R}$ is also equipped with the Euclidean norm in $\mathbb{R}^{2n+1}$. Let us denote the Euclidean norm by $| \cdot |_e$:

$$|g|_e = (|z|^2 + |t|^2)^{1/2}$$

when $g = (z, t) \in \mathbb{H}^n$.

Lemma 9.8.1 *For $|g|_e^2 \leq \frac{1}{2}$, we have*

$$|g|_e \leq |g|_h \leq |g|_e^{1/2}.$$

Proof: We see that

$$|g|_e = (|z|^2 + t^2)^{1/2} \leq (|z|^4 + t^2)^{1/4} = |g|_h$$

reduces to

$$(|z|^2 + t^2)^2 \leq |z|^4 + t^2,$$

or

$$2t^2|z|^2 + t^4 \leq t^2,$$

or

$$2|z|^2 + t^2 \leq 1.$$

Since we assumed that $|g|_e^2 = |z|^2 + t^2 \leq \frac{1}{2}$, we have $2|z|^2 + t^2 \leq 1$.

Furthermore,

$$|g|_h = (|z|^4 + t^2)^{1/4}$$
$$\leq (|z|^2 + t^2)^{1/4}$$
$$= |g|_e^{1/2}.$$

That completes the proof.   □

9.8.2 Polar Coordinates

Prelude: Of course the idea of polar coordinates is very basic and elementary. But this simple calculation shows us already that analysis on the Heisenberg group will be different from analysis on ordinary Euclidean space. In particular, the Heisenberg group polar coordinates begin to point to the idea of homogeneous dimension.

Let $g \in \mathbb{H}^n$, $g \neq 0$, and $r = |g|_h$. Then we may write $g = r \cdot \xi$, where $\xi = \frac{g}{|g|_h}$. Of course $|\xi|_h = 1$.

Lemma 9.8.2 *We have*

$$dV(g) = dx_1 dy_1 \cdots dx_n dy_n dt = r^{2n+1} dr \, d\sigma(\xi),$$

where $d\sigma$ is a smooth, positive measure on the Heisenberg unit sphere $\{\xi \in \mathbb{H}^n : |\xi|_h = 1\}$.

Proof: Write $g = (s_1, \ldots, s_{2n+1}) \in \mathbb{H}^n$. If we let $r = |g|_h$, then

$$g = (z, t) \approx (s_1, \ldots, s_{2n+1}) = r(\xi_1, \ldots, \xi_{2n+1}) = (r\xi_1, r\xi_2, \ldots, r\xi_{2n}, r^2 \xi_{2n+1}).$$

Since $|\xi|_h = 1$, we have

$$\xi_{2n+1}^2 = 1 - (\xi_1^2 + \cdots + \xi_{2n}^2)^2.$$

Therefore we may consider the coordinate transform

$$(s_1, \ldots, s_{2n+1}) \to (\xi_1, \ldots, \xi_{2n}, r).$$

Calculating the Jacobian matrix, we get

$$
\text{Jac} =
\begin{pmatrix}
\frac{\partial s_1}{\partial \xi_1} & \cdots & \frac{\partial s_1}{\partial \xi_{2n}} & \frac{\partial s_1}{\partial r} \\
\vdots & & & \vdots \\
\frac{\partial s_{2n+1}}{\partial \xi_1} & \cdots & \frac{\partial s_{2n+1}}{\partial \xi_{2n}} & \frac{\partial s_{2n+1}}{\partial r}
\end{pmatrix}
$$

$$
=
\begin{pmatrix}
r & 0 & \cdots & 0 & \xi_1 \\
0 & r & \cdots & 0 & \xi_2 \\
\vdots & & & & \vdots \\
0 & & \cdots & r & \xi_{2n} \\
r^2 \frac{\partial \xi_{2n+1}}{\partial \xi_1} & r^2 \frac{\partial \xi_{2n+1}}{\partial \xi_2} & \cdots & r^2 \frac{\partial \xi_{2n+1}}{\partial \xi_{2n}} & 2r\xi_{2n+1}
\end{pmatrix}
$$

Therefore

$$
|\det \text{Jac}| = r^{2n+1}\left(2\xi_{2n+1} - \xi_1 \frac{\partial \xi_{2n+1}}{\partial \xi_1} - \xi_2 \frac{\partial \xi_{2n+1}}{\partial \xi_2} - \cdots - \xi_{2n} \frac{\partial \xi_{2n+1}}{\partial \xi_{2n}} \right).
$$

Hence

$$
dV(z) = dz_1 \cdots dz_{2n+1} = r^{2n+1} dr\, d\sigma,
$$

where

$$
d\sigma = \left(2\xi_{2n+1} - \xi_1 \frac{\partial \xi_{2n+1}}{\partial \xi_1} - \xi_2 \frac{\partial \xi_{2n+1}}{\partial \xi_2} - \cdots - \xi_{2n} \frac{\partial \xi_{2n+1}}{\partial \xi_{2n}} \right) d\xi_1 \cdots d\xi_{2n}. \quad \square
$$

To repeat what we hinted at earlier, the factor of r^{2n+1} in the polar representation for volume suggests that the homogeneous dimension is $2n + 2$ (just as, in the classical Euclidean setting, the factor of r^{N-1} in the polar representation for volume suggests that the homogeneous dimension—which is of course the same as the topological dimension—is N).

9.8.3 Some Remarks about Hausdorff Measure

The work [FED] is a good reference for this material.

Let $S \subseteq \mathbb{R}^N$ be a set. Then for $\delta > 0$, $\lambda > 0$, we define

$$
\mathcal{H}^\lambda_\delta(S) = \inf_{S \subset \cup B_j} \sum_j [\text{diam}\,(B_j)]^\lambda,
$$

where the B_j's are Euclidean balls of radius less than δ. Note that if $0 < \delta_1 < \delta_2$, then

$$
\mathcal{H}^\lambda_{\delta_1}(S) \geq \mathcal{H}^\lambda_{\delta_2}(S).
$$

Therefore

$$
\mathcal{H}^\lambda(S) = \lim_{\delta \to 0} \mathcal{H}^\lambda_\delta
$$

exists and we call this limit the λ-dimensional Hausdorff measure of S.

Definition 9.8.3 (Hausdorff dimension) For any set S, $\exists \lambda_0 > 0$ such that

$$\mathcal{H}^\lambda(S) = \begin{cases} 0 & \text{if } \lambda > \lambda_0, \\ \infty & \text{if } \lambda < \lambda_0. \end{cases}$$

We call λ_0 the *Hausdorff dimension* of S.

In fact, the Hausdorff dimension of a given set S can be defined to be the infimum of all λ such that $\mathcal{H}^\lambda(S) = 0$. Or it can be defined to be the supremum of all λ such that $\mathcal{H}^\lambda(S) = +\infty$. It is a theorem that these two numbers are equal (see [FED]).

Example 9.8.4 Let $I = [0, 1] \subset \mathbb{R}$. If $I \subset \cup B_j$ and the B_j's are balls of radius less than $\delta > 0$, then $\mathcal{H}_\delta^1(I) \geq 1$ and $\mathcal{H}_\delta^1(I) \leq \frac{1}{2\delta} \cdot (2\delta) = 1$. Therefore $\mathcal{H}^1(I) = 1$. But for $\lambda = 1 + \epsilon, \epsilon > 0$, we get

$$\mathcal{H}_\delta^\lambda(I) \leq \frac{1}{2\delta}(2\delta)^{1+\epsilon} = (2\delta)^\epsilon \to 0, \quad \text{as } \delta \to 0.$$

Also, if $\lambda < 1$, then $\mathcal{H}^\lambda(I) = \infty$.

9.8.4 Integration in $\mathbb{H}^n$

We have defined polar coordinates in $\mathbb{H}^n$. Now we can calculate the volume of a ball in $\mathbb{H}^n$ using polar coodinates.

Let μ be the surface area of a unit sphere in $\mathbb{H}^n$:

$$\mu = \int_{|\xi|_h = 1} d\sigma(\xi).$$

Then the volume of the unit ball in $\mathbb{H}^n$ is

$$|B| = \int_{|z| \leq 1} dV(z) = \int_\Sigma \int_0^1 r^{2n+1} dr \, d\sigma = \frac{\mu}{2n+2}.$$

Hence the volume of a ball of radius ρ will be

$$|B(0, \rho)| = \int_{|z| \leq \rho} dV(z) = \int_{|z| \leq 1} dV(\rho z) = \rho^{2n+2} \int_{|z| \leq 1} dV(z) = \rho^{2n+2}|B|.$$

$$(9.8.4.1)$$

Observe the homogeneous dimension coming into play here.

Now the integration of characteristic functions of balls is well defined. Since characteristic functions of arbitrary sets can be approximated by aggregates of characteristic functions on balls and simple functions are just linear combinations of characteristic functions, the integration of simple functions is well defined. We define the integration of a function in $L^1(\mathbb{R}^{2n+1})$ as the limit of the integration of the simple functions that approximate it.

9.8.5 Distance in $\mathbb{H}^n$

For $x, y \in \mathbb{H}^n$, we define the distance $d(x, y)$ as follows:

$$d(x, y) \equiv |x^{-1} \cdot y|_h.$$

Then $d(x, y)$ satisfies the following properties:

1. $d(x, y) = 0 \iff x = y$;
2. $d(x, y) = d(y, x)$;
3. $\exists \gamma_0 > 0$ such that $d(x, y) \le \gamma_0[d(x, w) + d(w, y)]$.

Proof:

1. Obvious.
2. One may easily check that $x^{-1} = -x$. Thus

$$d(x, y) = |x^{-1} \cdot y|_h = |(-x) \cdot y|_h = |x \cdot (-y)|_h = d(y, x).$$

3. Let

$$\sup_{|x|_h, |y|_h \le 1} d(x, y) = C$$

$$\inf_{|x|_h, |y|_h, |w|_h \le 1} d(x, w) + d(w, y) = D.$$

Then $C \ge 1$ and $D > 0$. Therefore we get

$$d(x, y) \le C \le \frac{C}{D}(d(x, w) + d(w, y)), \quad \text{if } |x|_h, |y|_h, |w|_h \le 1.$$

Now, for general x, y, and w, let $r = \max\{|x|_h, |y|_h, |w|_h\}$. Then $x = rx'$, $y = ry'$, and $w = rw'$, where $|x'|_h, |y'|_h, |w'|_h \le 1$. Then we have

$$d(x, y) = d(rx', ry') = rd(x', y')$$

and

$$d(x, w) + d(w, y) = r(d(x', w') + d(w', y')).$$

Hence

$$d(x, y) \le \frac{C}{D}(d(x, w) + d(w, y)) \text{ for all } x, y, w. \qquad \square$$

9.8.6 $\mathbb{H}^n$ Is a Space of Homogeneous Type

Refer to Section 9.8.1. Define balls in $\mathbb{H}^n$ by $B(x, r) = \{y \in \mathbb{H}^n : d(x, y) < r\}$. Then, equipped with the Lebesgue measure on $\mathbb{R}^{2n+1}$, $\mathbb{H}^n$ is a space of homogeneous type. We need to check the following three conditions:

1. $0 < m(B(x, r)) < \infty$ for all $x \in \mathbb{H}^n$ and $r > 0$;
2. $\exists C_1 > 0$ such that $m(B(x, 2r)) \le C_1 m(B(x, r))$;
3. $\exists C_2 > 0$ such that if $B(x, r) \cap B(y, s) \ne 0$ and $s \ge r$, then $B(y, C_2 s) \supseteq B(x, r)$.

Proof:

1. From (9.8.5.1), we know that $m(B(x, r)) = r^{2n+2}|B|$, where $|B|$ is the volume of the unit ball.
2. $m(B(x, 2r)) = 2^{2n+2}m(B(x, r))$. Therefore $C_1 = 2^{2n+2}$.
3. This result follows because, equipped with the distance d, the Heisenberg group is a quasimetric space. In detail, let $w \in B(x, r) \cap B(y, s)$. Then $d(w, x) < r$ and $d(w, y) < s$. If $u \in B(x, r)$, then we obtain

$$d(y, u) \le \gamma_0[d(y, x) + d(x, u)]$$

$$< \gamma_0[\gamma_0 d(y, w) + \gamma_0 d(w, x) + d(x, u)]$$

$$\le \gamma_0[\gamma_0 s + \gamma_0 r + r]$$

$$\le \gamma_0(1 + 2\gamma_0)s.$$

Thus we may let $C_2 = \gamma_0 + 2\gamma_0^2$.   □

For $f \in L_{\text{loc}}^1(\mathbb{H}^n)$, define

$$Mf(z) = \sup_{r>0} \frac{1}{|B(z, r)|} \int_{B(z,r)} |f(t)|dt,$$

$$\widetilde{M}f(z) = \sup_{x \in B} \frac{1}{|B|} \int_B |f(t)|dt.$$

These two operators are closely related. In fact, $\widetilde{M}f \le C \cdot Mf$. If $B(z, r) \cap B(w, r) \ne \emptyset$, then $B(w, r) \subseteq B(z, C_2 r)$. Therefore

$$\frac{1}{|B(w, r)|} \int_{B(w,r)} |f(t)|dt \le \frac{1}{|B(w, r)|} \int_{B(z,C_2 r)} |f(t)|dt$$

$$= \frac{C_2^{2n+2}}{|B(z, C_2 r)|} \int_{B(z,C_2 r)} |f(t)|dt$$

$$\le C_2^{2n+2} Mf.$$

The reverse inequality is obvious.

Thus $M, \widetilde{M}$ are both bounded on L^∞ and both of weak-type $(1, 1)$ (just because $\mathbb{H}^n$ is a space of homogeneous type). Hence, by the Marcinkiewicz interpolation theorem, both are strong-type (p, p) for $1 < p < \infty$.

9.8.7 Homogeneous Functions

We say that a function $f : \mathbb{H}^n \to \mathbb{C}$ is *homogeneous of degree* $m \in \mathbb{R}$ if $f(\delta x) = \delta^m f(x)$.

The Schwartz space $\mathcal{S}$ of $\mathbb{H}^n$ is the Schwartz space of $\mathbb{R}^{2n+1}$:

$$\mathcal{S}(\mathbb{H}^n) = \left\{ \phi : \|\phi\|_{\alpha,\beta} \equiv \sup_{x \in \mathbb{H}^n} \left| x^\alpha \left(\frac{\partial}{\partial x} \right)^\beta f(x) \right| < \infty \right\}.$$

The norm $\| \cdot \|_{\alpha,\beta}$ is a seminorm and $\mathcal{S}$ is a Frechet space. The dual space of $\mathcal{S}$ is the space of Schwartz distributions. For $\phi \in \mathcal{S}$ and $\delta > 0$, set

$$\phi_\delta(x) = \phi(\delta x),$$

$$\phi^\delta(x) = \delta^{-2n-2} \phi \left(\frac{x}{\delta} \right).$$

Note the homogeneous dimension playing a role in the definition of ϕ^δ.

A Schwartz distribution τ is said to be *homogeneous of degree m* provided[6] that

$$\tau(\phi^\delta) = \delta^m \tau(\phi).$$

If it happens that the distribution τ is given by integration against a function K that is homogeneous of degree m, then the resulting distribution is homogeneous of degree m:

$$\tau(\phi^\delta) = \int K(x) \phi^\delta(x) \, dx$$

$$= \int K(\delta x) \phi(x) \, dx$$

$$= \int \delta^m K(x) \phi(x) \, dx$$

$$= \delta^m \tau(\phi).$$

Proposition 9.8.5 *Let f be a homogeneous function of degree $\lambda \in \mathbb{R}$. Assume that f is C^1 away from 0. Then $\exists C > 0$ such that*

$$|f(x) - f(y)| \leq C|x - y|_h \cdot |x|_h^{\lambda-1}, \quad \text{whenever } |x - y|_h \leq \frac{1}{2\gamma_0}|x|_h;$$

$$|f(x \cdot y) - f(x)| \leq C|y|_h|x|_h^{\lambda-1}, \quad \text{whenever } |y|_h \leq \frac{1}{2\gamma_0}|x|_h.$$

Remark: If f is homogeneous of degree λ, then Df (any first derivative of f with respect to X_j or Y_j) is homogeneous of degree $\lambda - 1$. We leave the proof as an easy exercise. What does this say about the homogeneity of Tf?

---

[6] A moment's thought reveals that the motivation for this definition is change of variables in the integral—see below.

Remark: In $\mathbb{H}^n$, the Dirac-δ mass is homogeneous of degree $-2n-2$.

Proof of Proposition 9.8.5:   Let us look at the first inequality. If we dilate x, y by $\alpha > 0$, then

$$LHS = |f(\alpha x) - f(\alpha y)| = \alpha^\lambda |f(x) - f(y)|,$$

$$RHS = C|\alpha x - \alpha y|_h |\alpha x|_h^{\lambda-1} = C\alpha^\lambda |x - y|_h |x|_h^{\lambda-1}.$$

Thus the inequality is invariant under dilation. So it is enough to prove the inequality when $|x|_h = 1$ and $|x - y|_h \leq \frac{1}{4\gamma_0}$. Then (assuming as we may that $\gamma_0 > 1$) y is bounded from 0:

$$d(x, 0) \leq \gamma_0(d(x, y) + d(y, 0)),$$

$$d(y, 0) \geq \frac{1}{\gamma_0} - d(x, y) \geq \frac{1}{\gamma_0} - \frac{1}{4\gamma_0} = \frac{3}{4\gamma_0} > 0.$$

Apply the classical Euclidean mean value theorem[7] to $f(x)$:

$$|f(x) - f(y)| \leq \sup |\nabla f| |x - y|_e.$$

Note that the supremum is taken on the segment connecting x and y. Since $|x|_h = 1$ and y is bounded from 0, we have

$$|f(x) - f(y)| \leq C|x - y|_e \leq C|x - y|_h.$$

The last inequality is by Lemma 9.8.1.

We use the same argument to prove the second inequality. $\square$

9.9 The Calderón–Zygmund Integral on $\mathbb{H}^n$ is Bounded on L^2

Capsule: This section is the payoff for our hard work. We can now define Calderón–Zygmund operators on the Heisenberg group. We can identify particular examples of such operators. And we can prove that they are bounded on L^p, $1 < p < \infty$. Some parts of the proof of this key result will be familiar. Other parts (such as the boundedness on L^2, which formerly depended critically on the Fourier transform) will require new techniques. The entire section is a tour de force of our new ideas and techniques.

---

[7] The fact is that there is no mean value theorem in higher dimensions. If f is a continuously differentiable function on $\mathbb{R}^N$ then we apply the "mean value theorem" at points P and Q in its domain by considering the one-variable function $[0, 1] \ni t \mapsto f((1-t)P + tQ)$ and invoking the one-dimensional calculus mean value theorem.

In $\mathbb{R}^N$, for a Calderón–Zygmund kernel $K(z)$, we know that $f \longmapsto f * K$ is bounded on L^2. We proved the result using Plancherel's theorem. Since K is homogeneous of degree $-N$, we know that $\widehat{K}$ is homogeneous of degree $-N - (-N) = 0$. Thus $\widehat{K}$ is bounded and

$$\| f * K \|_2 = \| \widehat{f * K} \|_2 = \| \widehat{f} \cdot \widehat{K} \|_2 \leq C \| \widehat{f} \|_2 = C \| f \|_2.$$

But we cannot use the same technique in $\mathbb{H}^n$ since we do not have the Fourier transform in $\mathbb{H}^n$ as a useful analytic tool. Instead we use the so-called Cotlar–Knapp–Stein lemma.

9.9.1 Cotlar–Knapp–Stein Lemma

Question. Let H be a Hilbert space. Suppose we have operators $T_j : H \to H$ that have uniformly bounded norm, $\| T_j \|_{op} = 1$. Then what can we say about $\| \sum_{j=1}^N T_j \|_{op}$?

Example 9.9.1 Of course if all the operators are just the identity then their sum has norm N. That is not a very interesting situation.

By contrast, let $H = L^2(\mathbb{T})$ and $T_j = f * e^{ijt}$ for $j \in \mathbb{Z}$. Then $\| T_j \|_{op} = 1$ and, by the Riesz–Fischer theorem, we get $\| \sum T_j \| = 1$.

The last example is a special circumstance; the kernels operate on orthogonal parts of the Hilbert space. It was Mischa Cotlar who first understood how to conceptualize this idea.

Prelude: The Cotlar lemma, now (in a more elaborate form) known as the Cotlar–Knapp–Stein lemma, has a long and colorful history. Certainly Misha Cotlar [COT] deserves full credit for coming up with the idea of summing operators that act on different parts of Hilbert space. Later on, Cotlar and Knapp–Stein [KNS2] nearly simultaneously came up with the more flexible version of Cotlar's idea that we use today. It is a matter of considerable interest to come up with a version of the Cotlar–Knapp–Stein theorem for operators on L^p spaces when $p \neq 2$. Some contributions in that direction appear in [COP].

Lemma 9.9.2 (Cotlar) *Let H be a Hilbert space. Let $T_j : H \to H, j = 1, \ldots, N$, be self-adjoint operators. Assume that*

1. *$\| T_j \| = 1 \ \forall j$;*
2. *$T_j T_k^* = 0, T_j^* T_k = 0, \forall j \neq k$.*

Then $\| \sum_{j=1}^N T_j \| \leq C$, where C is a universal constant (independent of N).

The proof of the lemma is a very complicated combinatorial argument.

Cotlar and Knapp–Stein independently found a much more flexible formulation of the result that has proved to be quite useful in the practice of harmonic analysis. We now formulate and prove a version of their theorem (see [COT] and [KNS2]).

Lemma 9.9.3 (Cotlar–Knapp–Stein) *Let H be a Hilbert space and $T_j : H \to H$ bounded operators, $j = 1, \ldots, N$. Suppose there exists a positive, bi-infinite sequence $\{a_j\}_{j=-\infty}^{\infty}$ of positive numbers such that $A = \sum_{-\infty}^{\infty} a_j < \infty$. Also assume that*

$$\|T_j T_k^*\|_{\mathrm{op}} \leq a_{j-k}^2, \quad \|T_j^* T_k\|_{\mathrm{op}} \leq a_{j-k}^2. \tag{9.9.3.1}$$

Then

$$\left\| \sum_{j=1}^{N} T_j \right\|_{\mathrm{op}} \leq A.$$

Remark: Note that $\|T_j\| = \sqrt{\|T_j T_j^*\|} \leq a_0$, for all j.

Proof: We will use the fact that $\|TT^*\| = \|T^*T\| = \|T\|^2 = \|T^*\|^2$. Also, since TT^* is self-adjoint, we have $\|(TT^*)^k\| = \|TT^*\|^k$. Let $T = \sum_{j=1}^{N} T_j$. Then we get

$$(TT^*)^m = \left[\left(\sum_{j=1}^{N} T_j\right)\left(\sum_{j=1}^{N} T_j^*\right)\right]^m = \sum_{1 \leq j_k \leq N} T_{j_1} T_{j_2}^* T_{j_3} T_{j_4}^* \cdots T_{j_{2m-1}} T_{j_{2m}}^*.$$

By (9.9.3.1), we get

$$\|T_{j_1} T_{j_2}^* \cdots T_{j_{2m-1}} T_{j_{2m}}^*\|$$
$$\leq \|T_{j_1} T_{j_2}^*\| \|T_{j_3} T_{j_4}^*\| \cdots \|T_{j_{2m-1}} T_{j_{2m}}^*\|$$
$$\leq a_{j_1-j_2}^2 \cdots a_{j_{2m-1}-j_{2m}}^2.$$

Also,

$$\|T_{j_1} T_{j_2}^* \cdots T_{j_{2m-1}} T_{j_{2m}}^*\| \leq \|T_{j_1}\| \|T_{j_2}^* T_{j_3}\| \|T_{j_4}^* T_{j_5}\| \cdots \|T_{j_{2m-2}}^* T_{j_{2m-1}}\| \|T_{j_{2m}}\|$$
$$\leq A a_{j_2-j_3}^2 \cdots a_{j_{2m-2}-j_{2m-1}}^2 A.$$

Therefore we may conclude that

$$\|T_{j_1} T_{j_2}^* \cdots T_{j_{2m-1}} T_{j_{2m}}^*\| = \|T_{j_1} T_{j_2}^* \cdots T_{j_{2m-1}} T_{j_{2m}}^*\|^{1/2}$$
$$\|T_{j_1} T_{j_2}^* \cdots T_{j_{2m-1}} T_{j_{2m}}^*\|^{1/2}$$
$$\leq A a_{j_1-j_2} a_{j_2-j_3} \cdots a_{j_{2m-2}-j_{2m-1}} a_{j_{2m-1}-j_{2m}}.$$

Hence

$$\|TT^*\|^m \leq \sum_{\substack{1 \leq j_k \leq N \\ k=1,2,\ldots,2m}} A a_{j_1-j_2} a_{j_2-j_3} \cdots a_{j_{2m-2}-j_{2m-1}} a_{j_{2m-1}-j_{2m}}.$$

First we sum over j_{2m} and then j_{2m-2} and so on. Then we get

$$\|TT^*\|^m \leq \sum_{\substack{1 \leq j_k \leq N \\ k=1,3,\dots,2m-1}} A^{2m} = \binom{N+m-1}{m} A^{2m}.$$

Taking the mth root, we get

$$\|TT^*\| \leq \left(\frac{N+m-1}{m}\right)^{\frac{1}{m}} \cdots \left(\frac{N}{1}\right)^{\frac{1}{m}} A^2.$$

Letting $m \to \infty$, we get $\|T\| \leq A$. $\square$

9.9.2 The Folland–Stein Theorem

Prelude: This next is the fundamental "singular integrals" result on the Heisenberg group. There are also theories of pseudodifferential operators on the Heisenberg group (for which see, for example, [NAS]). There is still much to be done to develop these ideas on general nilpotent Lie groups.

Theorem 9.9.4 (Folland, Stein. 1974) *Let K be a function on $\mathbb{H}^n$ that is smooth away from 0 and homogeneous of degree $-2n - 2$. Assume that*

$$\int_{|z|_h=1} K\, d\sigma = 0,$$

where $d\sigma$ is area measure (i.e., Hausdorff measure) on the unit sphere in the Heisenberg group. Define

$$Tf(z) = \mathrm{PV}(K * f) = \lim_{\epsilon \to 0} \int_{|t|_h > \epsilon} K(t) f(t^{-1}z)\, dt.$$

Then the limit exists pointwise and in norm and

$$\|Tf\|_2 \leq C\|f\|_2.$$

Remark: In fact, $T : L^p \to L^p$, for $1 < p < \infty$. We shall discuss the details of this assertion a bit later.

Proof: We start with an auxiliary function $\phi(z) = \phi_0(|z|_h)$, where $\phi_0 \in C^\infty(\mathbb{R}^1_+)$ and

$$\phi_0(t) = \begin{cases} 1 & \text{if } 0 \leq t \leq 1, \\ 0 & \text{if } 2 \leq t < \infty. \end{cases}$$

Let

$$\psi_j(g) = \phi(2^{-j}g) - \phi(2^{-j+1}g).$$

We stress here that the dilations are taking place in the Heisenberg group structure (the action of the Iwasawa subgroup A).

Note that

$$\phi(2^{-j}g) = \begin{cases} 1 & \text{if } |g|_h \le 2^j, \\ 0 & \text{if } |g|_h \ge 2^{j+1}, \end{cases}$$

and

$$\phi(2^{-j+1}g) = \begin{cases} 1 & \text{if } |g|_h \le 2^{j-1}, \\ 0 & \text{if } |g|_h \ge 2^j. \end{cases}$$

Therefore

$$\psi_j(g) = \phi(2^{-j}g) - \phi(2^{-j+1}g) = 0, \quad \text{if } |g|_h < 2^{j-1}, \text{ or } |g|_h > 2^{j+1},$$

i.e.,

$$\operatorname{supp} \psi_j(g) \subset \{2^{j-1} \le |g|_h \le 2^{j+1}\}. \tag{9.9.4.1}$$

Thus, for arbitrary g, there exist at most two ψ_j's such that $g \in \operatorname{supp} \psi_j$.
Observe that

$$\sum_{-N}^{N} \psi_j(g) = [\phi(2^N g) - \phi(2^{N+1}g)] + [\phi(2^{N-1}g) - \phi(2^N g)] + \cdots$$

$$+ [\phi(2^{-N+1}g) - \phi(2^{-N+2}g)] + [\phi(2^{-N}g) - \phi(2^{-N+1}g)]$$

$$= -\phi(2^{N+1}g) + \phi(2^{-N}g) = 1 \quad \text{if } 2^{-N} \le |g|_h \le 2^N. \tag{9.9.4.2}$$

Therefore

$$\sum_{-\infty}^{\infty} \psi_j(g) \equiv 1.$$

We let

$$K_j(g) = \psi_j(g)K(g)$$

and

$$T_j f = f * K_j.$$

Then

$$Tf = f * \operatorname{PV} K = f * \sum_{-\infty}^{\infty} K_j = \sum_{-\infty}^{\infty} T_j f.$$

Claim (i):

$$\|T_j\| \le C \quad \text{(independent of } j\text{)}, \tag{9.9.4.3}$$

$$\|T_j T_\ell^*\| \le C 2^{-|j-\ell|}, \tag{9.9.4.4}$$

$$\|T_\ell^* T_j\| \le C 2^{-|j-\ell|}. \tag{9.9.4.5}$$

Suppose the claim is proved. If we let $a_j = \sqrt{2^{-|j|}}$, then the hypothesis of Cotlar–Knapp–Stein is satisfied. We may conclude then that finite sums of the T_j have norm that is bounded by C. An additional argument (using Functional Analysis Principle I from Appendix 1) will be provided below to show that the same estimate holds for infinite sums.

Fact: If $Tg = g * M$, then $\|Tg\|_2 \leq \|M\|_1 \|g\|_2$ by the generalized Minkowski inequality (see [STE2]).

Therefore, to prove (9.9.4.3), we need to show that $\|K_j\|_1 \leq C$. Since K is homogeneous of degree $-2n - 2$, we have

$$K(2^{-j}g) = (2^{-j})^{-2n-2} K(g) = 2^{j(2n+2)} K(g).$$

Hence we get

$$K_j(g) = \psi_j(g) K(g)$$
$$= 2^{-j(2n+2)} K(2^{-j}g) \psi_j(g)$$
$$= 2^{-j(2n+2)} K(2^{-j}g) \left[\phi(2^{-j}g) - \phi(2^{-j+1}g) \right].$$

Thus

$$\|K_j\|_1 = \int_{\mathbb{H}^n} |K_j(g)| \, dg$$
$$= \int_{\mathbb{H}^n} 2^{-j(2n+2)} |K(2^{-j}g)| \left| \phi(2^{-j}g) - \phi(2^{-j+1}g) \right| dg$$
$$= \int_{\mathbb{H}^n} 2^{j(2n+2)} 2^{-j(2n+2)} |K(g)| |\phi(g) - \phi(2g)| \, dg$$
$$= \int_{\mathbb{H}^n} |K(g)| |\phi(g) - \phi(2g)| \, dg$$
$$= \int_{\mathbb{H}^n} |K_0(g)| \, dg$$
$$= C. \tag{9.9.4.6}$$

Hence (9.9.4.3) is proved.

Before proving (9.9.4.4), let us note the following:

Remark: If $S_1 f = f * L_1$ and $S_2 f = f * L_2$, then

$$S_1 S_2 f = S_1 (f * L_2) = (f * L_2) * L_1 = f * (L_2 * L_1).$$

Also, if $Tf = f * L$, then
$$T^* h = L^* * h,$$

where $L^*(g) = \overline{L}(g^{-1})$—here g^{-1} is the inverse of g in $\mathbb{H}^n$. We see this by calculating

$$\langle T^*h, f \rangle = \langle h, Tf \rangle$$

$$= \int_{\mathbb{H}^n} h(y) \overline{\left(\int_{\mathbb{H}^n} f(g) L(g^{-1}y) dg \right)} dy$$

$$= \int_{\mathbb{H}^n} \overline{f(g)} \int_{\mathbb{H}^n} h(y) \overline{L(g^{-1}y)} dy \, dg$$

$$= \langle L^* * h, f \rangle.$$

Let us assume that $j \geq l$ and prove (9.9.4.4), i.e.,

$$\| T_j T_\ell^* \| \leq C 2^{\ell - j}.$$

From the preceding remark, we know that $T_j T_\ell^* f = f * (K_\ell^* * K_j)$. Therefore, by the generalized Minkowski inequality, it is enough to show that

$$\| K_\ell^* * K_j \|_1 \leq C 2^{\ell - j}.$$

We can write $K_\ell^* * K_j$ as follows:

$$\int_{\mathbb{H}^n} K_j(y) K_\ell^*(y^{-1}x) dy$$

$$= \int_{\mathbb{H}^n} K_j(xy^{-1}) K_\ell^*(y)(-1)^{2n+2} dy$$

$$= \int_{\mathbb{H}^n} K_j(xy^{-1}) K_\ell^*(y) dy. \tag{9.9.4.7}$$

Claim (ii):

$$\int_{\mathbb{H}^n} K_j(x) dx = \int_{\mathbb{H}^n} K_\ell^*(x) dx = 0. \tag{9.9.4.8}$$

To see this, we calculate that

$$\int_{\mathbb{H}^n} K_j(x) dx = \int_{\mathbb{H}^n} K(x) [\phi(2^{-j}x) - \phi(2^{-j+1}x)] dx$$

$$= \int_{\Sigma} \int_0^\infty K(r\xi)(\phi_0(2^{-j}r) - \phi_0(2^{-j+1}r)) r^{2n+1} dr \, d\sigma(\xi)$$

$$= \int_0^\infty r^{-2n-2} r^{2n+1} (\phi_0(2^{-j}r) - \phi_0(2^{-j+1}r)) \int_{\Sigma} K(\xi) d\sigma(\xi) dr$$

$$= 0.$$

As a result,

$$\int_{\mathbb{H}^n} K_l^*(x) dx = \int_{\mathbb{H}^n} \overline{K_\ell(x^{-1})} dx = \int_{\mathbb{H}^n} \overline{K_\ell(x)}(-1)^{2n+2} dx = \overline{\int K_\ell(x) dx} = 0.$$

Thus, from (9.9.4.7) and (9.9.4.8), we can rewrite $K_\ell^* * K_j$ as follows:

$$K_\ell^* * K_j = \int_{\mathbb{H}^n} K_j(xy^{-1})K_\ell^*(y)dy = \int_{\mathbb{H}^n} [K_j(xy^{-1}) - K_j(x)]K_\ell^*(y)dy.$$

Claim (iii):

$$\int_{\mathbb{H}^n} \left| K_j(xy^{-1}) - K_j(x) \right| dx \leq C2^{-j}|y|_h. \tag{9.9.4.9}$$

To see this, recall that

$$K_j(x) = K(x)\psi_j(x) = K(x)[\phi(2^{-j}x) - \phi(2^{-j+1}x)]$$

and

$$K_0(x) = K(x)[\phi(x) - \phi(2x)].$$

Therefore

$$K_j(2^j x) = K(2^j x)[\phi(x) - \phi(2x)] = 2^{j(-2n-2)}K_0(x).$$

Hence we get

$$\int |K_0(xy^{-1}) - K_0(x)|\,dx \leq C|y|_h$$

$$\Longleftrightarrow \int |K_j(2^j(x \cdot y^{-1})) - K_j(2^j x)|\,dx \leq C2^{-j(2n+2)}|y|_h$$

$$\Longleftrightarrow \int 2^{-j(2n+2)}|K_j(x \cdot (2^j y)^{-1}) - K_j(x)|\,dx \leq C2^{-j(2n+2)}|y|_h$$

$$\Longleftrightarrow \int |K_j(x \cdot y^{-1}) - K_j(x)|\,dx \leq C\left|\frac{y}{2^j}\right|_h = C2^{-j}|y|_h.$$

Therefore, to prove (9.9.4.9), we only need to show that

$$\int \left| K_0(xy^{-1}) - K_0(x) \right| dx \leq C|y|_h.$$

First, suppose that $|y|_h \geq 1$. Then

$$\int |K_0(xy^{-1}) - K_0(x)|\,dx \leq \int |K_0(xy^{-1})|\,dx + \int |K_0(x)|\,dx$$

$$= 2\int |K_0(x)|\,dx$$

$$= 2\int_\Sigma \int_0^\infty |K_0(r\xi)|r^{2n+1}\,dr\,d\sigma(\xi)$$

$$= 2\int_\Sigma \int_0^\infty |K(r\xi)||\phi_0(r) - \phi_0(2r)|r^{2n+1}\,dr\,d\sigma(\xi)$$

$$= 2 \int_\Sigma |K(\xi)| \int_0^\infty |\phi_0(r) - \phi_0(2r)| r^{-2n-2} r^{2n+1} \, dr \, d\sigma(\xi)$$

$$\leq C \int_\Sigma |K(\xi)| \int_{\frac{1}{2}}^2 \frac{1}{r} \, dr \, d\sigma(\xi)$$

$$\leq C. \tag{9.9.4.10}$$

Thus we have

$$\int |K_0(xy^{-1}) - K_0(x)| \, dx \leq C|y|_h, \quad \text{if } |y|_h \geq 1.$$

Now suppose that $|y|_h < 1$. We may consider K_0 as a function on $\mathbb{R}^{2n+1}$. We use the notation $\widetilde{K}_0$ to denote such a function. Since y^{-1}, the inverse of y in $\mathbb{H}^n$, corresponds to $-y$ in $\mathbb{R}^{2n+1}$, we get

$$|K_0(x \cdot y^{-1}) - K_0(x)| = |\widetilde{K}_0(x - y) - \widetilde{K}_0(x)|. \tag{9.9.4.11}$$

Thus, using the mean value theorem, we have

$$|\widetilde{K}_0(x - y) - \widetilde{K}_0(x)| \leq C|y|_e.$$

Therefore, from Proposition 9.8.5, we have

$$|K_0(xy^{-1}) - K_0(x)| \leq C|y|_h.$$

Hence

$$\int_{\mathbb{H}^n} |K_0(xy^{-1}) - K_0(x)| \, dx \leq C|y|_h \int_{\mathbb{H}^n} \chi_{\text{supp}|K_0(xy^{-1}) - K_0(x)|} \, dx$$

$$\leq C|y|_h \left(\int_{\mathbb{H}^n} \chi_{\text{supp}K_0(xy^{-1})} dx + \int_{\mathbb{H}^n} \chi_{\text{supp}K_0(x)} dx \right). \tag{9.9.4.12}$$

We certainly have

$$\text{supp } K_0(xy^{-1}) \subset \left\{ \frac{1}{2} \leq |xy^{-1}| \leq 2 \right\} \quad \text{and} \quad \text{supp } K_0(x) \subset \left\{ \frac{1}{2} \leq |x|_h \leq 2 \right\}.$$

Since $|y|_h < 1$, if $x \in \text{supp } K_0(xy^{-1})$, we have

$$|x|_h \leq \gamma (|xy^{-1}|_h + |y|_h) \leq 3\gamma.$$

Therefore

$$\text{supp } K_0(xy^{-1}) \subset \{|x|_h \leq 3\gamma \}.$$

Hence we can rewrite (9.9.4.12) as follows:

$$\int_{\mathbb{H}^n} |K_0(xy^{-1}) - K_0(x)| dx \leq C|y|_h (3\gamma)^{2n+2} = C|y|_h.$$

Thus our claim is proved and we have established (9.9.4.4).

The proof of (9.9.4.5) is similar.

Now we invoke the Cotlar–Knapp–Stein lemma and get

$$\left\| \sum_{\ell=1}^{M} T_{j_\ell} \right\| \leq C, \quad \forall M \in \mathbb{N}.$$

We actually wish to consider

$$T_\epsilon^N f = \int_{\epsilon \leq |y|_h \leq N} f(zy^{-1}) K(y) dy$$

and let $\epsilon \to 0$, $N \to \infty$.

Claim (iv):

$$\| T_\epsilon^N f \|_2 \leq C \| f \|_2,$$

where C is independent of ϵ and N.

For the proof of claim (iv), let

$$K_\epsilon^N(y) = K(y) \chi_{[\epsilon, N]}(|y|_h).$$

Then

$$T_\epsilon^N f = \int_{\mathbb{H}^n} f(xy^{-1}) K_\epsilon^N(y) dy.$$

Therefore, to prove claim (iv), we will show that $\| K_\epsilon^N \|_1 \leq C$, where C is independent of ϵ and N.

We may find $j, \ell \in \mathbb{Z}$ such that

$$2^{j-1} \leq \epsilon < 2^j \quad \text{and} \quad 2^\ell \leq N < 2^{\ell+1}$$

and want to compare $\sum_{j \leq i \leq \ell} T_i$ and T_ϵ^N.

So, we look at

$$\left(\sum_{j \leq i \leq \ell} K_i \right) - K_\epsilon^N.$$

Note that

$$\sum_{j \leq i \leq \ell} K_i(x) = K(x)[\psi_j(x) + \cdots + \psi_\ell(x)]$$

$$= K(x)[\phi(2^{-j}x) - \phi(2^{-j+1}x) + \phi(2^{-j-1}x) - \phi(2^{-j}x) \pm \cdots$$

$$+ \phi(2^{-\ell+1}x) - \phi(2^{-\ell+2}x) + \phi(2^{-\ell}x) - \phi(2^{-\ell+1}x)]$$

$$= K(x)[\phi(2^{-j}x) - \phi(2^{-j+1}x)]. \tag{9.9.4.13}$$

Thus

$$\text{supp} \sum_{j \le i \le \ell} K_i \subset \{2^{j-1} \le |x_h| \le 2^{\ell+1}\}$$

and

$$\sum_{j \le i \le \ell} K_i(x) = K(x), \quad \text{if} 2^j \le |x|_h \le 2^\ell.$$

Hence

$$\text{supp} \left(\left[\sum_{j \le i \le \ell} K_i \right] - K_\epsilon^N \right) \subset [\{2^{j-1} \le |x|_h \le 2^j\} \cup \{2^\ell \le |x|_h \le 2^{\ell+1}\}]$$

$$\subset \left[\left\{ \frac{\epsilon}{2} \le |x|_h \le 2\epsilon \right\} \cup \left\{ \frac{N}{2} \le |x|_h \le 2N \right\} \right].$$

Therefore

$$\left\| \sum_{j \le i \le \ell} K_i - K_\epsilon^N \right\|_1 \lesssim \int_{\frac{\epsilon}{2} \le |x|_h \le 2\epsilon} |K(x)|\, dx + \int_{\frac{N}{2} \le |x|_h \le 2N} |K(x)|\, dx$$

$$= \int_\Sigma \int_{\frac{\epsilon}{2}}^{2\epsilon} |K(r\xi)| r^{2n+1}\, dr d\sigma(\xi)$$

$$+ \int_\Sigma \int_{\frac{N}{2}}^{2N} |K(r\xi)| r^{2n+1}\, dr d\sigma(\xi)$$

$$= \int_{\frac{\epsilon}{2}}^{2\epsilon} r^{2n+1} r^{-2n-2} \int_\Sigma |K(\xi)|\, d\sigma(\xi)\, dr$$

$$+ \int_{\frac{N}{2}}^{2N} r^{2n+1} r^{-2n-2} \int_\Sigma |K(\xi)|\, d\sigma(\xi) dr$$

$$= C(\log 4 + \log 4)$$

$$= C.$$

Therefore

$$\|K_\epsilon^N\|_1 \le \left\| \sum_{j \le i \le \ell} K_i \right\| + \left\| \sum_{j \le i \le \ell} K_i - K_\epsilon^N \right\| \le C.$$

Hence, applying Functional Analysis Principle I, the theorem is proved.

9.10 The Calderón–Zygmund Integral on $\mathbb{H}^n$ Is Bounded on L^p

In the last section we established L^2 boundedness of the Calderón–Zygmund operators on the Heisenberg group. Given the logical development that we have seen thus
far in the subject, the natural next step for us would be to prove a weak-type $(1, 1)$
estimate for these operators. Then one could apply the Marcinkiewicz interpolation
theorem to get strong L^p estimates for $1 < p < 2$. Finally, a simple duality argument would yield strong L^p estimates for $2 < p < \infty$. And that would complete the
picture for singular integrals on the Heisenberg group.

The fact is that a general paradigm for proving weak-type $(1, 1)$ estimates on a
space of homogeneous type has been worked out in [COIW1]. It follows roughly the
same lines as the arguments we gave in Section 9.7.3. There are a number of interesting new twists and turns in this treatment—for instance the geometry connected
with the Whitney decomposition is rather challenging—and we encourage readers to
consult this original source as interest dictates. But it would be somewhat repetitious
for us to present all the details here, and we shall not do so.

In the remainder of this book, we shall take it for granted that L^p boundedness
for Calderón–Zygmund operators has been established, $1 < p < \infty$, and we shall
use the result to good effect.

9.11 Remarks on H^1 and BMO

Capsule: As indicated in previous sections, H^1_{Re} and BMO are critical
spaces for the harmonic analysis of any particular context. These ideas were
originally discovered on classical Euclidean space. But today they are studied on arbitrary manifolds and fairly arbitrary Lie groups and other contexts.
Each function space has considerable intrinsic interest, but it is their interaction that is of greatest interest. The predual of H^1_{Re} is also known; it is the
space VMO of functions of *vanishing mean oscillation* (see [SAR]). There
is a great lore of Hardy spaces and functions of bounded mean oscillation.
We only scratch the surface here.

Refer to the discussion of atomic Hardy spaces in Section 8.10. That definition
transfers *grosso modo* to the Heisenberg group. Once one has balls and a measure—
with certain elementary compatibility conditions—then one can define the atomic
Hardy space H^1 for instance. In fact, all one needs is the structure of a space of
homogeneous type, and the Heisenberg group certainly possesses that feature.[8]

And it turns out that one can prove that a singular integral—such as we have
defined in Theorem 9.9.4—is bounded on H^1 and bounded on BMO. The proofs of

---

[8] It is a bit trickier to define H^p for p small on a space of homogeneous type. For the classical
atomic definition entails higher-order vanishing moment conditions, hence raises questions
of (i) smoothness of functions and (ii) the definition of "polynomials." Neither of these
ideas is a priori clear on a general space of homogeneous type. These delicate questions are
explored in the book [KRA10]. We cannot treat the details here.

these statements are straightforward, and are just as in the Euclidean case. We shall not provide the details. The book [STE2] is a good reference for the chapter and verse of these ideas.

If we push further and examine $H^p_{\text{Re}}(\mathbb{H}^n)$ for $p < 1$ but near to 1 (so that the moment condition on an atom is still $\int_B a(z)dx = 0$) then one can show that the dual of H^p is a certain nonisotropic Lipschitz space (see [FOST2] for a careful consideration of this point). This result was anticipated in the important paper [DRS]. Nonisotropic Lipschitz spaces on nilpotent Lie groups are developed, for example, in [KRA11].

In sum, virtually all the machinery developed in this book may be brought to bear on the Heisenberg group. As a result, the Heisenberg group has developed into a powerful tool of modern harmonic analysis. Because the Heisenberg group is naturally identified with the boundary of the unit ball in $\mathbb{C}^n$, it fits very naturally into complex function theory and explorations of the Cauchy–Riemann equations. There still remain many important avenues to explore in this important new byway of our field.

Analysis on the Heisenberg Group

Prologue: And now we reach the pinnacle of our work. We have spent quite a number of chapters laying the background and motivation for what we are doing. The last chapter set the stage for Heisenberg group analysis, and laid out the foundations for the subject. Now we are going to dive in and prove a big theorem.

Our goal in this chapter is to prove the L^p boundedness of the Szegő projection on the unit ball in $\mathbb{C}^n$ (we also make some remarks about the Poisson–Szegő integral). It must be emphasized here that this is a singular integral, but not in the traditional, classical sense. The singularity is *not* isotropic, and the mapping properties of the operator cannot simply be analyzed using traditional techniques. The anisotropic Heisenberg analysis that we have developed here is what is needed.

So our program has two steps. The first is to consider the Szegő kernel on the ball in some detail. We must calculate it explicitly, and render it in a form that is useful to us. In particular, we must see that it is a convolution operator on the Heisenberg group.

Second, we must see that the Szegő projection is in fact a Heisenberg convolution operator. Indeed, we want to see that it is a singular integral operator in the sense of this book. Then all of our analysis will come to bear and we can derive a significant theorem about mapping properties of the Szegő projection.

Along the way we shall also learn about the Poisson–Szegő operator. It is an essential feature of the analysis of the Heisenberg group. We shall be able to say something about its mapping properties as well.

We begin this chapter by reviewing a few of the key ideas about the Szegő and Poisson–Szegő kernels.

S.G. Krantz, *Explorations in Harmonic Analysis*, Applied and Numerical
Harmonic Analysis, DOI 10.1007/978-0-8176-4669-1_10,
© Birkhäuser Boston, a part of Springer Science+Business Media, LLC 2009

10.1 The Szegő Kernel on the Heisenberg Group

Capsule: We have met the Szegő kernel in earlier parts of this book (see Section 7.2). It is the canonical reproducing kernel for H^2 of any bounded domain. It fits the paradigm of Hilbert space with reproducing kernel (see [ARO]). And it is an important operator for complex function theory. But it does not have the invariance properties of the Bergman kernel. It is not quite as useful in geometric contexts. Nonetheless, the Szegő kernel is an important artifact of complex function theory and certainly worthy of our studies.

Let $\mathbb{B}$ be a unit ball in $\mathbb{C}^n$. Consider the Hardy space $H^2(\mathbb{B})$:

$$H^2(\mathbb{B}) = \left\{ f \text{ holomorphic on } \mathbb{B} : \right.$$

$$\left. \sup_{0<r<1} \left[\int_{\partial \mathbb{B}} |f(r\xi)|^2 d\sigma(\xi) \right]^{1/2} = \|f\|_{H^2(\mathbb{B})} < \infty \right\}$$

Note that $H^2(\mathbb{B})$ is a Hilbert space. The Szegő kernel is a canonical reproducing kernel for H^2. We have studied its properties earlier in the book—see Chapter 7. Now we turn to the particular properties of this kernel, and the allied Poisson–Szegő kernel, on the Siegel upper half-space and the Heisenberg group.

10.2 The Poisson–Szegő Kernel on the Heisenberg Group

Capsule: The Poisson–Szegő kernel, invented by Hua and Koranyi, is interesting because it is a positive kernel that reproduces H^2. It is quite useful in function-theoretic contexts. But it is not nearly so well-known as perhaps it should be. Like the Bergman and Szegő kernels, it is quite difficult in practice to compute. But various asymptotic expansions make it accessible in many contexts.

Let $\Omega \in \mathbb{C}^n$ be a bounded domain. We would like to construct a positive reproducing kernel. Let $S(z, \zeta)$ be the Szegő kernel on Ω.

Definition 10.2.1 We define the Poisson–Szegő kernel as follows:

$$P(z, \zeta) = \frac{|S(z, \zeta)|^2}{S(z, z)}.$$

Prelude: The Poisson–Szegő kernel is a fairly modern idea that grew perhaps out of representation-theoretic considerations (see [HUA]). It arises in a variety of contexts but is particularly useful in the harmonic analysis of several complex variables. Its chief virtue is that it is a positive kernel that reproduces H^2, and that has a "structure" (i.e., the shape of its singularity) that is closely tied to the complex structure of the domain in question. It happens that the Poisson–Szegő kernel also solves the Dirichlet problem for the invariant Laplacian on the ball in $\mathbb{C}^n$, but that seems to be a special effect true only for that domain and a very few other special domains. This circle of ideas has been explored in [GRA1], [GRA2]. See also [KRA3].

Lemma 10.2.2 *Let Ω be a bounded domain in $\mathbb{C}^n$. The Poisson–Szegő kernel on Ω is well defined and nonnegative.*

Proof: Note that

$$S(z, z) = \sum_{j=1}^{\infty} \phi_j(z)\overline{\phi_j(z)} = \sum_{j=1}^{\infty} |\phi_j(z)|^2 \geq 0.$$

Suppose that there exists a $z_0 \in \Omega$ such that $S(z_0, z_0) = 0$. Then $\phi_j(z_0) = 0$, for all $j \geq 1$. Thus

$$f(z_0) = \sum_{j=1}^{\infty} \alpha_j \phi_j(z_0) = 0, \quad \forall f \in H^2(\Omega).$$

But $f \equiv 1 \in H^2(\Omega)$. Contradiction. □

10.3 Various Kernels on the Siegel Upper Half-Space $\mathcal{U}$

Capsule: Now it is time to focus in on the Siegel half-space and the canonical kernels on that space. Both the Szegő and the Poisson–Szegő kernels will play a prominent role. Their mapping properties are of particular interest.

The Siegel upper half-space $\mathcal{U}$ is biholomorphically equivalent to the unit ball via the generalized Cayley map. Thus $\mathcal{U}$ has a Bergman and a Szegő kernel.

10.3.1 Sets of Determinacy

A set $S \subseteq U \subseteq \mathbb{C}^n$ is called a *set of determinacy* if any holomorphic function on U that vanishes on S must be identically zero on U.

Example 10.3.1 The set $S = \{(s+it, 0) : s, t \in \mathbb{R}\} \subset \mathbb{C}^2$ is *not* a set of determinacy on $\mathbb{C}^2$, because $f(z_1, z_2) = z_2^2$ is holomorphic and vanishes on S.

Example 10.3.2 The set $S = \{(s + i0, t + i0) : s, t \in \mathbb{R}\} \subset \mathbb{C}^2$ *is* a set of determinacy. This assertion follows from elementary one-variable power series considerations.

Remark: If $S \subset \mathbb{C}^n$ is a totally real[1] n-dimensional manifold, then S is a set of determinacy. See [WEL].

---

[1] A manifold $M \subseteq \mathbb{C}^n$ is said to be *totally real* if whenever α lies in the tangent space to M then $i\alpha$ does not lie in the tangent space.

Prelude: There is an important idea lurking in the background in the next lemma. Let S be a 2-dimensional linear subspace of $\mathbb{C}^2$. Let f be a holomorphic function. If $f|_S = 0$, does it follow that $f \equiv 0$? The answer depends on how S is positioned in space; put in other words, it depends on the complex structure of S.

Lemma 10.3.3 *Let $\Omega \subseteq \mathbb{C}^n$ be a domain and let $f(z, w)$ be defined on $\Omega \times \Omega$. Assume that f is holomorphic in z and conjugate holomorphic in w. If $f(z, z) = 0$ for all z, then $f(z, w) \equiv 0$ on $\Omega \times \Omega$ (i.e., the diagonal is a set of determinacy).*

Proof: Consider the mapping

$$\phi : (z, w) \to \left(\frac{z + \overline{w}}{2}, \frac{z - \overline{w}}{2i} \right).$$

Then

$$\phi^{-1} : (\alpha, \beta) \to (\alpha + i\beta, \overline{\alpha - i\beta}).$$

Let $\widetilde{f} = f \circ \phi^{-1}(\alpha, \beta)$. We can easily check that it is holomorphic in α and β:

$$\frac{\partial \widetilde{f}}{\partial \overline{\alpha}} = \frac{\partial f}{\partial z} \frac{\partial z}{\partial \overline{\alpha}} + \frac{\partial f}{\partial \overline{w}} \frac{\partial \overline{w}}{\partial \overline{\alpha}} = 0,$$

$$\frac{\partial \widetilde{f}}{\partial \overline{\beta}} = \frac{\partial f}{\partial z} \frac{\partial z}{\partial \overline{\beta}} + \frac{\partial f}{\partial \overline{w}} \frac{\partial \overline{w}}{\partial \overline{\beta}} = 0.$$

Since $f(z, z) = 0$ and

$$\phi : (z, z) \to (\operatorname{Re} z, \operatorname{Im} z),$$

we know that

$$\widetilde{f}(\operatorname{Re} z, \operatorname{Im} z) = f \circ \phi^{-1}(\operatorname{Re} z, \operatorname{Im} z) = f(z, z) = 0.$$

But $\{(\operatorname{Re} z, \operatorname{Im} z)\} \subset \mathbb{C}^{2n}$ is a totally real $2n$-dimensional manifold, thus a set of determinacy. Therefore, $f \circ \phi^{-1} \equiv 0$. Hence $f \equiv 0$. $\qquad\square$

10.3.2 The Szegő Kernel on the Siegel Upper Half-Space $\mathcal{U}$

Recall the height function (Section 9.4) ρ in $\mathcal{U}$:

$$\rho(w) = \operatorname{Im} w_1 - |w'|^2,$$

where $w' = (w_2, \ldots, w_n)$. We look at the almost analytic extension of ρ:

$$\rho(z, w) = \frac{i}{2}(\overline{w}_1 - z_1) - \sum_{k=2}^{n} z_k \overline{w}_k.$$

Note that ρ is holomorphic in z and conjugate holomorphic in w and $\rho(w, w) = \rho(w)$.

Prelude: The next theorem is key to the principal result of this chapter. The Szegő kernel is a Heisenberg singular integral, hence can be analyzed using the machinery that we have developed. Of course analogous results are true in one dimension as well.

Theorem 10.3.4 *On the Siegel upper half-space $\mathcal{U}$, the Szegő kernel $S(z, \zeta)$ is*

$$S(z, \zeta) = \frac{n!}{4\pi^n} \cdot \frac{1}{\rho(z, \zeta)^n}.$$

For $F \in H^2(\mathcal{U})$, we let

$$F_\rho(z', t) = F\big((z', t + i(|z'|^2 + \rho)\big).$$

Lurking in the background here is the map Ψ (originally defined in Section 9.4) that takes elements of $\partial\mathcal{U}$ to elements of the Heisenberg group. In the current discussion we will find it convenient *not* to mention Ψ explicitly. The role of Ψ will be understood from context.

We know that for $z \in \mathcal{U}$,

$$F(z) = \int_{\partial\mathcal{U}} F_0(w) S(z, w) d\sigma(w).$$

This is just the standard reproducing property of the Szegő kernel acting on H^2 of the Siegel upper half-space.

Corollary 10.3.5

$$F_\rho(z', t) = F_0 * K_\rho(z', t),$$

where $F_0 \in L^2(\partial\mathcal{U})$ is the L^2 boundary limit of F and

$$K_\rho(z', t) = 2^{n+1} C_n \cdot (|z'|^2 + it + \rho)^{-n-1}, \quad C_n = \frac{n!}{4\pi^{n+1}}.$$

Proof of Theorem 10.3.4: Let $z = (t + i(|z|^2 + \rho(z)), \zeta)$ and $\rho = \rho(z)$. Then

$$F(z) = F_\rho(z', t) = \int_{\mathbb{H}^n} S(z, w) F_0(w) d\sigma(w)$$

$$= \int_{\mathbb{H}^n} C_n \cdot \frac{1}{(\rho(z, w))^{n+1}} F_0(w) d\sigma(w).$$

Therefore, we need only show that

$$\rho(z, w)^{-n-1} = \frac{1}{2} K_\rho\big((z', t)^{-1} \cdot (w', u_1)\big), \quad u_1 = \operatorname{Re} w_1.$$

Now

$$\rho(z, w)^{-n-1} = \left[\frac{i}{2}(\overline{w}_1 - z_1) - z' \cdot \overline{w}' \right]^{-n-1}$$

$$= \left[\frac{i}{2}[u_1 - i|w'|^2 - t - i(|z'|^2 + \rho)] - \operatorname{Re} z' \cdot \overline{w}' - i\operatorname{Im} z' \cdot \overline{w}' \right]^{-n-1}$$

$$= \left[\frac{1}{2}[|z'|^2 + |w'|^2 - 2\mathrm{Re}\, z' \cdot \overline{w'} + \rho] + \frac{i}{2}[u_1 - t - 2\mathrm{Im}\, z' \cdot \overline{w'}]\right]^{-n-1}$$

$$= \left[\frac{1}{2}[|-z' + w'|^2 + \rho + i(-t + u_1 + 2\mathrm{Im}\,(-z' \cdot \overline{w'}))]\right]^{-n-1}$$

$$= \frac{1}{2^{n+1}} K_\rho\big((-z' + w', -t + u_1 + 2\mathrm{Im}\,(-z' \cdot \overline{w'}))\big)$$

$$= \frac{1}{2^{n+1}} K_\rho\big((z', t)^{-1} \cdot (w', u_1)\big). \qquad (10.3.4.1)$$

$\square$

Let us begin with the classical upper half-plane in $\mathbb{C}^1$, $U = \{x + iy : y > 0\}$, and its associated Hardy space

$$H^2(U) = \left\{ f : f \text{ is holomorphic on } \mathbb{R}_+^2, \sup_{y>0} \int |f(x + iy)|^2 dx < \infty \right\}.$$

This function space is a Hilbert space with norm given by

$$\|f\|_{H^2(U)} = \sup_{y>0} \left(\int_{\mathbb{R}^1} |f(x + iy)|^2 dx \right)^{\frac{1}{2}}.$$

The classical structure theorem for this Hardy space is given by the following result.

Prelude: The Paley–Wiener theorem is one of the big ideas of modern harmonic analysis. It dates back to the important book [PAW]. Today Paley–Wiener theory shows itself in signal processing, in wavelet theory, in partial differential equations, and many other parts of mathematics.

Theorem 10.3.6 (Paley–Wiener theorem) *The equation*

$$f \mapsto F(z) \equiv \int_0^\infty e^{2\pi i z \cdot \lambda} f(\lambda) d\lambda$$

yields an isomorphism between $\mathcal{L}^2(0, \infty)$ *and* $H^2(U)$.

This theorem is of particular value because the elements of $L^2(0, \infty)$ are easy to understand, whereas the elements of $H^2(\mathbb{R}_+^2)$ are less so—one cannot construct H^2 functions at will. Observe that one direction of the proof of this theorem is easy: given a function $f \in L^2(0, \infty)$, the integral above converges (absolutely) as soon as $\mathrm{Im}\, z > 0$. Furthermore, for any $y > 0$, we set $F_y(x) = F(x + iy)$ and see that

$$\|F_y\|_{L^2(\mathbb{R}^1)}^2 = \int_{\mathbb{R}^1} \left| \int_0^\infty e^{-2\pi y \lambda} \cdot e^{2\pi i x \lambda} f(\lambda) d\lambda \right|^2 dx$$

$$\leq \int_{\mathbb{R}^1} \left| \int_0^\infty e^{2\pi i x \lambda} f(\lambda) d\lambda \right|^2 dx$$

$$= \|\widehat{f}(x)\|^2_{L^2(\mathbb{R}^1)}$$

$$= \|f\|^2_{L^2(0,\infty)}.$$

It is also clear that

$$\|F\|_{H^2(\mathbb{R}^2_+)} = \sup_{y>0} \|F_y\|_{L^2(\mathbb{R}^1)} = \|f\|_{L^2(0,\infty)}.$$

The more difficult direction is the assertion that the map $f(\lambda) \mapsto F(z)$ is actually *onto* H^2. We shall not treat it in detail (although our proof below specializes down to the half-plane in $\mathbb{C}$), but refer the reader instead to [KAT].

We would like to develop an analogue for the Paley–Wiener theorem on the Siegel upper half-space $\mathcal{U}$. First we must discuss integration on $\mathbb{H}^n$. Recall that a measure $d\lambda$ on a topological group is the Haar measure (unique up to multiplication by a constant) if it is a Borel measure that is invariant under left translation. Our measure $d\zeta\, dt$ (ordinary Lebesgue measure) turns out to be both left and right invariant, i.e., it is *unimodular*. The proof is simply a matter of carrying out the integration:

$$\iint f([\xi, s] \cdot [\zeta, t]) d\zeta\, dt = \iint f(\zeta + \xi, t + s + 2\mathrm{Im}\,\xi \cdot \overline{\zeta}) d\zeta\, dt$$

$$= \iint f(\zeta + \xi, t + s + 2\mathrm{Im}\,\xi \cdot \overline{\zeta}) dt\, d\zeta$$

$$= \iint f(\zeta + \xi, t) dt\, d\zeta$$

$$= \iint f(\zeta, t) dt\, d\zeta.$$

Observe now that the map $[\zeta, t] \mapsto [-\zeta, -t]$ preserves the measure but also sends an element of $\mathbb{H}^n$ onto its inverse. Thus it sends left translation into right translation, and so the left invariance of the measure implies its right invariance.

With that preliminary step out of the way, we can make the following definition:

Definition 10.3.7 For f holomorphic on the Siegel upper half-space $\mathcal{U}$, we define

$$\|f\|_{H^2} = \sup_{\rho>0} \left(\iint |f(z', z_1 + i|z'|^2 + i\rho)|^2 \, d|z'| \, d|z_1| \right)^{\frac{1}{2}}.$$

Then we set

$$H^2(\mathcal{U}) = \{f :\ f \text{ is holomorphic on } \mathcal{U}, \|f\|_{H^2} < \infty\}.$$

Here ρ is the height function that we have introduced for $\mathcal{U}$. Just as in the case of $U \subset \mathbb{C}^1$, where we integrated over parallels to $\mathbb{R}^1 = \partial\mathbb{R}^2_+$, so here we integrate over parallels to $\mathbb{H}^n = \partial\mathcal{U}$.

We will now see that $H^2(\mathcal{U})$ is a Hilbert space, and we will develop a Fourier analysis for it. The substitute for $L^2(0, \infty)$ in the present context will be $\widehat{H^2}$, which consists of all functions $f = f(z', \lambda)$ with $z' \in \mathbb{C}^n$, $\lambda \in \mathbb{R}^+$ such that

(1) $\widehat{f}$ is jointly measurable in z' and λ.
(2) For almost every λ, the function $z' \mapsto \widehat{f}(z', \lambda)$ is entire on $\mathbb{C}^n$.
(3) $\|\widehat{f}\|_{\widehat{H^2}}^2 = \int_{\mathbb{C}^n} \int_{\mathbb{R}+} |\widehat{f}(z', \lambda)|^2 e^{-4\pi\lambda|z'|^2} \, d\lambda \, d|z'| < \infty\}$.

We have the following basic structure theorem:

Theorem 10.3.8 *Consider the equation*

$$F(z) = F(z_1, z') = \int_0^\infty e^{2\pi i \lambda z_1} f(z', \lambda) \, d\lambda. \tag{10.3.8.1}$$

(1) *Given an $f \in \widehat{H^2}$, the integral in (10.3.8.1) converges absolutely for $z \in \mathcal{U}$ and uniformly for $z \in K \subset\subset \mathcal{U}$. Thus we can interchange the order of differentiation and integration, and we see that the function F given by the integral is holomorphic.*
(2) *The function F defined in part (10.3.8.1) from an $f \in \widehat{H^2}$ is an element of H^2, and the resulting map $f \mapsto F$ is an isometry of $\widehat{H^2}$ onto H^2; i.e., it is an isomorphism of Hilbert spaces.*
(3) *Let $\tilde{\imath} = (0, 0, \ldots, 0, i) \in \mathbb{C}^{n+1}$ and let $f \in H^2$. Set $f_\varepsilon = f(z + \varepsilon \tilde{\imath})|_{\partial \mathcal{U}}$. Then f_ε is a function on $\mathbb{H}^n$, $f_\varepsilon \to f_0$ in $L^2(\mathbb{H}^n)$ as $\varepsilon \to 0$, and*

$$\|f_0\|_{L^2(\mathbb{H}^n)} = \|f\|_{H^2}.$$

The idea of the proof is to freeze z' and look at the Paley–Wiener representation of the half-space $\text{Im } z_1 > |z'|^2$. There are several nontrivial technical problems with this program, so we shall have to develop the proof in stages. First, we want $\widehat{H^2}$ to be complete so that it is a Hilbert space.

Lemma 10.3.9 $\widehat{H^2}$ *is a Hilbert space.*

Proof: Since $\widehat{H^2}$ is defined as $L^2(\mathbb{C}^n \times \mathbb{R}^+, d\mu)$ for a certain measure μ, its inner product is already determined. The troublesome part of the lemma is the completeness.

Since we are dealing with analytic functions, the L^2 convergence will lead to a very strong (i.e., uniformly on compact subsets) type of convergence on the interior of $\mathbb{C}^n \times \mathbb{R}^1$. Now suppose we are given a Cauchy sequence $\{f_j\}$ in $\widehat{H^2}$; we must show that some subsequence converges to an element of $\widehat{H^2}$. Since H^2 is an L^2 space, some subsequence converges in L^2 (L^2 being complete), and we can extract from that a subsequence converging both in L^2 and pointwise almost everywhere. Next take a compact set $K \subset\subset \mathbb{C}^n$ that is the closure of an open set and $L \subset\subset \mathbb{R}^+$, and a subsequence $\{f_{j_\ell}\}$ such that

$$\int_L \int_K |f_{j_\ell} - f_{j_{\ell+1}}|^2 \, d|z'| \, d\lambda \le \frac{1}{2^{2\ell}}.$$

Thus

$$\sum_\ell \|f_{j_\ell} - f_{j_{\ell+1}}\|^2_{K,L} = \sum_\ell \int_L \int_K |f_{j_\ell}(z',\lambda) - f_{j_{\ell+1}}(z',\lambda)|^2 \, d|z'| \, d\lambda < \infty.$$

If we set

$$\Delta_\ell(\lambda) = \int_K |f_{j_\ell}(z',\lambda) - f_{\ell+1}(z',\lambda)|^2 \, d|z'|,$$

then we have

$$\int_L \sum_\ell \Delta_\ell(\lambda) \, d\lambda < \infty.$$

Thus $\sum_\ell \Delta_\ell(\lambda) < \infty$ for almost every $\lambda \in L$. Passing to a set $K' \subset\subset K$ we find a number $r > 0$ such that $B(z',r) \subset\subset K$ for all $z' \in K'$. Since, for a fixed λ, the functions f_{j_ℓ} are holomorphic on K, they obey the mean value property. Therefore

$$|f_{j_\ell}(z',\lambda) - f_{j_{\ell+1}}(z',\lambda)| \leq \frac{1}{c_n r^{2n}} \cdot \int_{B(z',r)} |f_{j_\ell}(w',\lambda) - f_{j_{\ell+1}}(w',\lambda)| \, d|w'|$$

$$\leq \frac{1}{\tilde{c}_n r^n} \cdot \left(\int_{B(z',r)} |f_{j_\ell}(w',\lambda) - f_{j_{\ell+1}}(w',\lambda)|^2 \, d|w'| \right)^{\frac{1}{2}}$$

$$\leq C_r \cdot \Delta_\ell^{\frac{1}{2}}(\lambda)$$

for all $z' \in K'$ and λ fixed. Therefore the sequence $\{f_{j_\ell}(\cdot,\lambda)\}$ converges uniformly on compact subsets of $\mathbb{C}^n$ for almost every λ. Since for almost every λ the functions f_{j_ℓ} are holomorphic, the limit is then holomorphic. Since the functions f_{j_ℓ} already converge in $L^2(\mathbb{C}^n \times \mathbb{R}^+, d\mu)$ and pointwise almost everywhere, our limit is in $\widehat{H}^2$. $\qquad\square$

Next we prove a lemma:

Lemma 10.3.10 *If $f \in \widehat{H}^2$, then, for $(z',z_1) \in K \subset\subset \mathcal{U}$, we have that*

$$\int_0^\infty e^{2\pi i \lambda z_1} f(z',\lambda) \, d\lambda$$

converges absolutely. Its absolute value is $\leq C_K \|f\|_{\widehat{H}^2}$.

Proof: For $z = (z',z_1) \in K \subset\subset \mathcal{U}$ there is an $\epsilon > 0$ such that $\operatorname{Im} z_1 - |z'|^2 \geq \varepsilon > 0$. Also observe that for almost every λ (since $f(z',\lambda)$ is entire in z'), we have by the mean value property that

$$|f(z',\lambda)| \leq \frac{1}{\operatorname{Vol}[\mathbb{B}(z',\delta)]} \cdot \int_{\mathbb{B}(z',\delta)} |f(w',\lambda)| d|w'|. \qquad (10.3.10.1)$$

The number $\delta > 0$ will be selected later.

Since $-\operatorname{Im} z_1 \leq -\varepsilon - |z'|^2$, we calculate that

$$
I = \left| \int_0^\infty e^{2\pi i \lambda z_1} f(z', \lambda) \, d\lambda \right|
$$

$$
\leq \int_0^\infty e^{-2\pi \varepsilon \lambda} \cdot e^{-2\pi \lambda |z'|^2} \cdot |f(z', \lambda)| \, d\lambda
$$

$$
\leq \left(\int_0^\infty e^{-2\pi \varepsilon \lambda} \, d\lambda \right)^{\frac{1}{2}} \cdot \left(\int_0^\infty e^{-2\pi \varepsilon \lambda} \cdot e^{-4\pi \lambda |z'|^2} |f(z', \lambda)|^2 \, d\lambda \right)^{\frac{1}{2}}
$$

by Schwarz's inequality. Now set

$$
C_0 = \left(\int_0^\infty e^{-2\pi \varepsilon \lambda} \, d\lambda \right)^{\frac{1}{2}}
$$

and apply (10.3.10.1) to obtain

$$
I \leq \frac{C_0}{\operatorname{Vol}[\mathbb{B}(z', \delta)]} \cdot \left(\int_0^\infty e^{-2\pi \varepsilon \lambda} \cdot e^{-4\pi \lambda |z'|^2} \left(\int_{\mathbb{B}(z', \delta)} |f(w', \lambda)| d|w'| \right)^2 d\lambda \right)^{\frac{1}{2}}.
$$

But an application of the generalized Minkowski inequality to the w' integration yields

$$
I^2 \leq C_0^2 \cdot \int_0^\infty e^{-2\pi \varepsilon \lambda} \cdot e^{-4\pi \lambda |z'|^2} \int_{\mathbb{B}(z', \delta)} |f(w', \lambda)|^2 \, d|w'| \, d\lambda.
$$

Now we would like to replace the expression $e^{-4\pi \lambda |z'|^2}$ by $e^{-4\pi \lambda |w'|^2}$ and then apply condition (3) of the definition of $\widehat{H^2}$. Since $w' \in \mathbb{B}(z', \delta)$, we see that

$$
e^{-4\pi \lambda |w'|^2} \geq e^{-4\pi \lambda |z'|^2} \cdot e^{-4\pi \lambda \delta}.
$$

We choose $0 < \delta < \frac{\varepsilon}{2}$. It follows that

$$
e^{-2\pi \varepsilon \lambda} \cdot e^{-4\pi \lambda |z'|^2} = e^{-2\pi \varepsilon \lambda} \cdot e^{4\pi \delta \lambda} \cdot e^{-4\pi \delta \lambda} \cdot e^{-4\pi \lambda |z'|^2} \leq e^{-4\pi \lambda |w'|^2}
$$

and we find that

$$
I^2 \leq C_0^2 \cdot \int_0^\infty \int_{\mathbb{B}(z', \delta)} |f(w', \lambda)|^2 e^{-4\pi \lambda |w'|^2} \, d|w'| \, d\lambda.
$$

Hence we have

$$
I \leq C_0 \cdot \|f\|_{\widehat{H^2}}.
$$

This completes the proof of the lemma. $\square$

Now that we have the absolute convergence of our integral (and uniform convergence for $z \in K \subset\subset \mathcal{U}$), we are allowed to differentiate under the integral sign and it is clear that the F that is created from $f \in \widehat{H}^2$ is holomorphic.

Before continuing with the proof of Theorem 10.3.8 we make two observations:

(i) There would appear to be an ambiguity in the definition

$$f(z', \lambda) = \int_{-\infty}^{\infty} e^{-2\pi \lambda (x_1 + iy_1)} f(z', x_1 + iy_1) \, dx_1 \quad \text{with} \quad y_1 > |z'|^2.$$

After all, the right-hand side explicitly depends on y_1, and yet the left-hand side is independent of y_1. The fact is that the right-hand side is also independent of y_1. After all, f is holomorphic in the variable $x_1 + iy_1$, since it ranges over the half-plane $y_1 > |z'|^2$. Then our claim is simply that the integral of f over a line parallel to the z-axis is independent of the particular line we choose (as long as $y_1 > |z'|^2$). This statement is a consequence of Cauchy's integral theorem: the difference of the integral of f over two parallel horizontal lines is the limit of the integral of f over long horizontal rectangles—from $-N$ to N say. Now the integral of f over a rectangle is zero, and we will see that f has sufficiently rapid decrease at ∞ so that the integrals over the ends of the rectangle tend to 0 as $N \to +\infty$. Thus

$$\int_{-\infty}^{\infty} e^{-2\pi \lambda (x_1 + iy_1)} f(z', x_1 + iy_1) \, dx_1$$

$$= \int_{-\infty}^{\infty} e^{-2\pi \lambda (x_1 + i\tilde{y}_1)} f(z', x_1 + i\tilde{y}_1) \, dx_1 \qquad \text{for } 0 < y_1 < \tilde{y}_2.$$

(ii) Fix a point $(z_1, z') \in \mathcal{U}$. Consider the functional that sends $f \in \widehat{H^2}$ to $F(z', z_1)$, where F is the function created by the Fourier integral of f. Then this functional is continuous on $\widehat{H^2}$. However, the integral of f (which yields F) is taken over a 1-dimensional set, so how can the result be well defined pointwise as a function?

The answer is that for almost every λ, we are careful to pick an almost everywhere equivalent of $f(z', \lambda)$ that is entire in z', so that the resulting F is holomorphic. Thus the precise definition of our linear functional is "evaluation at the point (z_1, z') of the holomorphic function that is an almost everywhere equivalent of the function F arising from f."

We next prove another lemma:

Lemma 10.3.11 *Let $F \in H^2(\mathcal{U})$. Then, for a fixed z', $F_\varepsilon(z', \cdot) \in H^2(\{y_1 > |z'|^2\})$ (as a function of one complex variable), where*

$$F_\varepsilon(z', x_1 + iy_1) = f(z', x_1 + iy_1 + i\varepsilon), \quad \text{for } \varepsilon > 0.$$

Proof: We may assume $z' = 0$. Apply the mean value theorem to $F(0, x_1 + iy_1 + i\varepsilon)$ on $\mathbb{B}(0, \delta) \times D(x_1 + iy_1 + i\epsilon, \delta')$, where $\mathbb{B} \subset \mathbb{C}^n$ and D is a disk in the plane. We see that

$$|F(0, x_1 + iy_1 + i\varepsilon)|^2$$

$$\leq C_{\delta, \delta'} \cdot \int_{D(x_1 + iy_1 + i\varepsilon, \delta')} \int_{\mathbb{B}(0, \delta)} |F(z', x_1 + iy_1 + w + i\varepsilon)|^2 \, d|z'| \, d|w|.$$

Hence

$$\int_{-\infty}^{\infty} |F(0, x_1 + iy_1 + i\varepsilon)|^2 \, dx_1$$

$$\leq C_{\delta,\delta'} \cdot \int_{-\infty}^{\infty} \int_{D(x_1+iy_1+i\varepsilon,\delta')} \int_{\mathbb{B}(0,\delta)} |F|^2 \, d|z'| \, d|w| \, dx_1.$$

We write $w = u + iv$. Now

$$\int_{-\infty}^{\infty} \int_{D(w,\delta')} |G(z + w)| \, d|w| \, dx \leq \int_{-\infty}^{\infty} \int_{-\delta'}^{\delta'} |G(z + iv)| \, dv \, dx$$

for any G (use the fact that $D(w, \delta')$ lies in a box centered at w of side $2\delta'$ and sides parallel to the axes). Choose $\delta' = \frac{\varepsilon}{3}$ and set $\varepsilon' = \frac{2\varepsilon}{3}$, to obtain

$$\int_{-\delta'}^{\delta'} F_\varepsilon(z + iv) \, dv = \int_0^{2\delta'} F_{\varepsilon'}(z + iv) \, dv.$$

Thus

$$\int_{-\infty}^{\infty} |F_\varepsilon(0, x_1 + iy_1)|^2 \, dx_1$$

$$\leq C \cdot \int_{-\infty}^{\infty} \int_0^{2\delta'} \int_{\mathbb{B}(0,\delta)} |F(z', x_1 + iy_1 + iv + i\varepsilon')|^2 d|z'| \, dv \, dx_1.$$

Now $|z'| < \delta$; we choose $\delta = \sqrt{\frac{\varepsilon'}{2}}$ such that $|z'|^2 < \frac{\varepsilon'}{2}$. Therefore

$$\| F_\varepsilon(0, \cdot) \|_{H^2}^2$$

$$\leq C \cdot \int_{-\infty}^{\infty} \int_{\mathbb{B}(0,\delta)} \int_0^{2\delta'} |F(z', x_1 + i(|z'|^2 + y_1 + v + \varepsilon'/2 + (\varepsilon'/2 - |z'|^2)))|^2$$
$$dv \, d|z'| \, dx_1.$$

Next set $\tilde{v} = v + \frac{\varepsilon}{2} - |z'|^2$ and observe that

$$\int_0^{2\delta'} |F(0, v + \frac{\varepsilon'}{2} - |z'|^2)| \, dv \leq \int_0^{2\delta' + \frac{\varepsilon'}{2}} |F(0, \tilde{v})| \, d\tilde{v}.$$

But we know that $2\delta' + \frac{\varepsilon'}{2} = \varepsilon$ so we have

$$\| F_\varepsilon(0, \cdot) \|_{H^2}^2$$

$$\leq C \int_0^\varepsilon \int_{-\infty}^\infty \int_{\mathbb{B}(0,\delta)} \left| F(z', x_1 + i(|z'|^2 + y_1 + v + \varepsilon'/2)) \right|^2 d|z'| \, dx_1 \, dv$$

$$\leq C \int_0^\varepsilon \| F \|_{H^2(\mathcal{U})}^2 \, dv$$

$$= C \cdot \varepsilon \cdot \|F\|^2_{H^2(\mathcal{U})}$$

$$< \infty.$$

This completes the proof.   □

Notice that this lemma is not necessarily true for the boundary limit function $F(z', z_1 + i|z'|^2)$. For the constant $C_\delta \approx \delta^{-n}$, hence the right-hand side blows up as $\varepsilon \to 0^+$.

We are finally in a position to bring our calculations together and to prove Theorem 10.3.8. We have seen that from a given $f \in \widehat{H^2}$ we obtain a function $F(z_1, z')$, holomorphic in $\mathcal{U}$. We now show that this function is in H^2 and in fact that its H^2 norm equals $\|f\|_{\widehat{H^2}}$.

Now

$$\iint_{-\infty}^{\infty} |F(z', z_1 + i\rho + i|z'|^2)|^2 d|z_1| d|z'| = \iint_0^{\infty} |\widehat{F}(z', \lambda)|^2 e^{-4\pi\lambda(|z'|^2+\rho)} d\lambda d|z'|$$

and the integral on the right increases to $\int_{\mathbb{C}^n} \int_0^\infty |\widehat{F}(z', \lambda)|^2 e^{-4\pi\lambda|z'|^2} d\lambda d|z'|$ as $\rho \to 0^+$. Hence

$$\|F\|^2_{H^2(\mathcal{U})} = \sup_{\rho > 0} \iint |F(z', z_1 + i|z'|^2 + i\rho)|^2 d|z_1| d|z'|$$

$$= \iint |\widehat{F}(z', \lambda)|^2 e^{-4\pi\lambda|z'|^2} d\lambda d|z'|$$

$$= \|f\|^2_{\widehat{H^2}}.$$

Furthermore, this equality of norms implies that our map from $\widehat{H^2}$ to H^2 is injective. All that remains is to show that an arbitrary $F \in H^2$ has such a representation.

Given $F = F(z', z_1) \in H^2(\mathcal{U})$, Lemma 10.3.11 tells us that for any fixed $\varepsilon > 0$ and $z' \in \mathbb{C}^n$, the function $F_\varepsilon(z', z_1) \equiv F(z', z_1 + i\varepsilon)$ has a classical Paley–Wiener representation. We leave it as an exercise to check that the resulting function $F_\varepsilon(z', \lambda)$ is holomorphic in z'. Since we have the relation

$$F_\varepsilon(z', z_1) = \int_0^\infty \widehat{F}_\varepsilon(z', \lambda) e^{2\pi\lambda z_1} d\lambda$$

and since the functions $\{F_\varepsilon\}$ are uniformly bounded in $\widehat{H^2}$ as $\varepsilon \to 0$, it follows that the functions $\{\widehat{F}_\varepsilon\}$ are uniformly bounded in $\widehat{H^2}$. We can therefore extract a subsequence F_{ε_j} such that $\widehat{F}_{\varepsilon_j} \to f_0$ weakly as $j \to \infty$. Observe that since $f_0 \in \widehat{H^2}$, we can recover f from $F \in H^2(\mathcal{U})$.

Lemma 10.3.10 tells us that for $(z_1, z') \in K \subset\subset \mathcal{U}$, the (continuous) linear functional on $\widehat{H^2}$ given by Fourier inversion and then evaluation at the point (z_1, z') is uniformly bounded:

$$|F(z', z_1)| \leq C_K \cdot \|\widehat{F}\|_{\widehat{H^2}}.$$

Thus $F_{\varepsilon_j}(z', z_1) \to F_0(z', z_1)$ uniformly on compact subsets of $\mathcal{U}$. However,

$$F_{\varepsilon_j}(z', z_1) \equiv F(z', z_1 + i\varepsilon_j) \to F(z', z_1)$$

pointwise, so we know that $F_0 \equiv f$. Thus F has a representation in terms of a function in $\widehat{H^2}$ because F_ε does.

Finally we must show that if F_ε is defined as above, then F_ε converges to the function f in $L^2(\partial\mathcal{U})$. But we see that

$$F_\varepsilon(z) = F(z', z_1 + i\varepsilon) = \int_0^\infty e^{2\pi \lambda z_1} \cdot e^{-2\pi \lambda \varepsilon} \widehat{F}(z', \lambda) d\lambda,$$

so that

$$\int_{\mathbb{H}^n} |F(z', z_1 + i\varepsilon)|^2 d|z'| d|z_1| = \iint e^{-4\pi \varepsilon \lambda} |\widehat{F}(z', \lambda)|^2 e^{-4\pi \lambda |z'|^2} d|z'| d\lambda$$

and

$$\int_{\mathbb{H}^n} |F_{\varepsilon_1}(z) - F_{\varepsilon_2}(z)|^2 d|z'| d|z_1|$$
$$= \iint |e^{-2\pi \lambda \varepsilon_1} - e^{-2\pi \lambda \varepsilon_2}|^2 \cdot |\widehat{F}(z', \lambda)|^2 e^{-4\pi \lambda |z'|^2} d|z'| d\lambda.$$

Thus the dominated convergence theorem tells us that $\{F_\varepsilon\}$ is a Cauchy sequence in $L^2(\mathbb{H}^n)$. Therefore F has boundary values in $\mathcal{L}^2(\mathbb{H}^n)$.

As a direct consequence of Lemma 10.3.10, we have the following corollary:

Corollary 10.3.12 *The space $H^2(\mathcal{U})$ is a Hilbert space with reproducing kernel.*

The reproducing kernel for H^2 is the Cauchy–Szegő kernel; we shall see, by symmetry considerations, that it is uniquely determined up to a constant. We let $S(z, w)$ denote the reproducing kernel for $H^2(\mathcal{U})$.

Prelude: Although it may not be immediately apparent from the statement of the theorem, it will turn out that the Szegő kernel is a singular integral kernel on the Heisenberg group. This will be important for our applications.

Theorem 10.3.13 *We have that*

$$S(z, w) = C_n \cdot [\rho(z, w)]^{-n-1},$$

where

$$\rho(z, w) = \frac{i}{2}(\overline{w}_1 - z_1) - \sum_{k=1}^n z_k \overline{w}_k$$

and

$$C_n = \frac{n!}{4\pi^{n+1}}.$$

Observe that ρ is a polarization of our "height function" $\rho(z) = \operatorname{Im} z_1 - |z'|^2$, for $\rho(z, w)$ is holomorphic in z, antiholomorphic in w, and $\rho(z, z) = \rho(z)$. It is common to refer to the new function ρ as an "almost analytic continuation" of the old function ρ.

Before we prove the theorem we will formulate an important corollary. Since all our constructs are canonical, the Cauchy–Szegő representation ought to be modeled on a simple convolution operator on the Heisenberg group. Let us determine how to write the reproducing formula as a convolution.

A function F defined on $\mathcal{U}$ induces, for each value of the "height" ρ, a function on the Heisenberg group:

$$F_\rho(\zeta, t) = F\left(\zeta, t + i(|\zeta|^2 + \rho)\right).$$

Since $S(z, w)$ is the reproducing kernel, we know that

$$F(z) = \int_{\mathbb{H}^n} F(w)S(z, w)d\beta(w), \tag{10.3.14}$$

where $d\beta(w) = dw'du_1$ is the Haar measure on $\mathbb{H}^n$ with w written as $w = (u_1 + iv_1, w')$. Recall that part (3) of Theorem 10.3.8 guarantees the existence of L^2 boundary values for F, and the boundary of $\mathcal{U}$ is $\mathbb{H}^n$. Thus the integral (10.3.14) is well defined.

Observe that since the Heisenberg group is *not* commutative, we must be careful when discussing convolutions. We will deal with *right* convolutions, namely an integral of right translates of the given function F:

$$(F * K)(z) = \int_{\mathbb{H}^n} F(z \cdot y^{-1})K(y)dy = \int_{\mathbb{H}^n} F(y)K(y^{-1} \cdot z)dy.$$

The result we seek is the following:

Corollary 10.3.15 *We have that*

$$F_\rho(\zeta, t) = F_0 * K_\rho(\zeta, t),$$

where F_0 is the L^2 boundary limit of F, and

$$K_\rho(\zeta, t) = 2^{n+1}c_n(|\zeta|^2 - it + \rho)^{-n-1}.$$

Proof of the Corollary (assuming the Theorem): We write

$$F_\rho(\zeta, t) = \int_{\mathbb{H}^n} S\left((\zeta, t + i\rho + i|\zeta|^2), (w', u_1 + i|w'|^2)\right) F(w', u_1 + i|w'|^2)d\beta(w).$$

Therefore

$$F_\rho(\zeta, t)$$

$$= \int_{\mathbb{H}^n} c_n \left(\frac{i}{2}(u_1 - i|w'|^2) - t - i\rho - i|\zeta|^2) - \sum_{k=2}^n \zeta_k \overline{w}_k \right)^{-n-1}$$

$$\times F(w', u_1 + i|w'|^2) d\beta(w)$$

$$= \int_{\mathbb{H}^n} 2^{n+1} c_n \frac{F(w', u_1 + i|w'|^2)}{[|\zeta|^2 + |w'|^2 - 2\mathrm{Re}\, \zeta \cdot \overline{w}' + \rho - i(t - u_1 + 2\mathrm{Im}\, \zeta \cdot \overline{w}')]^{n+1}} d\beta(w)$$

$$= \int_{\mathbb{H}^n} F_0(w', u_1) K_\rho \left((\zeta, t)^{-1} \cdot (u_1, w') \right) d\beta(w).$$

That completes the proof. □

Proof of Theorem 10.3.13: First we need the following elementary uniqueness result from complex analysis:

We know that if $\rho(z, w)$ is holomorphic in z and antiholomorphic in w then it is uniquely determined by $\rho(z, z) = \rho(z)$. Next we demonstrate the following claim:

Claim (i): If g is an element of $\mathbb{H}^n$ then $S(gz, gw) \equiv S(z, w)$.

After all, if $F \in H^2(\mathcal{U})$ then the map $F \mapsto F_g$ (where $F_g(z) = F(g^{-1}z)$) is a unitary map of $H^2(\mathcal{U})$ to itself. Now

$$F(g^{-1}z) = \int_{\mathbb{H}^n} S(z, w) F(g^{-1}w) d\beta(w).$$

We make the change of variables $w' = gw$; since $d\beta$ is Haar measure, it follows that $d\beta(w') = d\beta(w)$. Thus

$$F(g^{-1}z) = \int_{\mathbb{H}^n} S(z, gw') F(w') d\beta(w'),$$

so

$$F(z) = \int_{\mathbb{H}^n} S(gz, gw) F(w) d\beta(w).$$

We conclude that $S(z, w)$ and $S(gz, gw)$ are *both* reproducing kernels for $H^2(\mathcal{U})$; hence they are equal.

Claim (ii): If δ is the natural dilation of $\mathcal{U}$ by $\delta(z_1, z') = (\delta^2 z_1, \delta z')$ then

$$S(\delta z, \delta w) = \delta^{-2n-2} S(z, w).$$

The proof is just as above:

$$F(\delta z) = \int_{\mathbb{H}^n} S(z, w) F(\delta w) d\beta(w)$$

$$= \int_{\mathbb{H}^n} S(z, \delta^{-1} w') F(w') \cdot \delta^{-2n-2} d\beta(w),$$

so that

$$F(z) = \delta^{-(2n+2)} \cdot \int_{\mathbb{H}^n} S(\delta^{-1}z, \delta^{-1}w)F(w)d\beta(w).$$

Then the uniqueness of the reproducing kernel yields

$$S(z, w) = \delta^{-(2n+2)} \cdot S(\delta^{-1}z, \delta^{-1}w) \text{ for } \delta > 0.$$

Now the uniqueness result following Theorem 10.3.6 shows that $S(z, w)$ will be completely determined if we can prove that

$$S(z, z) = c_n \cdot [\rho(z)]^{-n-1}.$$

However, $\rho(z)$ is invariant under translation of $\mathcal{U}$ by elements of the Heisenberg group (i.e., $\rho(gz) = \rho(z)$, for all $g \in \mathbb{H}^n$) and

$$\rho(\delta z) = \text{Im}\,(\delta^2 z_1) - \delta^2|z'|^2 = \delta^2\rho(z).$$

Therefore the function

$$S(z, z) \cdot [\rho(z)]^{n+1}$$

has homogeneity zero and is invariant under the action of the Heisenberg group. Since the Heisenberg group acts simply transitively on "parallels" to $\partial\mathcal{U}$, and since dilations enable us to move from any one parallel to another, any function with these two invariance properties must be constant. Hence we have

$$S(z, z) \equiv c_n[\rho(z)]^{-n-1}.$$

It follows that

$$S(z, w) = c_n[\rho(z, w)]^{-n-1}.$$

At long last we have proved Theorem 10.3.13. We have not taken the trouble to calculate the exact value of the constant in front of the canonical kernel. That value has no practical significance for us here.

The completes our presentation of the main results of this section.

Remark: We mention an alternative elegant method for demonstrating the reproducing property of the kernel $e^{-\pi z' \cdot \overline{w}'}$. For ease of calculation, let us assume that $\lambda = 1/4$. Thus we wish to show that

$$F(z') = \int_{\mathbb{C}^n} e^{\pi z' \cdot \overline{w}'} F(w')e^{-\pi|w'|^2}dw', \quad \text{for suitable } F.$$

First, the equality is true for $z' = 0$: the function F is entire, so

$$F(0) = \frac{1}{\omega_{n-1}r^{n-1}} \int_{\partial\mathbb{B}(0,r)} F(w')d\sigma(w'), \quad \text{for all } r > 0.$$

Also observe that

$$\frac{1}{\omega_{n-1}} = \int_0^\infty e^{-\pi r^2} r^{n-1} dr.$$

Thus

$$F(0) = \int_0^\infty e^{-\pi r^2} \cdot r^{n-1} \cdot \omega_{n-1} \cdot \frac{1}{\omega_{n-1} r^{n-1}} \cdot \int_{\partial \mathbb{B}(0,r)} F(w') d\sigma(w') dr$$

$$= \int_{\mathbb{C}^n} F(w') e^{-\pi |w'|^2} dw'.$$

We would like to translate this result to arbitrary z'. The naive translation T_ξ : $f(z') \mapsto f(z' + \xi)$ is not adequate for our purposes, in part because it is not unitary. The unitarized translation operator

$$U_\xi : f(z') \mapsto e^{-\pi \left(z' \cdot \bar{\xi} + \frac{|\xi|^2}{2} \right)} \cdot f(z' + \xi)$$

is better suited to the job.

After changing variables, we find that

$$f(\xi) = e^{\pi \frac{|\xi|^2}{2}} (U_\xi f)(0)$$

$$= e^{\pi \frac{|\xi|^2}{2}} \int e^{-\pi |w'|^2} \cdot (U_\xi f)(w') dw'$$

$$= e^{\pi \frac{|\xi|^2}{2}} \int e^{-\pi (|w'|^2 + w' \cdot \bar{\xi} + \frac{|\xi|^2}{2})} f(w' + \xi) dw'$$

$$= \int e^{-\pi |w'|^2} e^{-\bar{w}' \cdot \xi} f(w') dw'. \qquad \square$$

Exercise for the Reader: If $S(z, \zeta)$ is the Cauchy–Szegő kernel for a domain then its *Poisson–Szegő* kernel is defined to be $\mathcal{P}(z, \zeta) = |S(z, \zeta)^2|/S(z, z)$. Show that if Ω is a bounded domain, then the Poisson–Szegő kernel reproduces functions that are holomorphic on Ω and continuous on the closure. In case $\mathcal{U}$ is the Siegel upper half-space (which is equivalent to the ball, a bounded domain), show that the Poisson–Szegő integral is given by a convolution on the Heisenberg group. Give an explicit formula for the Poisson–Szegő kernel on $\mathcal{U}$. What sort of integral is this? Is it a singular integral? A fractional integral? Or some other sort?

One of the main points of our study here is to see that Theorem 10.3.13 tells us quite explicitly that the Szegő integral is a singular integral on the Heisenberg group. So it induces a bounded operator on $L^p(\partial \mathcal{U}), l < p < \infty$. This can be considered to be the main result of Chapter 9 and 10, and hence of the entire book. It is the pinnacle of our work.

It is the higher-dimensional analogue of the fact that the Cauchy integral on the boundary of the unit disc in the complex plane induces a classical singular integral (namely, the Hilbert transform).

A Coda on Domains of Finite Type

Prologue: The idea of finite type was first developed by J.J. Kohn in the study of subelliptic estimates for the $\overline{\partial}$-Neumann problem (see [KOH1]). It has grown and evolved into a fundamental idea in geometric function theory. It is an important geometric invariant, one that we can calculate. Properly viewed, it is the right generalization of strong pseudoconvexity. The book [DAN5] gives a comprehensive survey of the theory. The idea of finite type is fundamental both to the partial differential equations of several complex variables and also to a variety of mapping problems. It is considerably more complex in the n-variable setting than in the 2-variable setting. We give an indication of both aspects in our exposition here. And we shall put the idea of finite type into the context of this book. That will be the pinnacle of our studies and (we hope) an entree into further investigations for you.

11.1 Prefatory Remarks

We close our studies in this text by considering a broad context into which to place the analysis of strongly pseudoconvex domains (such as the ball B and the Siegel upper half-space $\mathcal{U}$). This is the realm of domains of finite type. An idea first developed by J.J. Kohn to study estimates for the $\overline{\partial}$-Neumann problem, this idea has become pervasive in much of the modern function theory of several complex variables. The present chapter may be considered to be a brief and self-contained introduction to these modern ideas.

Let us begin with the simplest domain in $\mathbb{C}^n$—the ball. Let $P \in \partial B$. It is elementary to see that no complex line (equivalently no affine analytic disk) can have geometric order of contact with ∂B at P exceeding 2. That is, a complex line may pass through P and also be tangent to ∂B at P, but it can do no better. The boundary of the ball has positive curvature and a complex line is flat. The differential geometric structures disagree at the level of second derivatives. Another way to say this is that if ℓ is a complex line tangent to ∂B at P, then for $z \in \ell$,

S.G. Krantz, *Explorations in Harmonic Analysis*, Applied and Numerical Harmonic Analysis, DOI 10.1007/978-0-8176-4669-1_11,
© Birkhäuser Boston, a part of Springer Science+Business Media, LLC 2009

$$\text{dist}\,(z, \partial B) = \mathcal{O}\big(|z - P|^2\big) \tag{11.1.1}$$

and the exponent 2 cannot be improved. The number 2 is called the "order of contact" of the complex line with ∂B.

The notion of strongly pseudoconvex point can be viewed as the correct biholomorphically invariant version of the phenomenon described in the second paragraph: no analytic disk can osculate to better than first-order tangency to a strongly pseudoconvex boundary point. In fact, the positive definiteness of the Levi form provides the obstruction that makes this statement true. Let us sketch a proof:

Suppose that $P \in \partial\Omega$ is a point of strong pseudoconvexity and that we have fixed a defining function ρ whose complex Hessian is positive definite near $\partial\Omega$. We may further suppose, by the proof of Narasimhan's lemma (Section 6.4), that the only second-order terms in the Taylor expansion of ρ about P are the mixed terms occurring in the complex Hessian.

Let $\phi : D \to \mathbb{C}^n$ be an analytic disk that is tangent to $\partial\Omega$ at P and such that $\phi(0) = P$, $\phi'(0) \neq 0$. The tangency means that

$$\text{Re}\left(\sum_{j=1}^{n} \frac{\partial\rho}{\partial z_j}(P)\phi_j'(0) \right) = 0.$$

It follows that if we expand $\rho \circ \phi(\zeta)$ in a Taylor expansion about $\zeta = 0$ then the zero and first-order terms vanish. As a result, for ζ small,

$$\rho \circ \phi(\zeta) = \left[\sum_{j,k=1}^{n} \frac{\partial^2\rho}{\partial z_j \partial \bar{z}_k}(P)\phi_j'(0)\overline{\phi}_k'(0) \right] |\zeta|^2 + o(|\zeta|^2).$$

But this last is at most

$$\geq C \cdot |\zeta|^2$$

for ζ small. This gives an explicit lower bound, in terms of the eigenvalues of the Levi form, for the order of contact of the image of ϕ with $\partial\Omega$.

11.1.1 The Role of the $\overline{\partial}$ Problem

Recall that we introduced the Cauchy–Riemann operator $\overline{\partial}$ and the inhomogeneous Cauchy–Riemann equations in Section 6.1. One of the most important things that we do in complex function theory is to construct holomorphic functions with specified properties. In one complex variable, there are many fundamental tools for achieving this goal: Blaschke products, Weierstrass products, integral formulas, series, canonical factorizations, function algebra techniques, and many more. Several complex variables does not have a number of these techniques—at least not in well-developed form. The most powerful method known today for constructing holomorphic functions in dimensions two and higher is the method of the inhomogeneous Cauchy–Riemann equations.

It is an important and profound fact that on a pseudoconvex domain, any equation of the form $\bar\partial u = f$ will always have a solution provided only that f satisfy the basic compatibilty condition $\bar\partial f = 0$. Furthermore, if f has smooth coefficients (measured in a variety of different topologies) then u will have correspondingly smooth coefficients. In particular, if $f \in C^\infty(\overline\Omega)$, then we may find a u that is in $C^\infty(\overline\Omega)$ (see [KOH2]). We cannot provide the details here, but see [KRA4, Ch. 4]. We now provide a simple but rather profound example of how the $\bar\partial$-technique works.

Example 11.1.2 Let $\Omega \subseteq \mathbb{C}^n$ be pseudoconvex. Let $\omega = \Omega \cap \{(z_1, \ldots, z_n) : z_n = 0\}$ and assume that this intersection is nonempty. Let $f : \omega \to \mathbb{C}$ satisfy the property that the map

$$(z_1, \ldots, z_{n-1}) \mapsto f(z_1, \ldots, z_{n-1}, 0)$$

is holomorphic on $\tilde\omega = \{(z_1, \ldots, z_{n-1}) \in \mathbb{C}^{n-1} : (z_1, \ldots, z_{n-1}, 0) \in \omega\}$. Then there is a holomorphic $F : \Omega \to \mathbb{C}$ such that $F|_\omega = f$. Indeed there is a linear operator

$$\mathcal{E}_{\omega,\Omega} : \{\text{holomorphic functions on } \omega\} \to \{\text{holomorphic functions on } \Omega\}$$

such that $(\mathcal{E}_{\omega,\Omega} f)\big|_\omega = f$. The operator is continuous in the topology of normal convergence.

We shall now prove this assertion, assuming the $\bar\partial$ solvability property that we enunciated a moment ago.

Let $\pi : \mathbb{C}^n \to \mathbb{C}^n$ be the Euclidean projection

$$(z_1, \ldots, z_n) \mapsto (z_1, \ldots, z_{n-1}, 0).$$

Let $\mathcal{B} = \{z \in \Omega : \pi z \notin \omega\}$. Then $\mathcal{B}$ and ω are *relatively closed* disjoint subsets of Ω. Hence there is a function $\Psi : \Omega \to [0, 1]$, $\Psi \in C^\infty(\Omega)$, such that $\Psi \equiv 1$ on a relative neighborhood of ω and $\Psi \equiv 0$ on $\mathcal{B}$. [This last assertion is intuitively nonobvious. It is a version of the C^∞ Urysohn lemma, for which see [HIR]. It is also a good exercise for the reader to construct Ψ by hand.] Set

$$F(z) = \Psi(z) \cdot f(\pi(z)) + z_n \cdot v(z),$$

where v is an unknown function to be determined.

Notice that $f(\pi(z))$ is well defined on supp Ψ. We wish to select $v \in C^\infty(\Omega)$ such that $\bar\partial F = 0$. Then the function F defined by the displayed equation will be the function that we seek.

Thus we require that

$$\bar\partial v(z) = \frac{\left(-\bar\partial\Psi(z)\right) \cdot f(\pi(z))}{z_n}. \tag{11.1.2.1}$$

Now the right-hand side of this equation is C^∞ since $\bar\partial\Psi \equiv 0$ on a neighborhood of ω. Also, by inspection, the right side is annihilated by the $\bar\partial$ operator (remember that $\bar\partial^2 = 0$). There exists a $v \in C^\infty(\Omega)$ that satisfies (11.1.2.1). Therefore the extension F exists and is holomorphic.

It is known that the solution operator for the $\bar{\partial}$ problem is linear; hence it follows that F depends linearly on f. The continuity of the solution operator follows from standard results about the $\bar{\partial}$ operator (see [KRA4]).

Now let us return to our discussion of contact geometry and relate it to ideas coming from the $\bar{\partial}$ equation. It turns out that the number 2, which we saw in discussions of order of contact in the last subsection, arises rather naturally from geometric considerations of a strongly pseudoconvex point; it has important analytic consequences. For instance, it is known [KRA12] that the optimal Λ_α regularity

$$\|u\|_{\Lambda_\alpha(\Omega)} \le C\|f\|_{L^\infty(\Omega)}$$

for solutions on a strongly pseudoconvex domain to the $\bar{\partial}$ equation $\bar{\partial}u = f$, f a bounded, $\bar{\partial}$-closed $(0, 1)$ form, occurs when $\alpha = 1/2$. No such inequality holds for $\alpha > 1/2$. We say that the $\bar{\partial}$ problem exhibits "a gain of $1/2$"—see [KRA13] for the details. Thus the best index is the reciprocal of the integer describing the optimal order of contact of varieties with the boundary of the domain in question.

It was J.J. Kohn [KOH1] who first appreciated the logical foundations of this geometric analysis. In [KOH1], Kohn studied the regularity of the $\bar{\partial}$ equation in a neighborhood of a point at which the maximal order of contact (to be defined precisely below) of one-dimensional complex curves is at most m (this work is in dimension 2 only). He proved (in the Sobolev topology rather than the Lipschitz topology) that the $\bar{\partial}$ problem near such a point exhibits a gain of $(1/m) - \epsilon$, any $\epsilon > 0$. He conjectured that the correct gain is $1/m$. P. Greiner [GRE] gave examples that showed that Kohn's conjecture was sharp (see also [KRA12] for a different approach and examples in other topologies). Folland and Stein [FOST1] showed that the statement is correct without the ϵ.

David Catlin [CAT1]–[CAT3] has shown that the $\bar{\partial}$-Neumann problem on a pseudoconvex domain in $\mathbb{C}^n$ exhibits a "gain" in regularity if and only if the boundary admits only finite order of contact of (possibly singular) varieties. This result was made possible by the work of D'Angelo, who laid the algebrogeometric foundations for the theory of order of contact of complex varieties with the boundary of a domain (see [DAN1]–[DAN5]).

Special to the theory of the $\bar{\partial}$ problem is the so-called $\bar{\partial}$-Neumann problem. This has to do with the canonical solution of $\bar{\partial}u = f$. Let us say a few words about this matter. First notice that the equation $\bar{\partial}u = f$ never has a unique solution. For if u is a solution to this equation on the domain Ω and if h is any holomorphic function on Ω then $u + h$ is also a solution. Thus the solution space of our partial differential equation is in fact a coset of the space of holomorphic functions on Ω.

It is desirable to have an explicit and canonical method for selecting a unique solution to $\bar{\partial}u = f$. One method is to take *any* solution u and to consider the auxiliary solution $v = u - \tilde{u}$, where $\tilde{u}$ is the projection of u into the space of holomorphic functions. No matter what solution (in a suitable Hilbert space) we use to begin, the resulting solution v will be the unique solution to $\bar{\partial}u = f$ that is orthogonal to holomorphic functions. It is called the *canonical solution* to the $\bar{\partial}$ problem.

Another approach to these matters is by way of the Hodge theory of $\bar{\partial}$. One considers the operator $\Box = \bar{\partial}\bar{\partial}^* + \bar{\partial}^*\bar{\partial}$. A special right inverse to $\Box$, called N, is constructed. It is called the $\bar{\partial}$-Neumann operator. Then the canonical solution is given by $v = \bar{\partial}^* N f$.

All these ideas are tied together in a nice way by an important formula of J.J. Kohn:

$$P = I - \bar{\partial}^* N \bar{\partial}.$$

Here P is the Bergman projection. Thus we can relate the Bergman kernel and projection to the central and important $\bar{\partial}$-Neumann operator (see [KRA3]).

A big idea in this subject is *Condition R*, which was first developed by S.R. Bell [BEL]. We say that a domain Ω satisfy *Condition R* if the Bergman projection P satisfies

$$P : C^\infty(\overline{\Omega}) \to C^\infty(\overline{\Omega}).$$

Note that this condition makes sense, because the domain space is a subspace of $L^2(\Omega)$, and P is naturally defined on $L^2(\Omega)$.

Bell has shown that if two pseudoconvex domains Ω_1, Ω_2 satisfy Condition R then any biholomorphic mapping $\Phi : \Omega_1 \to \Omega_2$ will extend to be a diffeomorphism of the closures. Thus we see that the $\bar{\partial}$-Neumann problem and the Bergman projection are intimately bound up with fundamental questions of biholomorphic mappings.

The purpose of the present chapter is to acquaint the reader with the circle of geometric ideas that was described in the preceding paragraphs and to indicate the applications of these ideas to the theory of holomorphic mappings.

11.2 Return to Finite Type

Example 11.2.1 Let m be a positive integer and define

$$\Omega = \Omega_m = \{(z_1, z_2) \in \mathbb{C}^2 : \rho(z_1, z_2) = -1 + |z_1|^2 + |z_2|^{2m} < 0\}.$$

Consider the boundary point $P = (1, 0)$. Let $\phi : D \to \mathbb{C}^2$ be an analytic disk that is tangent to $\partial\Omega$ at P and such that $\phi(0) = P, \phi'(0) \neq 0$. We may in fact assume, after a reparametrization, that $\phi'(0) = (0, 1)$. Then

$$\phi(\zeta) = (1 + 0\zeta + \mathcal{O}(\zeta^2), \zeta + \mathcal{O}(\zeta^2)). \qquad (11.2.1.1)$$

What is the greatest order of contact (measured in the sense of equation (11.2.1.1), with the exponent 2 replaced by some $\widetilde{m}$) that such a disk ϕ can have with $\partial\Omega$?

Obviously the disk $\phi(\zeta) = (1, \zeta)$ has order of contact $2m$ at $P = (1, 0)$, for

$$\rho \circ \phi(\zeta) = |\zeta|^{2m} = \mathcal{O}(|(1, \zeta) - (1, 0)|^{2m}).$$

The question is whether we can improve upon this estimate with a different curve ϕ. Since all curves ϕ under consideration must have the form (11.2.1.1), we calculate that

$$\rho \circ \phi(\zeta) = -1 + |1 + \mathcal{O}(\zeta^2)|^2 + |\zeta + \mathcal{O}(\zeta)^2|^{2m}$$

$$= -1 + [|1 + \mathcal{O}(\zeta^2)|^2] + [|\zeta|^{2m}|1 + \mathcal{O}(\zeta)|^{2m}].$$

The second expression in [ ] is essentially $|\zeta|^{2m}$, so if we wish to improve on the order of contact then the first term in [ ] must cancel it. But then the first term would have to be $|1 + c\zeta^m + \cdots|^{2m}$. The resulting term of order $2m$ would be positive and in fact would *not* cancel the second. We conclude that $2m$ is the optimal order of contact for complex curves with $\partial\Omega$ at P. Let us say that P is of "geometric type $2m$."

Now we examine the domain Ω_m from the analytic viewpoint. Consider the vector fields

$$L = \frac{\partial\rho}{\partial z_1}\frac{\partial}{\partial z_2} - \frac{\partial\rho}{\partial z_2}\frac{\partial}{\partial z_1} = \bar{z}_1\frac{\partial}{\partial z_2} - mz_2^{m-1}\bar{z}_2^m\frac{\partial}{\partial z_1}$$

and

$$\bar{L} = \frac{\partial\rho}{\partial\bar{z}_1}\frac{\partial}{\partial\bar{z}_2} - \frac{\partial\rho}{\partial\bar{z}_2}\frac{\partial}{\partial\bar{z}_1} = z_1\frac{\partial}{\partial\bar{z}_2} - m\bar{z}_2^{m-1}z_2^m\frac{\partial}{\partial\bar{z}_1}.$$

One can see from their very definition, or can compute directly, that both these vector fields are tangent to $\partial\Omega$. That is to say, $L\rho \equiv 0$ and $\bar{L}\rho \equiv 0$. It is elementary to verify that the commutator of two tangential vector fields must still be tangential *in the sense of real geometry.* That is, $[L, \bar{L}]$ must lie in the three-dimensional real tangent space to $\partial\Omega$ at each point of $\partial\Omega$. However, there is no a priori guarantee that this commutator must lie in the *complex* tangent space. And in general it will not. Take for example the case $m = 1$, when our domain is the ball. A calculation reveals that at the point $P = (1, 0)$,

$$[L, \bar{L}] \equiv L\bar{L} - \bar{L}L = -i\frac{\partial}{\partial y_1}.$$

This vector is indeed tangent to the boundary of the ball at P (it annihilates the defining function), but it is equal to the negative of the complex structure tensor J applied to the Euclidean normal $\partial/\partial x_1$; therefore it is what we call *complex normal.* [There is an excellent opportunity here for confusion. It is common in the literature to say that "the direction $\partial/\partial y_1$ is i times the direction ∂x_1" when what is meant is that when the complex structure tensor J is applied to $\partial/\partial x_1$ then one obtains $\partial/\partial y_1$. One must distinguish between the linear operator J and the tensoring of space with $\mathbb{C}$ that enables one to multiply by the scalar i. These matters are laid out in detail in [WEL].]

The reason that it takes only a commutator of order one to escape the complex tangent and have a component in the complex normal direction is that the ball is strongly pseudoconvex. One may see this using the invariant definition of the Levi form and Cartan's formula—see [KRA4, Ch. 5]. Calculate for yourself that on our domain Ω_m, at the point $P = (1, 0)$, it requires a commutator

$$[L, [\overline{L}, \dots [L, \overline{L}] \dots]]$$

of length $2m - 1$ (that is, a total of $2m$ L's and $\overline{L}$'s) to have a component in the complex normal direction. We say that P is of "analytic type $2m$."

Thus, in this simple example, a point of geometric type $2m$ is of analytic type $2m$.

Prelude: Finite type is one of the big ideas of the modern theory of several complex variables. Originally formulated as a means of determining whether the $\overline{\partial}$-Neumann problem satisfies subelliptic estimates on a domain, the finite type condition now plays a prominent role in mapping theory and the boundary behavior of holomorphic functions. D'Angelo's book [DAN5] is a fine introduction to the idea. D'Angelo has also written a number of important papers in the subject; some of these are listed in our bibliography.

It is noteworthy that the philosophy of finite type makes sense in the context of *convex* sets in $\mathbb{R}^N$. To our knowledge, this circle of ideas has never been fully developed.

Next we shall develop in full generality both the geometric and the analytic notions of "type" for domains in complex dimension 2. In this low-dimensional context, the whole idea of type is rather clean and simple (misleadingly so). In retrospect we shall see that the reason for this is that the varieties of maximal dimension that can be tangent to the boundary (that is, one-dimensional complex analytic varieties) have no interesting subvarieties (the subvarieties are all zero-dimensional). Put another way, any irreducible one-dimensional complex analytic variety V has a holomorphic parametrization $\phi : D \to V$. Nothing of the kind is true for higher-dimensional varieties.

11.3 Finite Type in Dimension Two

We begin with the formal definitions of geometric type and of analytic type for a point in the boundary of a smoothly bounded domain $\Omega \subseteq \mathbb{C}^2$. The main result of this section will be that the two notions are equivalent. We will then describe, but not prove, some sharp regularity results for the $\overline{\partial}$ problem on a finite type domain. Good references for this material are [KOH1], [BLG], and [KRA12].

Definition 11.3.3 A *first-order commutator* of vector fields is an expression of the form

$$[L, M] \equiv LM - ML.$$

Here the right-hand side is understood according to its action on C^∞ functions:

$$[L, M](\phi) \equiv (LM - ML)(\phi) \equiv L(M(\phi)) - M(L(\phi)).$$

Inductively, an mth-order commutator is the commutator of an $(m - 1)$st-order commutator and a vector field N. The commutator of two vector fields is again—as we proved in Chapter 9—a vector field.

Definition 11.3.4 A holomorphic vector field is any linear combination of the expressions

$$\frac{\partial}{\partial z_1}, \quad \frac{\partial}{\partial z_2},$$

with coefficients in the ring of C^∞ functions.

A conjugate holomorphic vector field is any linear combination of the expressions

$$\frac{\partial}{\partial \overline{z}_1}, \quad \frac{\partial}{\partial \overline{z}_2}$$

with coefficients in the ring of C^∞ functions.

Definition 11.3.5 Let M be a vector field defined on the boundary of $\Omega = \{z \in \mathbb{C}^2 : \rho(z) < 0\}$. We say that M is *tangential* if $M\rho = 0$ at each point of $\partial\Omega$.

Now we define a gradation of vector fields that will be the basis for our definition of analytic type. Throughout this section $\Omega = \{z \in \mathbb{C}^2 : \rho(z) < 0\}$ and ρ is C^∞. If $P \in \partial\Omega$ then we may make a change of coordinates so that $\partial\rho/\partial z_2(P) \neq 0$. Define the holomorphic vector field

$$L = \frac{\partial\rho}{\partial z_1}\frac{\partial}{\partial z_2} - \frac{\partial\rho}{\partial z_2}\frac{\partial}{\partial z_1}$$

and the conjugate holomorphic vector field

$$\overline{L} = \frac{\partial\rho}{\partial \overline{z}_1}\frac{\partial}{\partial \overline{z}_2} - \frac{\partial\rho}{\partial \overline{z}_2}\frac{\partial}{\partial \overline{z}_1}.$$

Both L and $\overline{L}$ are tangent to the boundary because $L\rho = 0$ and $\overline{L}\rho = 0$. They are both nonvanishing near P by our normalization of coordinates.

The real and imaginary parts of L (equivalently of $\overline{L}$) generate (over the ground field $\mathbb{R}$) the complex tangent space to $\partial\Omega$ at all points near P. The vector field L alone generates the space of all holomorphic tangent vector fields and $\overline{L}$ alone generates the space of all conjugate holomorphic vector fields.

Definition 11.3.6 Let L_1 denote the module, over the ring of C^∞ functions, generated by L and $\overline{L}$. Inductively, for μ an integer greater than 1, L_μ denotes the module generated by $L_{\mu-1}$ and all commutators of the form $[F, G]$ where $F \in L_0$ and $G \in L_{\mu-1}$.

Clearly $L_1 \subseteq L_2 \subseteq \cdots$. Each L_μ is closed under conjugation. *It is not generally the case that* $\cup_\mu L_\mu$ *is the entire three-dimensional tangent space at each point of the boundary.* A counterexample is provided by

$$\Omega = \{z \in \mathbb{C}^2 : |z_1|^2 + 2e^{-1/|z_2|^2} < 1\}$$

and the point $P = (1, 0)$. We invite the reader to supply details of this assertion.

Definition 11.3.7 Let $\Omega = \{\rho < 0\}$ be a smoothly bounded domain in $\mathbb{C}^2$ and let $P \in \partial\Omega$. Let m be a positive integer. We say that $\partial\Omega$ is of *finite analytic type m at P* if $\langle \partial\rho(P), F(P) \rangle = 0$ for all $F \in L_{m-1}$ while $\langle \partial\rho(P), G(P) \rangle \neq 0$ for some $G \in L_m$. In this circumstance we call P a point of type m.

Remark: A point is of finite analytic type m if it requires the commutation of m complex tangential vector fields to obtain a component in the complex normal direction. Such a commutator lies in L_m. This notation is different from that in our source [BLG], but is necessary for consistency with D'Angelo's ideas that will be presented below.

There is an important epistemological observation that needs to be made at this time. Complex tangential vector fields do not, after being commuted with each other finitely many times, suddenly "pop out" into the complex normal direction. What is really being discussed in this definition is an order of vanishing of coefficients.

For instance, suppose that at the point P, the complex normal direction is the z_1 direction. A vector field

$$F(z) = a(z)\frac{\partial}{\partial z_1} + b(z)\frac{\partial}{\partial z_2}$$

such that a vanishes to some finite positive order at P and $b(P) \neq 0$ will be tangential at P. But when we commute vector fields *we differentiate their coefficients*. Thus if F is commuted with the appropriate vector fields finitely many times then a will be differentiated (lowering the order of vanishing by one each time) until the coefficient of $\partial/\partial z_1$ vanishes to order 0. This means that after finitely many commutations, the coefficient of $\partial/\partial z_1$ does not vanish at P. In other words, after finitely many commutations, the resulting vector field has a component in the normal direction at P.

Notice that the condition $\langle \partial\rho(P), F(P) \rangle \neq 0$ is just an elegant way of saying that the vector $G(P)$ has nonzero component in the complex normal direction. As we explained earlier, any point of the boundary of the unit ball is of finite analytic type 2. Any point of the form $(e^{i\theta}, 0)$ in the boundary of $\{(z_1, z_2) : |z_1|^2 + |z_2|^{2m} < 1\}$ is of finite analytic type $2m$. Any point of the form $(e^{i\theta}, 0)$ in the boundary of $\Omega = \{z \in \mathbb{C}^2 : |z_1|^2 + 2e^{-1/|z_2|^2} < 1\}$ is *not* of finite analytic type. We say that such a point is of *infinite analytic type*.

Now we turn to a precise definition of finite geometric type. If P is a point in the boundary of a smoothly bounded domain then we say that an analytic disk $\phi : D \to \mathbb{C}^2$ is a *nonsingular disk tangent to $\partial\Omega$ at P* if $\phi(0) = P, \phi'(0) \neq 0$, and $(\rho \circ \phi)'(0) = 0$.

Definition 11.3.8 Let $\Omega = \{\rho < 0\}$ be a smoothly bounded domain and $P \in \partial\Omega$. Let m be a nonnegative integer. We say that $\partial\Omega$ is *of finite geometric type m at P* if the following condition holds: there is a nonsingular disk ϕ tangent to $\partial\Omega$ at P such that for small ζ,

$$|\rho \circ \phi(\zeta)| \leq C|\zeta|^m.$$

But there is no nonsingular disk ψ tangent to $\partial\Omega$ at P such that for small ζ,

$$|\rho \circ \phi(\zeta)| \leq C|\zeta|^{m+1}.$$

In this circumstance we call P a point of finite geometric type m.

We invite the reader to reformulate the definition of geometric finite type in terms of the order of vanishing of ρ restricted to the image of ϕ.

The principal result of this section is the following theorem:

Prelude: The next result was implicit in the important paper [KOH1]. But it was made explicit in [BLG]. The correspondence between the analytic and geometric theories of

Theorem 11.3.9 *Let* $\Omega = \{\rho < 0\} \subseteq \mathbb{C}^2$ *be smoothly bounded and* $P \in \partial\Omega$. *The point* P *is of finite geometric type* $m \geq 2$ *if and only if it is of finite analytic type* m.

Proof: We may assume that $P = 0$. Write ρ in the form

$$\rho(z) = 2\operatorname{Re} z_2 + f(z_1) + \mathcal{O}(|z_1 z_2| + |z_2|^2).$$

We do this of course by examining the Taylor expansion of ρ and using the theorem of E. Borel to manufacture f from the terms that depend on z_1 only. Notice that

$$L = \frac{\partial f}{\partial z_1}\frac{\partial}{\partial z_2} - \frac{\partial}{\partial z_1} + (\text{error terms}).$$

Here the error terms arise from differentiating $\mathcal{O}(|z_1 z_2| + |z_2|^2)$. Now it is a simple matter to notice that the best order of contact of a one-dimensional nonsingular complex variety with $\partial\Omega$ at 0 equals the order of contact of the variety $\zeta \mapsto (\zeta, 0)$ with $\partial\Omega$ at 0, which is just the order of vanishing of f at 0.

On the other hand,

$$[L, \bar{L}] = \left[-\frac{\partial^2 \bar{f}}{\partial z_1 \partial \bar{z}_1}\frac{\partial}{\partial z_2} \right] - \left[-\frac{\partial^2 f}{\partial \bar{z}_1 \partial z_1}\frac{\partial}{\partial z_2} \right] + (\text{error terms})$$

$$= 2i \operatorname{Im}\left[\frac{\partial^2 f}{\partial \bar{z}_1 \partial z_1}\frac{\partial}{\partial z_2} \right] + (\text{error terms}).$$

Inductively, one sees that a commutator of m vector fields chosen from $L, \bar{L}$ will consist of (real or imaginary parts of) mth order of derivatives of f times $\partial/\partial z_2$ plus the usual error terms. And the pairing of such a commutator with $\partial\rho$ at 0 is just the pairing of that commutator with dz_2; in other words it is just the coefficient of $\partial/\partial z_2$. We see that this number is nonvanishing as soon as the corresponding derivative of f is nonvanishing. Thus the analytic type of 0 is just the order of vanishing of f at 0.

Since both notions of type correspond to the order of vanishing of f, we are done.   □

From now on, when we say "finite type" (in dimension two), we can mean either the geometric or the analytic definition.

We say that a domain $\Omega \subseteq \mathbb{C}^2$ is of *finite type* if there is a number M such that every boundary point is of finite type not exceeding M. In fact, the semicontinuity of type (see Theorem 10.3.9) implies this statement immediately.

Analysis on finite type domains in $\mathbb{C}^2$ has recently become a matter of great interest. It has been proved, in the works [CNS], [FEK1]–[FEK2], [CHR1]–[CHR3], that the $\overline{\partial}$-Neumann problem on a domain $\Omega \subseteq \mathbb{C}^2$ of finite type m exhibits a gain of order $1/m$ in the Lipschitz space topology. In [KRA12] it was proved that this last result is sharp. Finally, the paper [KRA12] also provided a way to prove the nonexistence of certain biholomorphic equivalences by using sharp estimates for the $\overline{\partial}$-problem.

11.4 Finite Type in Higher Dimensions

The most obvious generalization of the notion of geometric finite type from dimension two to dimensions three and higher is to consider orders of contact of nonsingular $(n - 1)$-dimensional complex manifolds with the boundary of a domain Ω at a point P. The definition of analytic finite type generalizes to higher dimensions almost directly (one deals with tangent vector fields $L_1, \ldots, L_{n-1}$ and $\overline{L}_1, \ldots, \overline{L}_{n-1}$ instead of just L and $\overline{L}$). It is a theorem of [BLG] that with these definitions, geometric finite type and analytic finite type are the same in all dimensions.

This is an elegant result, and is entirely suited to questions of extension of CR functions and reflections of holomorphic mappings. However, it is not the correct indicator of when the $\overline{\partial}$-Neumann problem exhibits a gain. In the late 1970s and early 1980s, John D'Angelo realized that a correct understanding of finite type in all dimensions requires sophisticated ideas from algebraic geometry—particularly the intersection theory of analytic varieties. And he saw that *nonsingular varieties cannot tell the whole story.* An important sequence of papers, beginning with [DAN1], laid down the theory of domains of finite type in all dimensions. The complete story of this work, together with its broader mathematical context, appears in [DAN5]. David Catlin [CAT1]–[CAT3] validated the significance of D'Angelo's work by proving that the $\overline{\partial}$-Neumann problem has a gain in the Sobolev topology near a point $P \in \partial\Omega$ if and only if the point P is of finite type in the sense of D'Angelo. [It is interesting to note that there are partial differential operators that exhibit a gain in the Sobolev topology but not in the Lipschitz topology—see [GUA].]

The point is that analytic structure in the boundary of a domain is an obstruction to regularity for the $\overline{\partial}$ problem. It is known (see [KRA4, Ch. 4]) that if the boundary contains an analytic disk then it is possible for the equation $\overline{\partial}u = f$ to have data f that is C^∞ but no smooth solution u. What we now learn is that the order of contact of analytic varieties stratifies this insight into degrees, so that one may make precise statements about the "gain" of the $\overline{\partial}$ problem in terms of the order of contact of varieties at the boundary.

In higher dimensions, matters become technical rather quickly. Therefore we shall content ourselves with a primarily descriptive treatment of this material. One missing piece of the picture is the following: As of this writing, there is no "analytic" description of finite type using commutators of vector fields (but see recent progress by Lee and Fornæss [LEF]). We know that the notion of finite type that we are about to describe is the right one for the study of the $\overline{\partial}$-Neumann problem because **(i)** it enjoys certain important semicontinuity properties (to be discussed below) that other notions of finite type do not, and **(ii)** Catlin's theorem shows that the definition meshes perfectly with fundamental ideas like the $\overline{\partial}$-Neumann problem.

Let us begin by introducing some notation. Let $U \subseteq \mathbb{C}^n$ be an open set. A subset $V \subseteq U$ is called a *variety* if there are holomorphic functions $f_1, \ldots, f_k$ on U such that $V = \{z \in U : f_1(z) = \cdots = f_k(z) = 0\}$. A variety is called *irreducible* if it cannot be written as the union of proper nontrivial subvarieties. One-dimensional varieties are particularly easy to work with because they can be parametrized:

Proposition 11.4.1 *Let $V \subseteq \mathbb{C}^n$ be an irreducible one-dimensional complex analytic variety. Let $P \in V$. There is a neighborhood W of P and a holomorphic mapping $\phi : D \to \mathbb{C}^n$ such that $\phi(0) = P$ and the image of ϕ is $W \cap V$. When this parametrization is in place, we refer to the variety as a holomorphic curve.*

In general, we cannot hope that the parametrization ϕ will satisfy (nor can it be arranged that) $\phi'(0) \neq 0$. As a simple example, consider the variety

$$V = \{z \in \mathbb{C}^2 : z_1^2 - z_2^3 = 0\}.$$

Then the most natural parametrization for V is $\phi(\zeta) = (\zeta^3, \zeta^2)$. Notice that $\phi'(0) = 0$ and there is no way to reparametrize the curve to make the derivative nonvanishing. This is because the variety has a singularity—a cusp—at the point $P = 0$.

Definition 11.4.2 Let f be a scalar-valued holomorphic function of a complex variable and P a point of its domain. The *multiplicity* of f at P is defined to be the least positive integer k such that the kth derivative of f does not vanish at P. If m is that multiplicity then we write $v_P(f) = v(f) = m$.

If ϕ is instead a vector-valued holomorphic function of a complex variable then its multiplicity at P is defined to be the minimum of the multiplicities of its entries. If that minimum is m then we write $v_P(\phi) = v(\phi) = m$.

In this section we shall exclusively calculate the multiplicities of holomorphic curves $\phi(\zeta)$ at $\zeta = 0$.

For example, the function $\zeta \mapsto \zeta^2$ has multiplicity 2 at 0; the function $\zeta \mapsto \zeta^3$ has multiplicity 3 at 0. Therefore the curve $\zeta \mapsto (\zeta^2, \zeta^3)$ has multiplicity 2 at 0.

If ρ is the defining function for a domain Ω then of course the boundary of Ω is given by the equation $\rho = 0$. D'Angelo's idea is to consider the *pullback* of the function ρ under a curve ϕ:

Definition 11.4.3 Let $\phi : D \to \mathbb{C}^n$ be a holomorphic curve and ρ the defining function for a hypersurface M (usually, but not necessarily, the boundary of a domain). Then the pullback of ρ under ϕ is the function $\phi^* \rho(\zeta) = \rho \circ \phi(\zeta)$.

Definition 11.4.4 Let M be a real hypersurface and $P \in M$. Let ρ be a defining function for M in a neighborhoof of P. We say that P is a point of finite type (or finite 1-type) if there is a constant $C > 0$ such that

$$\frac{v(\phi^*\rho)}{v(\phi)} \leq C$$

whenever ϕ is a nonconstant one-dimensional holomorphic curve through P such that $\phi(0) = P$.

The infimum of all such constants C is called the type (or 1-type) of P. It is denoted by $\Delta(M, P) = \Delta_1(M, P)$.

This definition is algebrogeometric in nature. We now offer a more geometric condition that is equivalent to it.

Proposition 11.4.5 *Let P be a point of the hypersurface M. Let $\mathcal{E}_P$ be the collection of one-dimensional complex varieties passing through P. Then we have*

$$\Delta(M, P) = \sup_{V \in \mathcal{E}_P} \sup_{a>0} \left\{ a \in \mathbb{R}^+ : \lim_{V \ni z \to P} \frac{\text{dist}(z, M)}{|z - P|^a} \text{ exists} \right\}.$$

Notice that the statement is attractive in that it gives a characterization of finite type that makes no reference to a defining function. The proposition, together with the material in the first part of this section, motivates the following definition:

Definition 11.4.6 Let P be a point of the hypersurface M. Let $\mathcal{R}_P$ be the collection of nonsingular one-dimensional complex varieties passing through P (that is, we consider curves $\phi : D \to \mathbb{C}^n$, $\phi(0) = P$, $\phi'(0) \neq 0$). Then we define

$$\Delta^{\text{reg}}(M, P) = \Delta_1^{\text{reg}}(M, P)$$

$$= \sup_{V \in \mathcal{R}_P} \sup_{a>0} \left\{ a \in \mathbb{R}^+ : \lim_{V \ni z \to P} \frac{\text{dist}(z, M)}{|z - P|^a} \text{ exists} \right\}.$$

The number $\Delta_1^{\text{reg}}(M, P)$ measures order of contact of nonsingular complex curves (i.e., one-dimensional complex analytic *manifolds*) with M at P. By contrast, $\Delta_1(M, P)$ looks at all curves, both singular and nonsingular. Obviously $\Delta_1^{\text{reg}}(M, P) \leq \Delta_1(M, P)$. The following example of D'Angelo shows that the two concepts are truly different:

Example 11.4.7 Consider the hypersurface in $\mathbb{C}^3$ with defining function given by

$$\rho(z) = 2\text{Re}\, z_3 + |z_1^2 - z_2^3|^2.$$

Let the point P be the origin. Then we have the following facts:

- We may calculate that $\Delta_1^{\text{reg}}(M, P) = 6$. We determine this by noticing that the z_3 direction is the normal direction to M at P; hence any tangent curve must have the form

$$\zeta \mapsto (a(\zeta), b(\zeta), \mathcal{O}(\zeta^2)).$$

Since we are calculating the "regular type," one of the quantities $a'(0)$, $b'(0)$, must be nonzero. We see that if we let $a(\zeta) = \zeta + \cdots$ then the expression $|z_1^2 - z_2^3|^2$ in the definition of ρ provides the obstruction to the order of contact: the curve cannot have order of contact better than 4. Similar considerations show that if $b(\zeta) = \zeta + \cdots$ then the order of contact cannot be better than 6. Putting these ideas together, we see that a regular curve that exhibits maximum order of contact at $P = 0$ is $\phi(\zeta) = (0, \zeta, 0)$. Its order of contact with M at P is 6. Thus $\Delta_1^{\text{reg}}(M, P) = 6$.

- We may see that $\Delta_1(M, P) = \infty$ by considering the (singular) curve $\phi(\zeta) = (\zeta^3, \zeta^2, 0)$. This curve actually *lies in* M. □

To repeat: An appealing feature of the notion of analytic finite type (in dimension two) that we learned about above is that it is upper semicontinuous: if, at a point P, the expression $\langle \partial\rho, F \rangle$ is nonvanishing for some $F \in L_\mu$, then it will certainly be nonvanishing at nearby points. Therefore if P is a point of type m it follows that sufficiently nearby points will be of type at most m. It is considered reasonable that a viable notion of finite type should be upper semicontinuous. Unfortunately, this is not the case, as the following example of D'Angelo shows:

Example 11.4.8 Consider the hypersurface in $\mathbb{C}^3$ defined by the function

$$\rho(z_1, z_2, z_3) = \operatorname{Re}(z_3) + |z_1^2 - z_2 z_3|^2 + |z_2|^4.$$

Take $P = 0$. Then we may argue as in the last example to see that $\Delta_1(M, P) = \Delta_1^{\text{reg}}(M, P) = 4$. The curve $\zeta \mapsto (\zeta, \zeta, 0)$ gives best-possible order of contact.

But for a point of the form $P = (0, 0, ia)$, a a positive real number, let α be a square root of ia. Then the curve $\zeta \mapsto (\alpha\zeta, \zeta^2, ia)$ shows that $\Delta_1(M, P) = \Delta_1^{\text{reg}}(M, P)$ is at least 8 (in fact it equals 8—Exercise).

Thus we see that the number Δ_P is not an upper semicontinuous function of the point P.

It is proved in [DAN1] that the invariant Δ_1 can be compared with another invariant that comes from intersection-theoretic considerations; that is, the author compares Δ_1 with the dimension of the quotient of the ring $\mathcal{O}$ of germs of holomorphic functions by an ideal generated by the components of a special decomposition of the defining function. This latter *is* semicontinuous. The result gives essentially sharp bounds on how Δ_1 can change as the point P varies within M.

We give now a brief description of the algebraic invariant that is used in [DAN1]. Take $P \in M$ to be the origin. Let ρ be a defining function for M near 0. The first step is to prove that one can write the defining function in the form

$$\rho(z) = 2\operatorname{Re} h(z) + \sum_j |f_j(z)|^2 - \sum_j |g_j(z)|^2,$$

where h, f_j, g_j are holomorphic functions. In case ρ is a polynomial then each sum can be taken to be finite—say $j = 1, \ldots k$. Let us restrict attention to that case. [See [DAN5] for a thorough treatment of this decomposition and [KRA8] for auxiliary discussion.] Write $f = (f_1, \ldots, f_k)$ and $g = (g_1, \ldots, g_k)$.

Let U be a unitary matrix of dimension k. Define $\mathcal{I}(U, P)$ to be the ideal generated by h and $f - Ug$. We set $D(\mathcal{I}(U, P))$ equal to the dimension of $\mathcal{O}/\mathcal{I}(U, P)$. Finally, declare $B_1(M, P) = 2 \sup D(\mathcal{I}(U, P))$, where the supremum is taken over all possible unitary matrices of order k. Then we have the following results.

Prelude: The next two results are central to D'Angelo's theory of points of finite type. It is critical that we know that if $P \in \partial\Omega$ is of finite type then nearby boundary points are also of finite type. And we should be able to at least estimate their type. One upshot of these considerations is that if *every* boundary point of Ω is of finite type then there is a global upper bound on the type.

Theorem 11.4.9 With M, ρ, P as usual we have

$$\Delta_1(M, P) \leq B_1(M, P) \leq 2(\Delta_1(M, P))^{n-1}.$$

Theorem 11.4.10 The quantity $B_1(M, P)$ is upper semicontinuous as a function of P.

We learn from the two theorems that Δ_1 *is* locally finite in the sense that if it is finite at P then it is finite at nearby points. We also learn by how much it can change. Namely, for points Q near P we have

$$\Delta_1(M, Q) \leq 2(\Delta_1(M, P))^{n-1}.$$

In case the hypersurface M is pseudoconvex near P then the estimate can be sharpened. Assume that the Levi form is positive semidefinite near P and has rank q at P. Then we have

$$\Delta_1(M, Q) \leq \frac{(\Delta(M, P))^{n-1-q}}{2^{n-2-q}}.$$

We conclude this section with an informal statement of the theorem of [CAT3]:

Theorem 11.4.11 Let $\Omega \subseteq \mathbb{C}^n$ be a bounded pseudoconvex domain with smooth boundary. Let $P \in \partial\Omega$. Then the problem $\bar{\partial}u = f$, with f a $\bar{\partial}$-closed $(0, 1)$-form, enjoys a gain in regularity in the Sobolev topology if and only if P is a point of finite type in the sense that $\Delta_1(M, P)$ is finite.

It is not known how to determine the sharp "gain" in regularity of the $\bar{\partial}$-Neumann problem at a point of finite type in dimensions $n \geq 3$. There is considerable evidence (see [DAN5]) that our traditional notion of "gain" as described here will have to be refined in order to formulate a result.

It turns out that to study finite type, and concomitantly gains in Sobolev regularity for the problem $\bar{\partial}u = f$ when f is a $\bar{\partial}$-closed $(0, q)$-form, requires the study of order of contact of q-dimensional varieties with the boundary of the domain.

One develops an invariant $\Delta_q(M, P)$. The details of this theory have the same flavor as what has been presented here, but are considerably more complicated.

The next result summarizes many of the key ideas about finite type that have been developed in the past thirty years. Catlin, D'Angelo, and Kohn have been key players in this development.

Theorem 11.4.12 *Let Ω_1, Ω_2 be domains of finite type in $\mathbb{C}^n$. If $\Phi : \Omega_1 \rightarrow \Omega_2$ is a biholomorphic mapping then Φ extends to a C^∞ diffeomorphism of $\overline{\Omega}_1$ onto $\overline{\Omega}_2$.*

Heartening progress has been made in studying the singularities and mapping properties of the Bergman and Szegő kernels on domains of finite type both in dimension 2 and in higher dimensions. We mention particularly [NRSW], [CHR1]–[CHR3], and [MCN1]–[MCN2].

There is still a great deal of work to be done before we have reached a good working understanding of points of finite type.

Appendix 1: Rudiments of Fourier Series

This appendix provides basic background in the concepts of Fourier series. Some readers will be familiar with these ideas, and they can easily skip this material (or refer to it as needed). For ease of reading, proofs are placed at the end of the appendix.

This exposition follows familiar lines, and the reader may find similar expositions in [ZYG], [KRA5], [KAT].

A1.1 Fourier Series: Fundamental Ideas

In the present chapter we study the circle group, which is formally defined as $\mathbb{T} = \mathbb{R}/2\pi\mathbb{Z}$. When we think of a function f on $\mathbb{T}$, we transfer it naturally to a function on $[0, 2\pi)$. Then it is also useful to identify it with its 2π-periodic extension to the real line. Then, when we integrate f, we are free to integrate it from any real number b to $b + 2\pi$; the value will be independent of the choice of b.

It is also sometimes useful to identify elements of the circle group with the unit circle S in the complex plane. We do so with the mapping

$$[0, 2\pi) \ni x \longmapsto e^{ix} \in S. \tag{A1.1.1}$$

The circle group acts on itself naturally as a group. That is to say, if g is a fixed element of the circle group then it induces the map

$$\tau_g : [0, 2\pi) \ni x \longmapsto x + g \in [0, 2\pi),$$

where again we are performing addition modulo 2π. We concentrate on functions that transform naturally under this group action.

Of course if we were to require that a function f on the circle group literally *commute* with translation, then f would be constant. It turns out to be more natural to require that there exist a function ϕ, with $|\phi(x)| = 1$ for all x, such that

$$f(y + x) = \phi(x) \cdot f(y). \tag{A1.1.2}$$

Thus the *size* of f is preserved under the group action. Taking $y = 0$ in (A1.1.2) yields

$$f(x) = \phi(x) \cdot f(0),$$

so we see right away that f is completely determined by its value at 0 and by the factor function ϕ. In addition, we compute that

$$\phi(x + y) \cdot f(0) = f(x + y) = \phi(x) \cdot f(0 + y) = \phi(x) \cdot \phi(y) \cdot f(0).$$

Thus, as long as f is not the identically zero function, we see that

$$\phi(x + y) = \phi(x) \cdot \phi(y). \tag{A1.1.3}$$

Our conclusion is this: any function ϕ that satisfies the transformation law (A1.1.2) for some function f must have property (A1.1.3). If, in addition, $|\phi| = 1$, then (A1.1.3) says that ϕ must be a group homomorphism from the circle group into the unit circle S in the complex plane. [The calculations that we have just performed are taken from [FOL4].]

When studying the Fourier analysis of a locally compact abelian group G, one begins by classifying all the continuous homomorphisms $\phi : G \rightarrow \mathbb{C}^*$, where $\mathbb{C}^*$ is the group $\mathbb{C} \setminus \{0\}$ under multiplication. These mappings are called the *group characters*; the characters themselves form a group, and they are the building blocks of commutative Fourier analysis. The functions ϕ that we discovered in line (A1.1.3) are the characters of $\mathbb{T}$.

If our group G is compact, then it is easy to see that any character ϕ must have image lying in the unit circle. For the image of ϕ must be compact. If $\lambda = \phi(g)$ is in the image of ϕ and has modulus greater than 1, then $\lambda^k = \phi(g^k)$ will tend to ∞ as $\mathbb{N} \ni k \rightarrow +\infty$. That contradicts the compactness of the image. A similar contradiction results if $|\lambda| < 1$. It follows that the image of ϕ must lie in the unit circle.

It is a sophisticated exercise (see [KRA5]) to show that the characters of the circle group are the functions

$$\varphi_k(x) = e^{ikx}$$

for $k \in \mathbb{Z}$. All of Fourier analysis (on the circle group) is premised on the study of these special functions. They span L^2 in a natural sense, and serve as a Hilbert space basis. [Note that these statements are a special case of the Peter–Weyl theorem, for which see [FOL4] or [BAC].]

Now suppose that f is a function of the form

$$f(t) = \sum_{j=-N}^{N} a_j e^{ijt}.$$

We call such a function a *trigonometric polynomial*. Trigonometric polynomials are dense in $L^p(\mathbb{T})$, $1 \leq p < \infty$ (think about the Stone–Weierstrass theorem—see [RUD1]). In that sense they are "typical" functions. Notice that if $-N \leq k \leq N$ then

$$\frac{1}{2\pi} \int_0^{2\pi} f(t) e^{-ikt}\, dt = \sum_{j=-N}^{N} a_j \frac{1}{2\pi} \int_0^{2\pi} e^{ijt} e^{-ikt}\, dt = a_k.$$

This calculation shows that we may recover the kth *Fourier coefficient* of f by integrating f against the conjugate of the corresponding character. Note further that if $|k| > N$, then the preceding integral is equal to 0. These considerations lead us to the following definition:

Definition A1.1.4 Let f be an integrable function on $\mathbb{T}$. For $j \in \mathbb{Z}$, we define

$$\widehat{f}(j) = a_j \equiv \frac{1}{2\pi} \int_0^{2\pi} f(t) e^{-ijt}\, dt. \tag{A1.1.4.1}$$

We call $\widehat{f}(j) = a_j$ the jth *Fourier coefficient* of f.

In the subject of Fourier series, it is convenient to build a factor of $1/2\pi$ into our integrals. We have just seen this feature in the definition of the Fourier coefficients. But we will also let

$$\|f\|_{L^p(\mathbb{T})} \equiv \left[\frac{1}{2\pi} \int_0^{2\pi} |f(e^{it})|^p\, dt \right]^{1/p}, \qquad 1 \le p < \infty. \tag{A1.1.4.2}$$

The custom of building the factor of 2π or $1/(2\pi)$ into various expressions simplifies certain key formulas. It is a welcome convenience, though not one that is universally exploited.

The fundamental issue of Fourier analysis is this: For a given function f, we introduce the formal expression

$$Sf \sim \sum_{j=-\infty}^{\infty} \widehat{f}(j) e^{ijt}. \tag{A1.1.5}$$

Does this series converge to f? In what sense? Is the convergence pointwise, in the L^2 topology, or in some other sense? Is it appropriate to use some summation method to recover f from this series?

We call the expression (A1.1.5) *formal*, because we do not know whether the series converges; if it does converge, we do not know whether it converges to the function f.[1]

It will turn out that Fourier series are much more cooperative, and yield many more convergence results, than Taylor series and other types of series in analysis. For very broad classes of functions, the Fourier series is at least summable (a concept to be defined later) to f.

---

[1] Recall here the theory of Taylor series from calculus: the Taylor series for a typical C^∞ function g generally does not converge, and when it does converge it does not typically converge to the function g.

A1.2 Basics

The subject of basic Fourier analysis is so well trodden that we are loath to simply re-produce what has been covered in detail elsewhere (see, for instance, [ZYG], [KAT], [KRA5]). Thus we will adopt the following expeditious format. In each section we shall *state* results from Fourier analysis; we shall sometimes also provide brief discussion. The proofs will be put at the end of Appendix 1. Some readers may find it an instructive challenge to endeavor to produce their own proofs.

We begin this section with three basic results about the size of Fourier coefficients.

Proposition A1.2.1 *Let f be an integrable function on $\mathbb{T}$. Then, for each integer j,*

$$|\widehat{f}(j)| \leq \frac{1}{2\pi} \int |f(t)|dt.$$

In other words,

$$|\widehat{f}(j)| \leq \|f\|_{L^1}.$$

Proposition A1.2.2 (Riemann–Lebesgue) *Let f be an integrable function on $\mathbb{T}$. Then*

$$\lim_{j \to \pm\infty} |\widehat{f}(j)| = 0.$$

Proposition A1.2.3 *Let f be a k-times continuously differentiable 2π-periodic function. Then the Fourier coefficients of f satisfy*

$$|\widehat{f}(j)| \leq C_k \cdot (1 + |j|)^{-k}.$$

This last result has a sort of converse: if the Fourier coefficients of a function decay rapidly, then the function is smooth. Indeed, the more rapid the decay of the Fourier coefficients, the smoother the function. This circle of ideas continues to be an active area of research, and currently is being studied in the context of wavelet theory.

We next define the notion of a partial sum.

Definition A1.2.4 Let f be an integrable function on $\mathbb{T}$ and let the formal Fourier series of f be as in (A1.1.5). We define the Nth *partial sum* of f to be the expression

$$S_N f(x) = \sum_{j=-N}^{N} \widehat{f}(j)e^{ijx}.$$

We say that the Fourier series *converges* to f at the point x if

$$S_N f(x) \to f(x) \qquad \text{as } N \to \infty$$

in the sense of convergence of ordinary sequences of complex numbers.

We mention now, just for cultural purposes, that the theory of Fourier series would work just as well if we were to define

$$S_N f(x) = \sum_{j=-\phi(N)}^{\psi(N)} \widehat{f}(j)e^{ijx}$$

for functions $\phi(N)$, $\psi(N)$ that tend strictly monotonically to $+\infty$. This assertion follows from the theory of the Hilbert transform and its connection to summability of Fourier series, to be explicated below. We leave the details as an open-ended exercise for the reader.

It is most expedient to begin our study of summation of Fourier series by finding an integral formula for $S_N f$. Thus we write

$$S_N f(x) = \sum_{j=-N}^{N} \widehat{f}(j)e^{ijx}$$

$$= \sum_{j=-N}^{N} \frac{1}{2\pi} \int_0^{2\pi} f(t)e^{-ijt}\, dt\, e^{ijx}$$

$$= \frac{1}{2\pi} \int_0^{2\pi} \left[\sum_{j=-N}^{N} e^{ij(x-t)} \right] f(t)dt. \qquad (A1.2.5)$$

We need to calculate the sum in brackets; for that will be a universal object associated to the summation process S_N, and unrelated to the particular function f that we are considering.

Now

$$\sum_{j=-N}^{N} e^{ijs} = e^{-iNs} \sum_{j=0}^{2N} e^{ijs} = e^{-iNs} \sum_{j=0}^{2N} [e^{is}]^j. \qquad (A1.2.6)$$

The sum on the right is a geometric sum, and we may instantly write a formula for it (as long as $s \neq 0 \bmod 2\pi$):

$$\sum_{0}^{2N} [e^{is}]^j = \frac{e^{i(2N+1)s} - 1}{e^{is} - 1}.$$

Substituting this expression into (A1.2.6) yields

$$\sum_{j=-N}^{N} e^{ijs} = e^{-iNs} \frac{e^{i(2N+1)s} - 1}{e^{is} - 1}$$

$$= \frac{e^{i(N+1)s} - e^{-iNs}}{e^{is} - 1}$$

$$= \frac{e^{i(N+1)s} - e^{-iNs}}{e^{is} - 1} \cdot \frac{e^{-is/2}}{e^{-is/2}}$$

$$= \frac{e^{i(N+1/2)s} - e^{-i(N+1/2)s}}{e^{is/2} - e^{-is/2}}$$

$$= \frac{\sin(N + \frac{1}{2})s}{\sin \frac{1}{2}s}.$$

We see that we have derived a closed formula (no summation signs) for the relevant sum. In other words, using (A1.2.5), we now know that

$$S_N f(x) = \frac{1}{2\pi} \int_0^{2\pi} \frac{\sin\left[N + \frac{1}{2}\right](x - t)}{\sin \frac{x-t}{2}} f(t)dt.$$

The expression

$$D_N(s) = \frac{\sin\left[N + \frac{1}{2}\right]s}{\sin \frac{s}{2}}$$

is called the *Dirichlet kernel*. It is the fundamental object in any study of the summation of Fourier series. In summary, our formula is

$$S_N f(x) = \frac{1}{2\pi} \int_0^{2\pi} D_N(x - t) f(t)dt,$$

or (after a change of variable—where we exploit the fact that our functions are periodic to retain the same limits of integration)

$$S_N f(x) = \frac{1}{2\pi} \int_0^{2\pi} D_N(t) f(x - t)dt.$$

For now, we notice that

$$\frac{1}{2\pi} \int_0^{2\pi} D_N(t)dt = \frac{1}{2\pi} \int_0^{2\pi} \sum_{j=-N}^{N} e^{ijt} \, dt$$

$$= \sum_{j=-N}^{N} \frac{1}{2\pi} \int_0^{2\pi} e^{ijt} \, dt$$

$$= \frac{1}{2\pi} \int_0^{2\pi} e^{i0t} \, dt$$

$$= 1. \tag{A1.2.7}$$

We begin our treatment of summation of Fourier series with Dirichlet's theorem (1828):

Theorem A1.2.8 *Let f be an integrable function on $\mathbb{T}$ and suppose that f is differentiable at x. Then $S_N f(x) \to f(x)$.*

This result, even though not as well-known as it should be, is foundational. It tells us that most reasonable functions—certainly all "calculus-style" functions—have convergent Fourier series. That is certainly useful information.

A1.3 Summability Methods

For many practical applications, the result presented in Theorem A1.2.8 is sufficient. Many "calculus-style" functions that we encounter in practice are differentiable except at perhaps finitely many points (we call these *piecewise differentiable
functions*). The theorem guarantees that the Fourier series of such a function will
converge back to the original function except perhaps at those finitely many singular
points. A standard—and very useful—theorem of Fejér provides the further information that if f is continuous except at finitely many jump discontinuities, then the
Fourier series of f is "summable" to $\frac{1}{2}[f(x+) + f(x-)]$ at every point x. Thus, in
particular, Fejér summation works at points of nondifferentiability of a continuous
function, and also at discontinuities of the first kind (see [RUD1] for this terminology). Another refinement is this: the conclusion of Theorem A1.2.8 holds if the
function f satisfies only a suitable Lipschitz condition at the point x.

However, for other purposes, one wishes to treat an entire Banach space of
functions—for instance L^2 or L^p. Pointwise convergence for the Fourier series of a
function in one of these spaces is the famous Carleson–Hunt theorem [CAR], [HUN],
one of the deepest results in all of modern analysis. We certainly cannot treat it here.
Recent advances [LAT] provide more accessible treatments of these ideas.

"Summability"—the idea of averaging the partial sums in a plausible manner—
is much easier, and in practice is just as useful. Zygmund himself—the great avatar
of twentieth-century Fourier analysis—said in the introduction to the new edition
of his definitive monograph [ZYG] that we had illspent our time worrying about
convergence of Fourier series. Summability was clearly the way to go. We can indeed
explain the basic ideas of summability in this brief treatment.

In order to obtain a unified approach to various summability methods, we shall
introduce some ancillary machinery. This will involve some calculation, and some
functional analysis. Our approach is inspired by [KAT]. We begin with two concrete
examples of summability methods, and explain what they are.

As we noted previously, one establishes ordinary convergence of a Fourier series by examining the sequence of partial-summation operators $\{S_N\}$. Figure A1.1
exhibits the "profile" of the operator S_N. In technical language, this figure exhibits
the *Fourier multiplier* associated to the operator S_N. More generally, let f be an in-
tegrable function and $\sum_{-\infty}^{\infty} \widehat{f}(j)e^{ijx}$ its (formal) Fourier series. If $\Lambda = \{\lambda_j\}_{j=-\infty}^{\infty}$
is a sequence of complex numbers, then Λ acts as a Fourier multiplier according to
the rule

$$\mathcal{M}_\Lambda : f \longmapsto \sum_j \lambda_j \widehat{f}(j)e^{ijx}.$$

In this language, the multiplier

$$\lambda_j = \begin{cases} 1 \text{ if } |j| \leq N; \\ 0 \text{ if } |j| > N \end{cases}$$

corresponds to the partial-summation operator S_N. The picture of the multiplier,
shown in Figure A1.1, enables us to see that the multiplier exhibits a precipitous

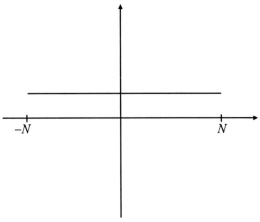

Figure A1.1. "Profile" of the Nth partial-summation multiplier.

"drop" at $\pm N$: the multiplier has a sudden change of value from 1 to 0. According to the philosophy of the Marcinkiewicz multiplier theorem (see [STE1]), this causes difficulties for summability of Fourier series.

The spirit of summability methods, as we now understand them, is to average the partial-summation operators in such a way as to mollify the sharp drop of the multiplier corresponding to the operators S_N. Fejér's method for achieving this effect—now known as a special case of the Cesàro summation method—is as follows. For f an integrable function, we define

$$\sigma_N f(x) = \frac{1}{N+1} \sum_{j=0}^{N} S_j f(x).$$

Notice that we are simply averaging the first $N+1$ partial-summation operators for f. Just as we calculated a closed formula for $S_N f$, let us now calculate a closed formula for $\sigma_N f$.

If we let K_N denote the kernel of σ_N, then we find that

$$K_N(x) = \frac{1}{N+1} \sum_{j=0}^{N} D_j(x)$$

$$= \frac{1}{N+1} \sum_{j=0}^{N} \frac{\sin\left[j + \frac{1}{2}\right]x}{\sin \frac{x}{2}}$$

$$= \frac{1}{N+1} \sum_{j=0}^{N} \frac{\cos jx - \cos(j+1)x}{2 \sin^2 \frac{x}{2}}$$

(since $\sin a \sin b = \frac{1}{2}[\cos(a-b) - \cos(a+b)]$). Of course the sum collapses and we find that

$$K_N(x) = \frac{1}{N+1} \frac{1 - \cos(N+1)x}{2\sin^2\frac{x}{2}}$$

$$= \frac{1}{N+1} \frac{1 - [\cos^2(\frac{(N+1)x}{2}) - \sin^2(\frac{(N+1)x}{2})]}{2\sin^2\frac{x}{2}}$$

$$= \frac{1}{N+1} \frac{2\sin^2(\frac{(N+1)x}{2})}{2\sin^2\frac{x}{2}}$$

$$= \frac{1}{N+1} \left(\frac{\sin(\frac{(N+1)x}{2})}{\sin\frac{x}{2}} \right)^2.$$

Notice that the Fourier multiplier associated to Fejér's summation method is

$$\lambda_j = \begin{cases} \frac{N+1-|j|}{N+1} & \text{if } |j| \leq N, \\ 0 & \text{if } |j| > N. \end{cases}$$

We can see that this multiplier effects the transition from 1 to 0 gradually, over the range $|j| \leq N$. Contrast this with the multiplier associated to ordinary partial summation—again view Figure A1.1 and also Figure A1.2.

On the surface, it may not be apparent why K_N is a more useful and accessible kernel than D_N, but we shall attend to those details shortly. Before we do so, let us look at another summability method for Fourier series. This method is due to Poisson, and is now understood to be a special instance of the summability method of Abel.

For f an integrable function and $0 < r < 1$ we set

$$P_r f(x) = \sum_{j=-\infty}^{\infty} r^{|j|} \widehat{f}(j) e^{ijx}.$$

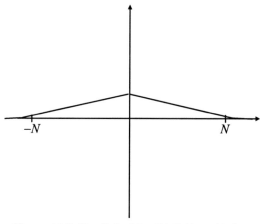

Figure A1.2. "Profile" of the Nth Fejér multiplier.

Notice now that the Fourier multiplier is $\Lambda = \{r^{|j|}\}$. Again the Fourier multiplier exhibits a smooth transition from 1 to 0—but now over the entire range of positive (or negative) integers—in contrast to the multiplier for the partial-summation operator S_N.

Let us calculate the kernel associated to the Poisson summation method. It will be given by

$$P_r(t) = \sum_{j=-\infty}^{\infty} r^{|j|} e^{ijt}$$

$$= \sum_{j=0}^{\infty} [re^{it}]^j + \sum_{j=0}^{\infty} [re^{-it}]^j - 1$$

$$= \frac{1}{1 - re^{it}} + \frac{1}{1 - re^{-it}} - 1$$

$$= \frac{2 - 2r\cos t}{|1 - re^{it}|^2} - 1$$

$$= \frac{1 - r^2}{1 - 2r\cos t + r^2}.$$

We see that we have rediscovered the familiar Poisson kernel of complex function theory and harmonic analysis. Observe that for fixed r (or, more generally, for r in a compact subinterval of $[0, 1)$), the series converges uniformly to the Poisson kernel.

Now let us summarize what we have learned, or are about to learn. Let f be an integrable function on the group $\mathbb{T}$.

A. Ordinary partial summation of the Fourier series for f, which is the operation

$$S_N f(x) = \sum_{j=-N}^{N} \widehat{f}(j) e^{ijx},$$

is given by the integral formulas

$$S_N f(x) = \frac{1}{2\pi} \int_{-\pi}^{\pi} f(t) D_N(x - t)\,dt = \frac{1}{2\pi} \int_{-\pi}^{\pi} f(x - t) D_N(t)\,dt,$$

where

$$D_N(t) = \frac{\sin\left[N + \frac{1}{2}\right]t}{\sin\frac{1}{2}t}.$$

B. Fejér summation of the Fourier series for f, a special case of Cesàro summation, is given by

$$\sigma_N f(x) = \frac{1}{N+1} \sum_{j=0}^{N} S_j f(x).$$

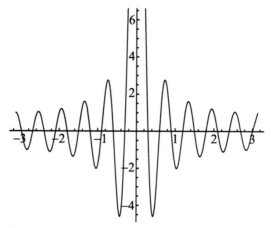

Figure A1.3. Graph of the Dirichlet kernel for $N = 10$.

It is also given by the integral formulas

$$\sigma_N f(x) = \frac{1}{2\pi} \int_{-\pi}^{\pi} f(t) K_N(x - t)\,dt = \frac{1}{2\pi} \int_{-\pi}^{\pi} f(x - t) K_N(t)\,dt,$$

where

$$K_N(t) = \frac{1}{N + 1} \left(\frac{\sin \frac{(N+1)}{2} t}{\sin \frac{t}{2}} \right)^2.$$

C. Poisson summation of the Fourier series for f, a special case of Abel summation, is given by

$$P_r f(x) = \frac{1}{2\pi} \int_{-\pi}^{\pi} f(t) P_r(x - t)\,dt = \frac{1}{2\pi} \int_{-\pi}^{\pi} f(x - t) P_r(t)\,dt,$$

where

$$P_r(t) = \frac{1 - r^2}{1 - 2r \cos t + r^2}, \quad 0 \le r < 1.$$

Figures A1.2, A1.3, and A1.4 show the graphs of the Dirichlet, Fejér, and Poisson kernels.

In the next section we shall isolate properties of the summability kernels P_r and K_N that make their study direct and efficient.

A1.4 Ideas from Elementary Functional Analysis

In this section we formulate and prove two principles of functional analysis that will serve us well in the sequel—particularly in our studies of summability of Fourier series. In fact, they will come up repeatedly throughout the book, in many different contexts.

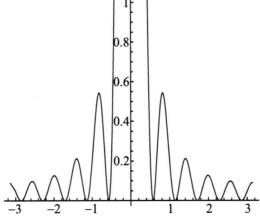

Figure A1.4. Graph of the Fejér kernel for $N = 10$.

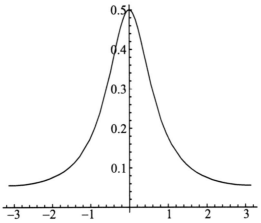

Figure A1.5. Graph of the Poisson kernel for $r = 1/2$.

Theorem A1.4.1 (Functional Analysis Principle I (FAPI)) *Let X be a Banach space and S a dense subset. Let $T_j : X \to X$ be linear operators. Suppose that*

(A1.4.1.1) *For each $s \in S$, $\lim_{j \to \infty} T_j s$ exists in the Banach space norm;*

(A1.4.1.2) *There is a finite constant $C > 0$, independent of x, such that*

$$\|T_j x\|_X \le C \cdot \|x\|$$

for all $z \in X$ and all indices j.
     Then $\lim_{j \to \infty} T_j x$ exists for every $x \in X$.

For the second functional analysis principle, we need a new notion. If T_j : $L^p \to L^p$ are linear operators on L^p (of some Euclidean space $\mathbb{R}^N$), then we let

$$T^* f(x) \equiv \sup_j |T_j f(x)| \text{ for any } x \in \mathbb{R}^N.$$

We call T^* the *maximal function* associated to the T_j.

Theorem A1.4.2 (Functional Analysis Principle II (FAPII)) *Let* $1 \leq p < \infty$ *and suppose that* $T_j : L^p \to L^p$ *are linear operators (these could be* L^p *on the circle, or the line, or* $\mathbb{R}^N$, *or some other Euclidean setting). Let* $\mathcal{S} \subseteq L^p$ *be a dense subset. Assume that*

(A1.4.2.1) *For each* $s \in \mathcal{S}$, $\lim_{j \to \infty} T_j s(x)$ *exists in* $\mathbb{C}$ *for almost every* x;

(A1.4.2.2) *There is a universal constant* $0 \leq C < \infty$ *such that for each* $\alpha > 0$,

$$m\{x : T^* f(x) > \alpha\} \leq \frac{C}{\alpha^p} \|f\|_{L^p}^p.$$

[Here m *denotes Lebesgue measure.]*

Then, for each $f \in L^p$,

$$\lim_{j \to \infty} T_j f(x)$$

exists for almost every x.

The inequality hypothesized in condition **(A1.4.2.2)** is called a *weak-type* (p, p) *inequality* for the maximal operator T^*. Weak-type inequalities have been fundamental tools in harmonic analysis ever since M. Riesz's proof of the boundedness of the Hilbert transform (see Section 2.1 and the treatment of singular integrals in Chapter 3). A classical L^p estimate of the form

$$\|Tf\|_{L^p} \leq C \cdot \|f\|_{L^p}$$

is sometimes called a *strong-type* estimate.

We shall use FAPI (Functional Analysis Principle I) primarily as a tool to prove norm convergence of Fourier series and other Fourier-analytic entities. We shall use FAPII (Functional Analysis Principle II) primarily as a tool to prove pointwise convergence of Fourier series and other Fourier entities.

A1.5 Summability Kernels

It is an interesting diversion to calculate

$$\frac{1}{2\pi} \int_{-\pi}^{\pi} |D_N(t)| dt = \|D_N\|_{L^1}.$$

As an exercise, endeavor to do so by either (i) using `Mathematica` or `Maple` or (ii) breaking up the interval $[0, \pi]$ into subintervals on which $\sin[N + \frac{1}{2}]t$ is essentially constant. By either method, you will see that the value of the integral is approximately equal to the partial sum for the harmonic series.

You will find in particular that

$$\|D_N\|_{L^1} \approx c \cdot \log N$$

(here the notation $\approx$ means "is of the size"). This nonuniform integrability of the Dirichlet kernel is, for the moment, a roadblock to our understanding of the partial-summation process. The theory of singular integral operators will give us a method for handling integral kernels like D_N. We shall say more about these in Chapter 5.

Meanwhile, let us isolate the properties of summability kernels that distinguish them from D_N and make them most useful. Give the kernels the generic name $\{k_N\}_{N\in\mathbb{Z}^+}$. We will consider asymptotic properties of these kernels as $N \to +\infty$. [Note that in most of the examples we shall present, the indexing space for the summability kernels will be $\mathbb{Z}^+$, the nonnegative integers—just as we have stated. But in some examples, such as the Poisson kernel, it will be more convenient to let the parameter be $r \in [0, 1)$. In this last case we shall consider asymptotic properties of the kernels as $r \to 1^-$. We urge the reader to be flexible about this notation.] There are three properties that are desirable for a family of summability kernels:

(A1.5.1) $\frac{1}{2\pi} \int_{-\pi}^{\pi} k_N(x)\,dx = 1 \qquad \forall N$;

(A1.5.2) $\frac{1}{2\pi} \int_{-\pi}^{\pi} |k_N(x)|\,dx \leq C \qquad \forall N$, some finite $C > 0$;

(A1.5.3) If $\delta > 0$, then $\lim_{N\to\infty} k_N(x) = 0$, uniformly for $\pi \geq |z| \geq \delta$.

Let us call any family of summability kernels *standard* if it possesses these three properties. [See [KAT] for the genesis of some of these ideas.]

It is worth noting that condition (A1.5.1) plus positivity of the kernel automatically give condition (A1.5.2). Positive kernels will prove to be "friendly" in other respects as well. Both the Fejér and the Poisson kernels are positive.

Now we will check that the family of Fejér kernels and the family of Poisson kernels both possess these three properties.

A1.5.1. The Fejér Kernels

Notice that since $K_N > 0$,

$$\frac{1}{2\pi} \int_{-\pi}^{\pi} |K_N(t)|\,dt = \frac{1}{2\pi} \int_{-\pi}^{\pi} K_N(t)\,dt$$

$$= \frac{1}{N+1} \sum_{j=0}^{N} \frac{1}{2\pi} \int_{-\pi}^{\pi} D_j(t)\,dt$$

$$= \frac{1}{N+1} \sum_{j=0}^{N} 1$$

$$= \frac{1}{N+1} \cdot (N+1) = 1.$$

This takes care of **(A1.5.1)** and **(A1.5.2)**. For **(A1.5.3)**, notice that $|\sin(t/2)| \geq |\sin(\delta/2)| > 0$ when $\pi \geq |t| \geq \delta > 0$. Thus, for such t,

$$|K_N(t)| \leq \frac{1}{N+1} \cdot \frac{1}{|\sin(\delta/2)|} \to 0,$$

uniformly in t as $N \to \infty$. Thus the Fejér kernels form a standard family of summability kernels.

A1.5.2. The Poisson Kernels

First we observe that since $P_r > 0$,

$$\frac{1}{2\pi} \int_{-\pi}^{\pi} |P_r(t)| dt = \frac{1}{2\pi} \int_{-\pi}^{\pi} P_r(t)\, dt$$

$$= \sum_{j=-\infty}^{\infty} \frac{1}{2\pi} \int_{-\pi}^{\pi} r^{|j|} e^{ijt}\, dt$$

$$= \frac{1}{2\pi} \int_{-\pi}^{\pi} r^0 e^{i0t}\, dt = 1.$$

This takes care of **(A1.5.1)** and **(A1.5.2)**. For **(A1.5.3)**, notice that

$$|1 - 2r\cos t + r^2| = (r - \cos t)^2 + (1 - \cos^2 t)$$

$$\geq 1 - \cos^2 t$$

$$= \sin^2 t$$

$$\geq \left(\frac{2}{\pi} t\right)^2$$

$$\geq \frac{4}{\pi^2} \delta^2$$

if $\pi/2 \geq |t| \geq \delta > 0$. [The estimate for $\pi \geq |t| > \pi/2$ is even easier.] Thus, for such t,

$$|P_r(t)| \leq \frac{\pi^2}{4} \cdot \frac{1 - r^2}{\delta^2} \to 0$$

as $r \to 1^-$. Thus the Poisson kernels form a standard family of summability kernels.

Now let us enunciate a rather general theorem about convergence-inducing properties of families of summability kernels:

Theorem A1.5.4 *Suppose that* $\{k_N\}$ *is a standard family of summability kernels. If* f *is any continuous function on* $\mathbb{T}$, *then*

$$\lim_{N \to \infty} \frac{1}{2\pi} \int_{-\pi}^{\pi} f(x - t)k_N(t)dt = f(x),$$

with the limit existing uniformly for $z \in \mathbb{T}$.

We see that if we use a summability method such as Cesàro's or Abel's to assimilate the Fourier data of a function f, then we may recover *any* continuous function f as the uniform limit of the trigonometric "sums" coming from the Fourier coefficients of f. Contrast this last theorem—especially its proof—with the situation for the ordinary partial sums of the Fourier series. Note that it is property **(A1.5.2)** that fails for the kernels D_N—see the beginning of this section—and, in fact, makes the theorem false for partial summation. [Property **(A1.5.3)** fails as well, but it is **(A1.5.2)** that will be the focus of our attention.]

For each N, let $\phi_N(t)$ be the function that equals $+1$ when $D_N(t) \geq 0$ and equals -1 when $D_N(t) < 0$. Of course ϕ_N is discontinuous. But now let $\psi_N(t)$ be a continuous function, bounded in absolute value by 1, that agrees with ϕ_N except in a very small interval about each point where ϕ_N changes sign. Integrate D_N against ψ_N. The calculation alluded to at the start of the section then shows that the value of the integral is about $c \cdot \log N$, even though ψ_N has supremum norm 1. The uniform boundedness principle now tells us that convergence for partial summation in norm fails dramatically for continuous functions on the circle group.

The next lemma, due to I. Schur, is key to a number of our elementary norm convergence results:

Lemma A1.5.5 (Schur's lemma) *Let* X *and* Y *be measure spaces equipped with the measures* μ *and* v, *respectively. Let* $K(x, y)$ *be a measurable kernel on* $X \times Y$. *Assume that there is a finite constant* $M > 0$ *such that for almost every* z,

$$\int_{y \in Y} |K(x, y)| \, dv(y) \leq M$$

and, for almost every y,

$$\int_{x \in X} |K(x, y)| \, d\mu(x) \leq M.$$

Then the operator

$$T : f \longmapsto \int_{y \in Y} K(x, y)f(y) \, dv(y)$$

is bounded from L^p *to* L^p, $1 \leq p \leq \infty$. *Moreover, the operator norm does not exceed* M.

Lemma A1.5.5 gives us a straightforward device for applying Functional Analysis Principle I (FAPI). If our operators are given by integration against kernels— $T_j f(x) = \int k_j(x, y)f(y) \, dy$—then, in order to confirm property **(A1.4.2)**, it suffices for us to check that $\int_{x \in \mathbb{T}} |k_j(x, y)| \, dx \leq C$ and $\int_{y \in \mathbb{T}} |k_j(x, y)| \, dy \leq C$.

Now we turn to the topic of "norm convergence" of the summability methods. The question is this: Let $\{k_N\}$ be a family of kernels. Fix $1 \leq p \leq \infty$. Is it true that for all $f \in L^p$, we have

$$\lim_{N \to \infty} \left\| \left[\frac{1}{2\pi} \int_{-\pi}^{\pi} k_N(t) f(\cdot - t) dt \right] - f(\cdot) \right\|_{L^p} = 0 \,?$$

This is a question about recovering f "in the mean," rather than pointwise, from the Fourier data. In fact, we have the following theorem.

Theorem A1.5.6 *Let* $1 \leq p < \infty$. *Let* $\{k_N\}$ *be a standard family of summability kernels. If* $f \in L^p$, *then*

$$\lim_{N \to \infty} \left\| \frac{1}{2\pi} \int_{-\pi}^{\pi} k_N(t) f(\cdot - t) dt - f(\cdot) \right\|_{L^p} = 0.$$

Remark: We shall present two proofs of this fundamental result. The first is traditional, and relies on basic techniques from real analysis. The second is more modern, and uses FAPI. We record first a couple of preliminary facts that are pervasive in this subject.

Lemma A1.5.7 *Let* $f \in L^p(\mathbb{R}^N)$, $1 \leq p < \infty$. *Then*

$$\lim_{t \to 0} \|f(\cdot - t) - f(\cdot)\|_{L^p} = 0.$$

It is easy to see that the lemma fails for L^∞—simply let f be the characteristic function of the unit ball.

Lemma A1.5.8 (Generalized Minkowski's inequality) *Let* $f(x, y)$ *be a measurable function on the product measure space* $(X, \mu) \times (Y, \nu)$. *Let* $1 \leq p < \infty$. *Then*

$$\int_X \left| \int_Y f(x, y) \, d\nu(y) \right|^p d\mu(x)^{1/p} \leq \int_Y \int_X |f(x, y)|^p \, d\mu(x)^{1/p} \, d\mu(y).$$

Remark: It is informative to think of Minkowski's inequality as saying that "the norm of the sum is less than or equal to the sum of the norms." For with the norm being the L^p norm, the left-hand side is the norm of an integral; the right-hand side is the integral of the norm—we think of the integral as a generalized sum. We shall not prove Minkowski's inequality, but refer the reader to [STG1] or [SAD] or leave the matter as an exercise.

Theorem A1.5.6 tells us in particular that the Fejér means of an L^p function f, $1 \leq p < \infty$, converge in norm back to f. It also tells us that the Poisson means converge to f for the same range of p. Finally, Theorem A1.5.4 says that both the Fejér and the Poisson means of a continuous function g converge uniformly to g.

A1.6 Pointwise Convergence

Pointwise convergence for ordinary summation of Fourier series is a very difficult and technical subject. We cannot treat it here. Pointwise convergence for the standard summability methods is by no means trivial, but it is certainly something that we can discuss here. We do so by way of the Hardy–Littlewood maximal function, an important tool in classical analysis.

Definition A1.6.1 For f an integrable function on $\mathbb{T}$, $z \in \mathbb{T}$, we set

$$Mf(x) = \sup_{R>0} \frac{1}{2R} \int_{x-R}^{x+R} |f(t)| dt.$$

The operator M is called the *Hardy–Littlewood maximal operator*.

One can see that Mf is measurable by using the following reasoning. The definition of Mf does not change if $\sup_{R>0}$ is replaced by $\sup_{R>0, R \text{ rational}}$. But each average is measurable, and the supremum of countably many measurable functions is measurable. [In fact, one can reason a bit differently as follows: Each average is continuous, and the supremum of continuous functions is lower semicontinuous—see [RUD2] or [KRA4].]

A priori, it is not even clear that Mf will be finite almost everywhere. We will show that in fact it is, and furthermore obtain an estimate of its relative size.

Lemma A1.6.2 *Let K be a compact set in $\mathbb{T}$. Let $\{U_\alpha\}_{\alpha \in A}$ be a covering of K by open intervals. Then there is a finite subcollection $\{U_{\alpha_j}\}_{j=1}^{M}$ with the following properties:*

(A1.6.2.1) *The intervals U_{α_j} are pairwise disjoint.*
(A1.6.2.2) *If we take $3U_\alpha$ to be the interval with the same center as U_α but with three times the length, then $\cup_j 3U_{\alpha_j} \supseteq K$.*

Now we may present our boundedness statement for the Hardy–Littlewood maximal function:

Proposition A1.6.3 *If f is an integrable function on $\mathbb{T}$, then for any $\lambda > 0$,*

$$m\{x \in \mathbb{T} : Mf(x) > \lambda\} \le \frac{6\pi \|f\|_{L^1}}{\lambda}. \qquad (A1.6.3.1)$$

Here m stands for the Lebesgue measure of the indicated set. [The displayed estimate is called a weak-type $(1, 1)$ bound for the operator M.]

Of course the maximal operator is trivially bounded on L^∞. It then follows from the Marcinkiewicz interpolation theorem (see [STG1]) that M is bounded on L^p for $1 < p \le \infty$.

The maximal operator is certainly unbounded on L^1. To see this, calculate the maximal function of

$$f_N(x) = \begin{cases} N & \text{if } |x| \leq \frac{1}{2N}, \\ 0 & \text{if } |x| > \frac{1}{2N}. \end{cases}$$

Each of the functions f_N has L^1 norm 1, but their associated maximal functions have L^1 norms that blow up. [Remember that we are working on the circle group $\mathbb{T}$; the argument is even easier on the real line, for Mf_1 is already not integrable at infinity.]

And now the key fact is that, in an appropriate sense, each of our families of standard summability kernels is majorized by the Hardy–Littlewood maximal function. This assertion must be checked in detail, and by a separate argument, for each particular family of summability kernels. To illustrate the ideas, we will treat the Poisson family at this time.

Proposition A1.6.4 *There is a constant finite $C > 0$ such that if $f \in L^1(\mathbb{T})$, then*

$$P^* f(e^{i\theta}) \equiv \sup_{0 < r < 1} |P_r f(e^{i\theta})| \leq CM f(e^{i\theta})$$

for all $\theta \in [0, 2\pi)$.

Corollary A1.6.5 *The operator $P^* f \equiv \sup_{0 < r < 1} P_r f$ is weak-type $(1, 1)$.*

Now we will invoke our second functional analysis principle to derive a pointwise convergence result.

Theorem A1.6.6 *Let f be an integrable function on $\mathbb{T}$. Then, for almost every $z \in \mathbb{T}$, we have that*

$$\lim_{r \to 1^-} \frac{1}{2\pi} \int_{-\pi}^{\pi} f(t) P_r(x - t) dt = f(x).$$

Of course it must be noted that $L^p(\mathbb{T}) \subseteq L^1(\mathbb{T})$ for $1 \leq p < \infty$. So Theorem A1.6.6 applies *a fortiori* to these L^p spaces.

The reader should take special note that this last theorem, whose proof appears to be a rather abstract manipulation of operators, says that a fairly "arbitrary" function f may be recovered, pointwise almost everywhere, by the Poisson summation method from the Fourier data of f.

A similar theorem holds for the Fejér summation method. We invite the interested reader to show that the maximal Fejér means of an L^1 function are bounded above (pointwise) by a multiple of the Hardy–Littlewood maximal function. The rest is then automatic from the machinery that we have set up.

A1.7 Square Integrals

We begin with the theory of L^2 convergence. This is an exercise in "soft" analysis,[2] for it consists in interpreting some elementary Hilbert space ideas for the particular Hilbert space $L^2(\mathbb{T})$. As usual, we define

---

[2] Analysts call an argument "soft" if it does not use estimates, particularly if it does not use ϵ's and δ's.

$$L^2(\mathbb{T}) \equiv \left\{ f \text{ measurable on } \mathbb{T} : \|f\|_{L^2} \equiv \left[\frac{1}{2\pi} \int_0^{2\pi} |f(e^{it})|^2 \, dt \right]^{1/2} < \infty \right\}.$$

Recall that L^2 is equipped with the inner product

$$\langle f, g \rangle = \frac{1}{2\pi} \int f(e^{it}) \overline{g(e^{it})} \, dt$$

and the induced metric

$$\mathbf{d}(f, g) = \|f - g\|_{L^2} = \sqrt{\frac{1}{2\pi} \int |f(e^{it}) - g(e^{it})|^2 \, dt}$$

(under which L^2 is complete).

Observe that the sequence of functions $\mathcal{F} \equiv \{e^{ijt}\}_{j=-\infty}^{\infty}$ is a complete orthonormal basis for L^2. [It will sometimes be useful to write $e_j(t) = e^{ijt}$.] The orthonormality is obvious, and the completeness can be seen by noting that the algebra generated by the exponential functions satisfies the hypotheses of the Stone–Weierstrass theorem (see [RUD1]) on the circle group $\mathbb{T}$. Thus the trigonometric polynomials[3] are uniformly dense in $C(\mathbb{T})$, the continuous functions on $\mathbb{T}$. If f is an L^2 function that is orthogonal to every e^{ijx}, then it is orthogonal to all trigonometric polynomials and hence to all continuous functions on $\mathbb{T}$. But the continuous functions are dense in L^2. So it must be that $f \equiv 0$ and the family $\mathcal{F}$ is complete. This fact, that the group characters for the circle group $\mathbb{T}$ also form a complete orthonormal system for the Hilbert space $L^2(\mathbb{T})$ (which is a special case of the Peter–Weyl theorem—see [FOL4]), will play a crucial role in what follows.

In fact, the quadratic theory of Fourier integrals is treated in considerable detail in Chapter 3—specifically Section 3.1. We shall take this opportunity to summarize some of the key facts about Fourier series of L^2 functions. Their proofs are transcriptions either of general facts about Hilbert space or of particular arguments presented in Chapter 3 for the Fourier integral.

Proposition A1.7.1 (Bessel's inequality) Let $f \in L^2(\mathbb{T})$. Let N be a positive integer and let $a_j = \widehat{f}(j)$, for each j. Then

$$\sum_{j=-N}^{N} |a_j|^2 \leq \|f\|_{L^2}^2.$$

Theorem A1.7.2 (Riesz–Fischer) Let $\{a_j\}_{j=-\infty}^{\infty}$ be a square-summable sequence (i.e., $\sum_j |a_j|^2$ is finite). Then the series

$$\sum_{j=-\infty}^{\infty} a_j e^{ijx}$$

---

[3] Recall that a trigonometric polynomial is simply a finite linear combination of exponential functions.

converges in the L^2 topology (i.e., the sequence of partial sums converges in L^2). It defines a function $f \in L^2(\mathbb{T})$. Moreover, for each n,

$$\widehat{f}(n) = a_n.$$

Theorem A1.7.3 (Parseval's formula) Let $f \in L^2(\mathbb{T})$. Then the sequence $\{\widehat{f}(j)\}$ is square-summable and

$$\frac{1}{2\pi} \int_{-\pi}^{\pi} |f(x)|^2 \, dx = \sum_{j=-\infty}^{\infty} |\widehat{f}(j)|^2.$$

Exercise for the Reader: Apply Parseval's formula to the function $f(x) = x$ on the interval $[0, 2\pi]$. Actually calculate the integral on the left, and write out the terms of the series on the right. Conclude that

$$\frac{\pi^2}{6} = \sum_{j=1}^{\infty} \frac{1}{j^2}.$$

The next result is the key to our treatment of the L^2 theory of norm convergence. It is also a paradigm for the more elaborate L^p theory that we treat afterward.

Proposition A1.7.4 Let $\Lambda = \{\lambda_j\}$ be a sequence of complex numbers. Then the multiplier operator $\mathcal{M}_\Lambda$ is bounded on L^2 if and only if Λ is a bounded sequence. Moreover, the supremum norm of the sequence is equal to the operator norm of the multiplier operator.

Of course the multiplier corresponding to the partial-summation operator S_N is just the sequence $\Lambda^N \equiv \{\lambda_j\}$ given by

$$\lambda_j = \begin{cases} 1 \text{ if } |j| \leq N, \\ 0 \text{ if } |j| > N. \end{cases}$$

[In what follows, we will often denote this particular multiplier by $\chi_{[-N,N]}$. It should be clearly understood that the domain of $\chi_{[-N,N]}$ is the set of integers $\mathbb{Z}$.] This sequence is bounded by 1, so the proposition tells us that S_N is bounded in norm on L^2—with operator norm 1. But in fact more is true. We know that $\|S_N\|_{op} = 1$ for every N. So the operators S_N are uniformly bounded in norm. In addition, the trigonometric polynomials are dense in L^2, and if p is such a polynomial, then when N exceeds the degree of p, $S_N(p) = p$; therefore norm convergence obtains for p. By Functional Analysis Principle I, we conclude that norm convergence is valid in L^2. More precisely:

Theorem A1.7.5 Let $f \in L^2(\mathbb{T})$. Then $\|S_N f - f\|_{L^2} \to 0$ as $N \to \infty$. Explicitly,

$$\lim_{N \to \infty} \left[\int_{\mathbb{T}} |S_N f(x) - f(x)|^2 \, dx \right]^{1/2} = 0.$$

Proofs of the Results in Appendix 1

Proof of Proposition A1.2.1: We observe that

$$|\widehat{f}(j)| = \left| \frac{1}{2\pi} \int_0^{2\pi} f(t)e^{-ijt}\, dt \right| \leq \frac{1}{2\pi} \int_{-\pi}^{\pi} |f(t)|\, dt,$$

as was to be proved.   □

Proof of Proposition A1.2.2: First consider the case in which f is a trigonometric polynomial. Say that

$$f(t) = \sum_{j=-M}^{M} a_j e^{ijt}.$$

Then $\widehat{f}(j) = 0$ as soon as $|j| \geq M$. That proves the result for trigonometric polynomials.

Now let f be any integrable function. Let $\epsilon > 0$. Choose a trigonometric polynomial p such that $\|f - p\|_{L^1} < \epsilon$. Let N be the degree of the trigonometric polynomial p and let $|j| > N$. Then

$$|\widehat{f}(j)| \leq \left| [f - p]\widehat{}(j) \right| + |\widehat{p}(j)|$$

$$\leq \|f - p\|_{L^1} + 0$$

$$< \epsilon.$$

which proves the result.   □

Sketch of Proof of Proposition A1.2.3: We know that for $j \neq 0$,

$$\widehat{f}(j) = \frac{1}{2\pi} \int_0^{2\pi} f(t)e^{-ijt}\, dt$$

$$\overset{\text{(parts)}}{=} \frac{1}{2\pi} \frac{1}{ij} \int_0^{2\pi} f'(t)e^{-ijt}\, dt$$

$$= \cdots =$$

$$\overset{\text{(parts)}}{=} \frac{1}{2\pi} \frac{1}{(ij)^k} \int_0^{2\pi} f^{(k)}(t)e^{-ijt}\, dt.$$

Notice that the boundary terms vanish by the periodicity of f. Thus

$$|\widehat{f}(j)| \leq C \cdot \frac{|\widehat{f^{(k)}}(j)|}{|j|^k} \leq \frac{C'}{|j|^k} \leq \frac{2^k C'}{(1 + |j|)^k},$$

where

$$C' = C \cdot \frac{1}{2\pi} \int |f^{(k)}(x)| \, dx = C \cdot \|f^{(k)}\|_{L^1}.$$

This is the desired result.   □

Proof of Theorem A1.2.8: We examine the expression $S_N f(x) - f(x)$:

$$|S_N f(x) - f(x)| = \left| \frac{1}{2\pi} \int_0^{2\pi} D_N(t) f(x-t) \, dt - f(x) \right|$$

$$= \left| \frac{1}{2\pi} \int_0^{2\pi} D_N(t) f(x-t) \, dt - \frac{1}{2\pi} \int_{-\pi}^{\pi} D_N(t) f(x) \, dt \right|.$$

Notice that something very important has transpired in the last step: we used the fact that $\frac{1}{2\pi} \int_{-\pi}^{\pi} D_N(t) \, dt = 1$ to rewrite the simple expression $f(x)$ (which is constant with respect to the variable t) in an interesting fashion; this step will allow us to combine the two expressions inside the absolute value signs.

Thus we have

$$|S_N f(x) - f(x)| = \left| \frac{1}{2\pi} \int_{-\pi}^{\pi} D_N(t)[f(x-t) - f(x)] \, dt \right|.$$

We may translate f so that $z = 0$, and (by periodicity) we may perform the integration from $-\pi$ to π (instead of from 0 to 2π). Thus our integral is

$$P_N \equiv \left| \frac{1}{2\pi} \int_{-\pi}^{\pi} D_N(t)[f(t) - f(0)] \, dt \right|.$$

Note that another change of variable has allowed us to replace $-t$ by t. Now fix $\epsilon > 0$ and write

$$P_N \le \left\{ \left| \frac{1}{2\pi} \int_{-\pi}^{-\epsilon} D_N(t)[f(t) - f(0)] \, dt \right| + \left| \frac{1}{2\pi} \int_{\epsilon}^{\pi} D_N(t)[f(t) - f(0)] \, dt \right| \right\}$$

$$+ \left| \frac{1}{2\pi} \int_{-\epsilon}^{\epsilon} D_N(t)[f(t) - f(0)] \, dt \right|$$

$$\equiv I + II.$$

We may note that

$$\sin(N + 1/2)t = \sin Nt \cos t/2 + \cos Nt \sin t/2$$

and thus rewrite I as

$$\left| \frac{1}{2\pi} \int_{-\pi}^{-\epsilon} \sin Nt \left[\cos \frac{1}{2}t \cdot \frac{f(t) - f(0)}{\sin \frac{t}{2}} \right] dt \right.$$

$$+ \frac{1}{2\pi} \int_{-\pi}^{-\epsilon} \cos Nt \left[\sin \frac{1}{2}t \cdot \frac{f(t) - f(0)}{\sin \frac{t}{2}} \right] dt \Bigg|$$

$$+ \left| \frac{1}{2\pi} \int_{\epsilon}^{\pi} \sin Nt \left[\cos \frac{1}{2}t \cdot \frac{f(t) - f(0)}{\sin \frac{t}{2}} \right] dt \right.$$

$$+ \frac{1}{2\pi} \int_{\epsilon}^{\pi} \cos Nt \left[\sin \frac{1}{2}t \cdot \frac{f(t) - f(0)}{\sin \frac{t}{2}} \right] dt \Bigg|.$$

These four expressions are all analyzed in the same way, so let us look at the first of them. The expression

$$\left[\cos \frac{1}{2}t \cdot \frac{f(t) - f(0)}{\sin \frac{t}{2}} \right] \chi_{[-\pi, -\epsilon]}(t)$$

is an integrable function (because t is bounded from zero on its support). Call it $g(t)$. Then our first integral may be written as

$$\left| \frac{1}{2\pi} \frac{1}{2i} \int_{-\pi}^{\pi} e^{iNt} g(t) dt - \frac{1}{2\pi} \frac{1}{2i} \int_{-\pi}^{\pi} e^{-iNt} g(t) dt \right|.$$

Each of these last two expressions is ($1/2i$ times) the $\pm N$th Fourier coefficient of the integrable function g. The Riemann–Lebesgue lemma tells us that as $N \to \infty$, they tend to zero. That takes care of I.

The analysis of II is similar, but slightly more delicate. First observe that

$$f(t) - f(0) = \mathcal{O}(t).$$

[Here $\mathcal{O}(t)$ is Landau's notation for an expression that is not greater than $C \cdot |t|$.] More precisely, the differentiability of f at 0 means that $[f(t) - f(0)]/t \to f'(0)$, hence $|f(t) - f(0)| \leq C \cdot |t|$ for t small.

Thus

$$II = \left| \frac{1}{2\pi} \int_{-\epsilon}^{\epsilon} \frac{\sin \left[N + \frac{1}{2} \right] t}{\sin \frac{t}{2}} \cdot \mathcal{O}(t) dt \right|.$$

Regrouping terms, as we did in our estimate of I, we see that

$$II = \left| \frac{1}{2\pi} \int_{-\epsilon}^{\epsilon} \sin Nt \left[\cos \frac{t}{2} \cdot \frac{\mathcal{O}(t)}{\sin \frac{t}{2}} \right] dt \right| + \left| \frac{1}{2\pi} \int_{-\epsilon}^{\epsilon} \cos Nt \left[\sin \frac{t}{2} \cdot \frac{\mathcal{O}(t)}{\sin \frac{t}{2}} \right] dt \right|.$$

The expressions in brackets are integrable functions (in the first instance, because $\mathcal{O}(t)$ cancels the singularity that would be induced by $\sin[t/2]$), and (as before) integration against $\cos Nt$ or $\sin Nt$ amounts to calculating a $\pm N$th Fourier coefficient. As $N \to \infty$, these tend to zero by the Riemann–Lebesgue lemma.

To summarize, our expression P_N tends to 0 as $N \to \infty$. That is what we wished to prove. $\qquad\square$

Proof of Theorem A1.4.1:   Let $f \in X$ and suppose that $\epsilon > 0$. There is an element $s \in S$ such that $\| f - s \| < \epsilon/3(C+1)$. Now select $J > 0$ so large that if $j, k \geq J$ then $\| T_j s - T_k s \| < \epsilon/3$. We calculate, for such j, k, that

$$\| T_j f - T_k f \| \leq \| T_j f - T_j s \| + \| T_j s - T_k s \| + \| T_k s - T_k f \|$$

$$\leq \| T_j \|_{\mathrm{op}} \| f - s \| + \frac{\epsilon}{3} + \| T_k \|_{\mathrm{op}} \| f - s \|$$

$$\leq C \cdot \| f - s \| + \frac{\epsilon}{3} + C \cdot \| f - s \|$$

$$< \frac{\epsilon}{3} + \frac{\epsilon}{3} + \frac{\epsilon}{3}$$

$$= \epsilon.$$

This establishes the result. Note that the converse holds by the uniform boundedness principle. $\square$

Proof of Theorem A1.4.2:   This proof parallels that of FAPI, but it is more technical.

Let $f \in L^p$ and suppose that $\delta > 0$ is given. Then there is an element $s \in S$ such that $\| f - s \|_{L^p}^p < \delta$. Let us assume for simplicity that f and the $T_j f$ are real-valued (the complex-valued case then follows from linearity). Fix $\epsilon > 0$ (independent of δ). Then

$$m\{x : \left| \limsup_{j \to \infty} T_j f(x) - \liminf_{j \to \infty} T_j f(x) \right| > \epsilon\}$$

$$\leq m\{x : \left| \limsup_{j \to \infty} [T_j (f - s)](x) \right| > \epsilon/3\}$$

$$+ m\{x : \left| \limsup_{j \to \infty} (T_j s)(x) - \liminf_{j \to \infty} (T_j s)(x) \right| > \epsilon/3\}$$

$$+ m\{x : \left| \limsup_{j \to \infty} [T_j (s - f)](x) \right| > \epsilon/3\}$$

$$\leq m\{x : \sup_{j} \left| [T_j (f - s)](x) \right| > \epsilon/3\}$$

$$+ 0$$

$$+ m\{x : \sup_{j} \left| [T_j (s - f)](x) \right| > \epsilon/3\}$$

$$= m\{x : T^*(f - s)(x) > \epsilon/3\}$$

$$+ 0$$

$$+ \{x : T^*(s - f)(x) > \epsilon/3\}$$

$$\leq C \cdot \frac{\|f-s\|_{L^p}^p}{[\epsilon/3]^p} + 0 + C \cdot \frac{\|f-s\|_{L^p}^p}{[\epsilon/3]^p}$$

$$< \frac{2C\delta}{\epsilon/3}.$$

Since this estimate holds no matter how small δ is, we conclude that

$$m\left\{x : \left|\limsup_{j\to\infty} T_j f(x) - \liminf_{j\to\infty} T_j f(x)\right| > \epsilon\right\} = 0.$$

This completes the proof of FAPII, for it shows that the desired limit exists almost everywhere. $\square$

Proof of Theorem A1.5.4:   Let $\epsilon > 0$. Since f is a continuous function on the compact set $\mathbb{T}$, it is uniformly continuous. Let $\epsilon > 0$. Choose $\delta > 0$ such that if $|t| < \delta$ then $|f(x) - f(x-t)| < \epsilon$ for all $z \in \mathbb{T}$. Further, let M be the maximum value of $|f|$ on $\mathbb{T}$.

Using Property (**A1.5.1**) of a standard family, we write

$$\left|\frac{1}{2\pi} \int_{-\pi}^{\pi} f(x-t) k_N(t) dt - f(x)\right|$$

$$= \left|\frac{1}{2\pi} \int_{-\pi}^{\pi} f(x-t) k_N(t) dt - \frac{1}{2\pi} \int_{-\pi}^{\pi} f(x) k_N(t) dt\right|$$

$$= \left|\frac{1}{2\pi} \int_{-\pi}^{\pi} [f(x-t) - f(x)] k_N(t) dt\right|$$

$$\leq \frac{1}{2\pi} \int_{-\pi}^{\pi} |f(x-t) - f(x)||k_N(t)| dt$$

$$= \frac{1}{2\pi} \int_{\{t:|t|<\delta\}} + \frac{1}{2\pi} \int_{\{t:|t|\geq\delta\}}$$

$$\equiv I + II.$$

We notice, by property (**A1.5.2**) and the choice of δ, that

$$I \leq \frac{1}{2\pi} \int_{-\delta}^{\delta} \epsilon \cdot |k_N(t)| dt \leq \epsilon \cdot \left(\frac{1}{2\pi} \int_{-\pi}^{\pi} |k_N(t)| dt\right) \leq \epsilon \cdot C.$$

That takes care of I.

For II we use property (**A1.5.3**). If N is large enough, we see that

$$II \leq \frac{1}{2\pi} \int_{\{t:|t|\geq\delta\}} 2M \cdot |K_N(t)| dt \leq 2M \cdot \epsilon.$$

In summary, if N is sufficiently large, then

$$\left| \frac{1}{2\pi} \int_{-\pi}^{\pi} f(x - t) k_N(t) dt - f(x) \right| \leq [C + 2M] \cdot \epsilon$$

for all $z \in \mathbb{T}$. This is what we wished to prove. $\qquad \square$

Proof of Lemma A1.5.5: For $p = \infty$ the result is immediate; we leave it as an exercise (or the reader may derive it as a limiting case of $p < \infty$, which we now treat). For $p < \infty$, we use Hölder's inequality to estimate

$$|Tf(x)| \leq \int_Y |K(x, y)| \cdot |f(y)| \, d\nu(y)$$

$$= \int_Y |K(x, y)|^{1/p'} \cdot |K(x, y)|^{1/p} |f(y)| \, d\nu(y)$$

$$\leq \left(\int_Y |K(x, y)| \, d\nu(y) \right)^{1/p'} \cdot \left(\int_Y |K(x, y)| \cdot |f(y)|^p \, d\nu(y) \right)^{1/p}$$

$$\leq M^{1/p'} \left(\int_Y |K(x, y)| \cdot |f(y)|^p \, d\nu(y) \right)^{1/p}.$$

We use the last estimate to determine the size of $\|Tf\|_{L^p}$:

$$\|Tf\|_{L^p} \leq M^{1/p'} \left(\int_{z \in X} \int_{y \in Y} |K(x, y)| \cdot |f(y)|^p \, d\nu(y) d\mu(x) \right)^{1/p}$$

$$= M^{1/p'} \left(\int_{y \in Y} \int_{z \in X} |K(x, y)| \, d\mu(x) \, |f(y)|^p \, d\nu(y) \right)^{1/p}$$

$$\leq M^{1/p'+1/p} \left(\int_Y |f(y)|^p \, d\nu(y) \right)^{1/p}$$

$$= M \cdot \|f\|_{L^p},$$

which completes the proof. $\qquad \square$

Proof of Lemma A1.5.7: If $\varphi \in C_c(\mathbb{R}^N)$ (the continuous functions with compact support) then obviously

$$\lim_{t \to 0} \|\varphi(\cdot - t) - \varphi(\cdot)\|_{L^p} = 0.$$

If now $f \in L^p$ and $\epsilon > 0$, choose $\varphi \in C_c(\mathbb{R}^N)$ such that $\|f - \varphi\|_{L^p(\mathbb{R}^N)} < \epsilon$. Choose $\delta > 0$ such that $|t| < \delta$ implies $\|\varphi(\cdot - t) - \varphi(\cdot)\|_{L^p} < \epsilon$. Then, for such t,

$$\|f(\cdot - t) - f(\cdot)\|_{L^p} \leq \|f(\cdot - t) - \varphi(\cdot - t)\|_{L^p} + \|\varphi(\cdot - t) - \varphi(\cdot)\|_{L^p}$$

$$+ \|\varphi(\cdot) - f(\cdot)\|_{L^p}$$

$$< \epsilon + \epsilon + \epsilon = 3\epsilon.$$

which gives the result. $\qquad \square$

First Proof of Theorem A1.5.6: Fix $f \in L^p$. Let $\epsilon > 0$. Choose $\delta > 0$ so small that if $|t| < \delta$, then $\|f(\cdot - t) - f(\cdot)\|_{L^p} < \epsilon$. Then

$$\left\| \left[\frac{1}{2\pi} \int_{-\pi}^{\pi} k_N(t) f(\cdot - t) dt \right] - f(\cdot) \right\|_{L^p}$$

$$= \left[\frac{1}{2\pi} \int_{-\pi}^{\pi} \left| \frac{1}{2\pi} \int_{-\pi}^{\pi} k_N(t) [f(x - t) dt - f(x)] dt \right|^p dx \right]^{1/p}$$

$$\leq \left[\frac{1}{2\pi} \int_{-\pi}^{\pi} \left(\frac{1}{2\pi} \int_{-\pi}^{\pi} |k_N(t) [f(x - t) - f(x)]| dt \right)^p dx \right]^{1/p}$$

$$\overset{\text{(Minkowski)}}{\leq} \frac{1}{2\pi} \int_{-\pi}^{\pi} \left[\frac{1}{2\pi} \int_{-\pi}^{\pi} |f(x - t) - f(x)|^p dx \right]^{1/p} |k_N(t)| dt$$

$$= \frac{1}{2\pi} \int_{-\pi}^{\pi} |k_N(t)| \|f(\cdot - t) - f(\cdot)\|_{L^p} dt$$

$$= \frac{1}{2\pi} \int_{|t| < \delta} + \frac{1}{2\pi} \int_{|t| \geq \delta}$$

$$\equiv I + II.$$

Now

$$I \leq \frac{1}{2\pi} \int_{|t| < \delta} |k_N(t)| \cdot \epsilon \, dt \leq C \cdot \epsilon.$$

For II, we know that if N is sufficiently large, then $|k_N|$ is uniformly small (less than ϵ) on $\{t : |t| \geq \delta\}$. Moreover, we have the easy estimate $\|f(\cdot - t) - f(\cdot)\|_{L^p} \leq 2\|f\|_{L^p}$. Thus

$$II \leq \int_{|t| \geq \delta} \epsilon \cdot 2\|f\|_{L^p} dt.$$

This last does not exceed $C' \cdot \epsilon$.

In summary, for all sufficiently large N,

$$\left\| \frac{1}{2\pi} \int_{0}^{2\pi} k_N(t) f(\cdot - t) dt - f(\cdot) \right\|_{L^p} < C'' \cdot \epsilon,$$

which is what we wished to prove.   □

Second Proof of Theorem A1.5.6: We know from Theorem A1.5.4 that the desired conclusion is true for continuous functions, and these are certainly dense in L^p. Secondly, we know from Schur's lemma (Lemma A1.5.5) that the operators

$$T_N : f \longmapsto k_N * f$$

are bounded from L^p to L^p with norms bounded by a constant C, independent of N (here we use property **(A1.5.2)** of a standard family of summability kernels). Now FAPI tells us that the conclusion of the theorem follows. $\qquad\square$

Proof of Lemma A1.6.2: Since K is compact, we may suppose at the outset that the original open cover is finite. So let us call it $\{U_\ell\}_{\ell=1}^p$. Now let U_{ℓ_1} be the open interval among these that has greatest length; if there are several of these longest intervals, then choose one of them. Let U_{ℓ_2} be the open interval, chosen from those remaining, that is disjoint from U_{ℓ_1} and has greatest length—again choose just one if there are several. Continue in this fashion. The process must cease, since we began with only finitely many intervals.

This subcollection does the job. The subcollection chosen is pairwise disjoint by design. To see that the threefold dilates cover K, it is enough to see that the threefold dilates cover the original open cover $\{U_\ell\}_{\ell=1}^p$. Now let U_i be some element of the original open cover. If it is in fact one of the selected intervals, then of course it is covered. If it is *not* one of the selected intervals, then let U_{ℓ_k} be the *first* in the list of selected intervals that intersects U_i (by the selection process, one such must exist). Then, by design, U_{ℓ_k} is at least as long as U_i. Thus, by the triangle inequality, the threefold dilate of U_{ℓ_k} will certainly cover U_i. That is what we wished to prove. $\qquad\square$

Proof of Proposition A1.6.3: By the inner regularity of the measure, it is enough to estimate $m(K)$, where K is any compact subset of $\{x \in \mathbb{T} : Mf(x) > \lambda\}$. Fix such a K, and let $k \in K$. Then, by definition, there is an open interval I_k centered at k such that

$$\frac{1}{m(I_k)} \int_{I_k} |f(t)|dt > \lambda.$$

It is useful to rewrite this as

$$m(I_k) < \frac{1}{\lambda} \int_{I_k} |f(t)|dt.$$

Now the intervals $\{I_k\}_{k \in K}$ certainly cover K. By the lemma, we may extract a pairwise disjoint subcollection $\{I_{k_j}\}_{j=1}^M$ whose threefold dilates cover K. Putting these ideas together, we see that

$$m(K) \leq m\left[\bigcup_{j=1}^M 3I_{k_j}\right]$$

$$\leq 3\sum_{j=1}^M m(I_{k_j})$$

$$\leq 3\sum_{j=1}^M \frac{1}{\lambda} \int_{I_{k_j}} |f(t)|dt.$$

Note that the intervals over which we are integrating in the last sum on the right are pairwise disjoint. So we may majorize the sum of these integrals by (2π times) the L^1 norm of f. In other words,

$$m(K) \leq \frac{3 \cdot 2\pi}{\lambda} \|f\|_{L^1}.$$

This is what we wished to prove. $\square$

Proof of Proposition A1.6.4:   The estimate for $0 < r \leq 1/2$ is easy (do it as an exercise—either using the maximum principle or by imitating the proof that we now present). Thus we concentrate on $1/2 < r < 1$. We estimate

$$|P_r f(e^{i\theta})| = \left| \frac{1}{2\pi} \int_0^{2\pi} f(e^{i(\theta - \psi)}) \frac{1 - r^2}{1 - 2r \cos \psi + r^2} \, d\psi \right|$$

$$= \left| \frac{1}{2\pi} \int_{-\pi}^{\pi} f(e^{i(\theta - \psi)}) \frac{1 - r^2}{(1 - r)^2 + 2r(1 - \cos \psi)} \, d\psi \right|$$

$$\leq \frac{1}{2\pi} \sum_{j=0}^{[\log_2(\pi/(1-r))]} \int_{S_j} |f(e^{i(\theta - \psi)})| \frac{1 - r^2}{(1 - r)^2 + 2r(2^{j-2}(1 - r))^2} \, d\psi$$

$$+ \frac{1}{2\pi} \int_{|\psi| < 1-r} |f(e^{i(\theta - \psi)})| \frac{1 - r^2}{(1 - r)^2} \, d\psi,$$

where $S_j = \{\psi : 2^j(1 - r) \leq |\psi| < 2^{j+1}(1 - r)\}$. Now this last expression is (since $1 + r < 2$)

$$\leq \frac{1}{2\pi} \sum_{j=0}^{\infty} \frac{1}{2^{2j-4}(1 - r)} \int_{|\psi| < 2^{j+1}(1-r)} |f(e^{i(\theta - \psi)})| \, d\psi$$

$$+ \frac{1}{\pi} \frac{1}{1 - r} \int_{|\psi| < 1-r} |f(e^{i(\theta - \psi)})| \, d\psi$$

$$\leq \frac{64}{2\pi} \sum_{j=0}^{\infty} 2^{-j} \left[\frac{1}{2 \cdot 2^{j+1}(1 - r)} \int_{|\psi| < 2^{j+1}(1-r)} |f(e^{i(\theta - \psi)})| \, d\psi \right]$$

$$+ \frac{2}{\pi} \left[\frac{1}{2(1 - r)} \int_{|\psi| < 1-r} |f(e^{i(\theta - \psi)})| \, d\psi \right]$$

$$\leq \frac{64}{2\pi} \cdot \sum_{j=0}^{\infty} 2^{-j} M f(e^{i\theta}) + \frac{2}{\pi} M f(e^{i\theta})$$

$$\leq \frac{128}{2\pi} Mf(e^{i\theta}) + \frac{2}{\pi} Mf(e^{i\theta})$$

$$\equiv C \cdot Mf(e^{i\theta}).$$

This is the desired estimate.   □

Proof of Corollary A1.6.5: We know that $P^* f$ is majorized by a constant times the Hardy–Littlewood maximal operator of f. Since (by (A1.6.3.1)) the latter is weak-type $(1, 1)$, the result follows.   □

Proof of Theorem A1.6.6: The remarkable thing to notice is that this result now follows with virtually no additional work: First observe that the continuous functions are dense in $L^1(\mathbb{T})$. Second, Theorem A1.5.4 tells us that the desired conclusion holds for functions in this dense set. Finally, Proposition A1.6.5 gives that P^* is weak-type $(1, 1)$. Thus we may apply FAPII and obtain the result.   □

Proof of Proposition A1.7.4: This is a direct application of Parseval's formula A1.7.3. To wit, let $f \in L^2(\mathbb{T})$ and let Λ be as in the statement of the proposition. Then

$$\|\mathcal{M}_\Lambda f\|_{L^2(\mathbb{T})}^2 = \sum_{j=-\infty}^{\infty} |\widehat{\mathcal{M}_\Lambda f}(j)|^2$$

$$= \sum_{j=-\infty}^{\infty} |\lambda_j \widehat{f}(j)|^2$$

$$\leq (\sup_j |\lambda_j|^2) \cdot \sum_{j=-\infty}^{\infty} |\widehat{f}(j)|^2$$

$$= (\sup_j |\lambda_j|^2) \cdot \|f\|_{L^2(\mathbb{T})}^2.$$

This shows that

$$\|\mathcal{M}_\Lambda\|_{\mathrm{op}} \leq \sup_j |\lambda_j| = \|\{\lambda_j\}\|_{\ell^\infty}.$$

For the reverse inequality, fix an integer j_0 and let $e_{j_0}(x) = e^{ij_0 x}$. Then $\|e_{j_0}\|_{L^2(\mathbb{T})} = 1$ and

$$\|\mathcal{M}_\Lambda e_{j_0}\|_{L^2(\mathbb{T})} = \|\lambda_{j_0} e^{ij_0 x}\|_{L^2(\mathbb{T})} = |\lambda_{j_0}|,$$

showing that $\|\mathcal{M}_\Lambda\|_{\mathrm{op}} \geq |\lambda_{j_0}|$ for each j_0. The result is that $\|\mathcal{M}_\Lambda\|_{\mathrm{op}} = \|\{\lambda_j\}\|_{\ell^\infty}$.   □

Appendix 2: The Fourier Transform

This appendix provides basic background in the concepts of the Fourier transform. Some readers will be familiar with these ideas, and they can easily skip this material (or refer to it as needed). For ease of reading, proofs are placed at the end of the appendix.

This exposition follows familiar lines, and the reader may find similar expositions in [ZYG], [KRA5], [KAT].

A2.1 Fundamental Properties of the Fourier Transform

Capsule: The basic properties of the Fourier transform are quite analogous (and proved very similarly) to the basic properties of Fourier series. The two theories begin to diverge when we discuss continuity, differentiability, and other subtle features of the Fourier transform function. Perhaps more significantly, the question of invariance with respect to group actions makes more sense, and is more natural, for the Fourier transform. We cannot say much about this matter here, but see [STE2].

The full-blown theory of the Fourier transform on N-dimensional Euclidean space is a relatively modern idea (developed mainly in the last sixty years). A good modern source is [STG1]. We present here a sketch of the main ideas, but there is no claim of originality either in content or in presentation.

Most of the results parallel facts that we have already seen in the context of Fourier series on the circle. But some, such as the invariance properties of the Fourier transform under the groups that act on Euclidean space (Section 3.2), will be new. One of the main points of this discussion of the Fourier transform is that it is valid in *any* Euclidean space $\mathbb{R}^N$.

We again follow the paradigm of Chapter 2 by stating theorems without proof, and providing the proof at the end of the chapter. We hope that this makes for a brisk transit for the reader, with sufficient details where needed. The energetic reader may attempt to supply his or her own proofs.

If $t, \xi \in \mathbb{R}^N$ then we let

$$t \cdot \xi \equiv t_1 \xi_1 + \cdots + t_N \xi_N.$$

We define the *Fourier transform* of a function $f \in L^1(\mathbb{R}^N)$ by

$$\widehat{f}(\xi) = \int_{\mathbb{R}^N} f(t) e^{it \cdot \xi} \, dt.$$

Here dt denotes Lebesgue N-dimensional measure. Many references will insert a factor of 2π in the exponential or in the measure. Others will insert a minus sign in the exponent. There is no agreement on this matter. We have opted for this definition because of its simplicity.

We note that the significance of the exponentials $e^{it \cdot \xi}$ is that the only continuous multiplicative homomorphisms of $\mathbb{R}^N$ into the circle group are the functions $\phi_\xi(t) = e^{it \cdot \xi}$, $\xi \in \mathbb{R}^N$. [We leave it to the reader to observe that the argument mentioned in Section 2.1 for the circle group works nearly verbatim here. See also [FOL4].] These functions are called the *characters* of the additive group $\mathbb{R}^N$.

Prelude: Together with the fact that the Fourier transform maps L^2 to L^2 (as an isometry), the boundedness from L^1 to L^∞ forms part of the bedrock of the subject. Interpolation then gives that $\widehat{}$ maps L^p to $L^{p/(p-1)}$ for $1 < p < 2$. William Beckner [BEC] has calculated the actual norm of the operator on each of these spaces. The behavior of $\widehat{}$ on L^p for $p > 2$ is problematic. It is difficult to make useful quantitative statements.

Proposition A2.1.1 *If* $f \in L^1(\mathbb{R}^N)$, *then*

$$\|\widehat{f}\|_{L^\infty(\mathbb{R}^N)} \le \|f\|_{L^1(\mathbb{R}^N)}.$$

In other words, $\widehat{}$ is a bounded operator from L^1 to L^∞. We sometimes denote the operator by $\mathcal{F}$.

Prelude: The next two results are the reason that the Fourier transform is so useful in the theory of partial differential equations. The Fourier transform turns an equation involving partial derivatives into an algebraic equation involving monomials. Thus a variety of algebraic and analytic techniques may be invoked in finding and estimating solutions.

Proposition A2.1.2 *If* $f \in L^1(\mathbb{R}^N)$, f *is differentiable, and* $\partial f / \partial x_j \in L^1(\mathbb{R}^N)$, *then*

$$\left(\frac{\partial f}{\partial x_j} \right)^{\widehat{}}(\xi) = -i\xi_j \widehat{f}(\xi).$$

Proposition A2.1.3 *If* $f \in L^1(\mathbb{R}^N)$ *and* $i x_j f \in L^1(\mathbb{R}^N)$, *then*

$$\left(i x_j f \right)^{\widehat{}} = \frac{\partial}{\partial \xi_j} \widehat{f}.$$

Prelude: The next result has intuitive appeal. For if the function is smooth (and, after all, smooth functions are dense), then the function is locally nearly constant. And clearly a trigonometric function of high frequency will nearly integrate to zero against a constant function. That is what the proposition is telling us. The proposition says then that the Fourier transform maps L^1 to C_0, the space of continuous functions that vanish at ∞ (the continuity follows from Lebesgue dominated convergence). It can be shown (see below), using the open mapping theorem, that this mapping is *not* onto. It is not possible to say exactly what the image of the Fourier transform is (although a number of necessary and sufficient conditions are known).

Proposition A2.1.4 (The Riemann–Lebesgue lemma)
If $f \in L^1(\mathbb{R}^N)$, then

$$\lim_{\xi \to \infty} |\widehat{f}(\xi)| = 0.$$

Proposition A2.1.5 *Let $f \in L^1(\mathbb{R}^N)$. Then $\widehat{f}$ is uniformly continuous and vanishes at ∞.*

Let $C_0(\mathbb{R}^N)$ denote the continuous functions on $\mathbb{R}^N$ that vanish at ∞. Equip this space with the supremum norm. Then our results show that the Fourier transform maps L^1 to C_0 continuously, with operator norm 1 (Propositions A2.1.1, A2.1.4). We show in Proposition 3.3.8, using a clever functional analysis argument, that it is *not* onto. It is rather difficult—see [KAT]—to give an explicit example of a C_0 function that is not in the image of the Fourier transform.

Later in this chapter (Section 3.3), we examine the action of the Fourier transform on L^2.

A2.2 Invariance and Symmetry Properties of the Fourier Transform

Capsule: The symmetry properties that we consider here fit very naturally into the structure of Euclidean space, and thus suit the Fourier transform very naturally. They would not arise in such a fashion in the study of Fourier series. Symmetry is of course one of the big ideas of modern mathematics. It certainly plays a central role in the development of the Fourier transform.

The three Euclidean groups that act naturally on $\mathbb{R}^N$ are

- rotations
- dilations
- translations

Certainly a large part of the utility of the Fourier transform is that it has natural invariance properties under the actions of these three groups. We shall now explicitly describe those properties.

We begin with the orthogonal group $O(N)$; an $N \times N$ matrix is *orthogonal* if it has real entries and its rows form an orthonormal system of vectors. Orthogonal matrices M are characterized by the property that $M^{-1} = {}^t M$. A *rotation* is an orthogonal matrix with determinant 1 (also called a *special orthogonal* matrix).

Prelude: The next result should not be a big surprise, for the rotation of an exponential is just another exponential (exercise—remember that the adjoint of a rotation is its inverse, which is another rotation). This observation is intimately bound up with the fact that the Laplacian, and of course its fundamental solution Γ, is rotation-invariant.

Proposition A2.2.1 *Let ρ be a rotation of $\mathbb{R}^N$. Let $f \in L^1(\mathbb{R}^N)$. We define $\rho f(x) = f(\rho(x))$. Then we have the formula*

$$\widehat{\rho f} = \rho \widehat{f}.$$

Definition A2.2.2 For $\delta > 0$ and $f \in L^1(\mathbb{R}^N)$ we set $\alpha_\delta f(x) = f(\delta x)$ and $\alpha^\delta f(x) = \delta^{-N} f(x/\delta)$. These are the dual *dilation operators* of Euclidean analysis.

Prelude: Dilation operators are a means of bringing the part of space near infinity near the finite part of space (near the origin) and vice versa. While this last statement may sound a bit dreamy, it is in fact an important principle in harmonic analysis. The way that the Fourier transform interacts with dilations is fundamental.

Proposition A2.2.3 *The dilation operators interact with the Fourier transform as follows:*

$$(\alpha_\delta f)\widehat{} = \alpha^\delta(\widehat{f}),$$

$$\widehat{\alpha^\delta f} = \alpha_\delta(\widehat{f}).$$

For any function f on $\mathbb{R}^N$ and $a \in \mathbb{R}^N$ we define $\tau_a f(x) = f(x - a)$. Clearly τ_a is a *translation operator*.

Prelude: One of the principal objects of study in classical harmonic analysis is the so-called translation-invariant operator. Thus the invariance of the Fourier transform under translations is essential for us.

A modern thrust in harmonic analysis is to develop tools for doing analysis where there is less structure—for instance on the boundary of a domain. In such a setting there is no notion of translation-invariance. The relatively new field of non-linear analysis also looks at matters from quite a different point of view.

Proposition A2.2.4 *If $f \in L^1(\mathbb{R}^N)$ then*

$$\widehat{\tau_a f}(\xi) = e^{ia \cdot \xi} \widehat{f}(\xi)$$

and

$$\left[\tau_a\{\widehat{f}\}\right](\xi) = \left[e^{-ia \cdot t} f(t)\right]\widehat{}(\xi).$$

Much of classical harmonic analysis—especially in the last century—concentrated on what we call translation-invariant operators. An operator T on functions is called *translation-invariant* if

$$T(\tau_a f)(x) = (\tau_a T f)(x)$$

for every x.[4] It is a basic fact that any translation-invariant integral operator is given by convolution with a kernel k (see also Lemma 2.1.3). We shall not prove this general fact (but see [STG1]), because the kernels that we need in the sequel will be explicitly calculated in context.

Prelude: The next two results are not profound, but they are occasionally useful—as in the proof of Plancherel's theorem.

Proposition A2.2.5 *For $f \in L^1(\mathbb{R}^N)$ we let $\widetilde{f}(x) = f(-x)$. Then $\widehat{\widetilde{f}} = \widetilde{\widehat{f}}$.*

Proposition A2.2.6 *We have*

$$\widehat{\overline{f}} = \overline{\widetilde{\widehat{f}}},$$

where the overbar denotes complex conjugation.

Prelude: The proposition that follows lies at the heart of Schwartz's distribution theory. It is obviously a means of dualizing the Fourier transform. It is extremely useful in the theory of partial differential equations, and of course in the study of pseudodifferential operators.

Proposition A2.2.7 *If $f, g \in L^1$, then*

$$\int \widehat{f}(\xi) g(\xi) \, d\xi = \int f(\xi) \widehat{g}(\xi) \, d\xi.$$

We conclude this section with a brief discussion of homogeneity. Recall the definition of the basic dilations of Fourier analysis given in Definition A2.2.2. Now let $\beta \in \mathbb{R}$. We say that a function f on $\mathbb{R}^N$ (or, sometimes, on $\mathbb{R}^N \setminus \{0\}$) is *homogeneous of degree β* if for each $x \in \mathbb{R}^N$ and each $\delta > 0$,

$$\alpha_\delta f(x) = \delta^\beta f(x).$$

The typical example of a function homogeneous of degree β is $f(x) = |x|^\beta$, but this is not the only example. In fact, let ϕ be any function on the unit sphere of $\mathbb{R}^N$. Now set

$$f(x) = |x|^\beta \cdot \phi\left(\frac{x}{|x|}\right), \qquad x \neq 0.$$

Then f is homogeneous of degree β.

There is a slight technical difficulty for Fourier analysis in this context because no function that is homogeneous of any degree β lies in any L^p class. If we are going

---

[4] It is perhaps more accurate to say that such an operator *commutes with translations*. However, the terminology "translation-invariant" is standard.

to apply the Fourier transform to a homogeneous function then some additional idea will be required. Thus we must pass to the theory of distributions.

We take a moment now for very quick digression on Schwartz functions and the Fourier transform. On $\mathbb{R}^N$, a *multi-index* α is an N-tuple $(a_1, \ldots, a_N)$ of nonnegative integers (see also Section 1.1). Then we let

$$x^\alpha \equiv x_1^{a_1} \cdot x_2^{a_2} \cdots x_N^{a_N}$$

and

$$\frac{\partial^\alpha}{\partial x^\alpha} = \frac{\partial^{a_1}}{\partial x_1^{a_1}} \frac{\partial^{a_2}}{\partial x_2^{a_2}} \cdots \frac{\partial^{a_N}}{\partial x_N^{a_N}}.$$

If α, β are multi-indices then we define the seminorm $\rho_{\alpha,\beta}$ on C^∞ functions by

$$\rho_{\alpha,\beta}(f) \equiv \sup_{x \in \mathbb{R}^N} \left| x^\alpha \cdot \frac{\partial^\beta}{\partial x^\beta} f(x) \right|.$$

We call $\rho_{\alpha,\beta}$ a seminorm because it will annihilate constants and certain low-degree polynomials. The *Schwartz space* S is the collection of all C^∞ functions f on $\mathbb{R}^N$ such that $\rho_{\alpha,\beta}(f) < \infty$ for all choices of α and β. A simple example of a Schwartz function is $f(x) = e^{-x^2}$. Of course any derivative of f is also a Schwartz function; and the product of f with any polynomial is a Schwartz function.

Now S, equipped with the family of seminorms $\rho_{\alpha,\beta}$, is a topological vector space. Its dual is the space of *Schwartz distributions* S'. See [STG1] or [KRA3] for a more thorough treatment of these matters. It is easy to see that any function φ in any L^p class, $1 \leq p \leq \infty$, induces a Schwartz distribution by

$$S \ni f \longmapsto \int_{\mathbb{R}^N} f(x) \cdot \varphi(x) \, dx \, ;$$

just use Hölder's inequality to verify this assertion.

It is easy to use Propositions A2.1.2 and A2.1.3 to see that the Fourier transform maps S to S. In fact, the mapping is one-to-one and onto, as one can see with Fourier inversion (see Section 3.3).

Let f be a function that is homogeneous of degree β on $\mathbb{R}^N$ (or on $\mathbb{R}^N \setminus \{0\}$). Then f will not be in any L^p class, so we may not consider its Fourier transform in the usual sense. We need a different, more general, definition of the Fourier transform.

Let λ be any Schwartz distribution. We define the *distribution Fourier transform* of λ by the condition

$$\langle \widehat{\lambda}, \varphi \rangle = \langle \lambda, \widehat{\varphi} \rangle$$

for any Schwartz testing function φ. The reader may check (see Proposition A2.2.7) that this gives a well-defined definition of $\widehat{\lambda}$ as a distribution (Fourier inversion, discussed in the next section, may prove useful here). In particular, if f is a function that induces a Schwartz distribution (by integration, as usual), then

$$\int_{\mathbb{R}^N} \widehat{f}(\xi) \cdot \varphi(\xi)\, d\xi = \int_{\mathbb{R}^N} f(\xi) \cdot \widehat{\varphi}(\xi)\, d\xi \qquad (\text{A2.2.8})$$

for any Schwartz function φ. In other words, $\widehat{f}$ is defined by this equality. [The reader familiar with the theory of differential equations will note the analogy with the definition of "weak solution."]

We have defined dilations of functions, and the corresponding notion of homogeneity, earlier in this chapter. If λ is a Schwartz distribution and φ a Schwartz testing function and $\delta > 0$ then we define the dilations

$$[\alpha_\delta \lambda](\varphi) = \lambda(\alpha^\delta \varphi)$$

and

$$[\alpha^\delta \lambda](\varphi) = \lambda(\alpha_\delta \varphi).$$

In the next proposition, it is not a priori clear that the Fourier transform of the function f is a function (after all, f will certainly *not* be L^1). So the proposition should be interpreted in the sense of distributions.

Prelude: Homogeneity has been one of the key ideas in modern analysis. It has been known since the time of Riesz that kernels homogeneous of degree $-N + \alpha$, $0 < \alpha < N$, are relatively tame, whereas kernels homogeneous of degree $-N$ are critical in their behavior (these ideas are treated in detail in our discussion of fractional and singular integrals). As a result of these considerations, the behavior of the Fourier transform vis à vis homogeneity is vital information for us.

Proposition A2.2.9 Let f be a function or distribution that is homogeneous of degree β. Then $\widehat{f}$ is homogeneous of degree $-\beta - N$.

Remark: Technically speaking, (A2.2.8) defines $\widehat{f}$ only as a distribution, or generalized function—even when f itself is a function. Extra considerations are necessary to determine that $\widehat{f}$ is a function when f is a function.

A2.3 Convolution and Fourier Inversion

Capsule: From our point of view, the natural interaction of the Fourier transform with convolution is a consequence of symmetry (particularly, the commutation with translations). The formula $\widehat{(f * g)} = \widehat{f} \cdot \widehat{g}$ is of preeminent importance. And also convolution (again, it is the translation-invariance that is the key) makes the elegant form of Fourier inversion possible. Fourier inversion, in turn, tells us that the Fourier transform is injective, and that opens up an array of powerful ideas.

Now we study how the Fourier transform respects convolution. Recall that if f, g are in $L^1(\mathbb{R}^N)$ then we define the convolution

$$f * g(x) = \int_{\mathbb{R}^N} f(x - t)g(t)dt = \int_{\mathbb{R}^N} f(t)g(x - t)dt.$$

The second equality follows by change of variable. The function $f * g$ is automatically in L^1 just by an application of the Fubini–Tonelli theorem.

Prelude: The next result is one manifestation of the translation-invariance of our ideas. In particular, this formula shows that a convolution kernel corresponds in a natural way to a Fourier multiplier.

Proposition A2.3.1 *If* $f, g \in L^1$, *then*

$$\widehat{f * g} = \widehat{f} \cdot \widehat{g}.$$

A2.3.1 The Inverse Fourier Transform

Our goal is to be able to recover f from $\widehat{f}$. The process is of interest for several reasons:

(a) We wish to be able to invert the Fourier transform.
(b) We wish to know that the Fourier transform is one-to-one.
(c) We wish to establish certain useful formulas that involve the inversion concept.

The program just described entails several technical difficulties. First, we need to know that the Fourier transform is one-to-one in order to have any hope of success. Second, we would like to say that

$$f(t) = c \cdot \int \widehat{f}(\xi)e^{-it \cdot \xi} \, d\xi. \tag{A2.3.1}$$

But in general the Fourier transform $\widehat{f}$ of an L^1 function f is not integrable (just calculate the Fourier transform of $\chi_{[0,1]}$)—so the expression on the right of (A2.3.1) does not necessarily make any sense.

To handle this situation, we will construct a family of *summability kernels* G_ϵ having the following properties:

(A2.3.2) $G_\epsilon * f \to f$ in the L^1 topology as $\epsilon \to 0$;
(A2.3.3) $\widehat{G_\epsilon}(\xi) = e^{-\epsilon|\xi|^2/2}$;
(A2.3.4) $G_\epsilon * f$ and $\widehat{G_\epsilon * f}$ are both integrable.

It will be useful to prove formulas about $G_\epsilon * f$ and then pass to the limit as $\epsilon \to 0^+$. Notice that **(A2.3.3)** mandates that the Fourier transform of G_ϵ be the Gaussian. The Gaussian (suitably normalized) is known to be an eigenfunction of the Fourier transform.

Prelude: It is a fundamental fact of Fourier analysis that the Fourier transform of the Gaussian e^{-x^2} is (up to certain adjustments by constants) the Gaussian itself. This tells us right away that the Gaussian is an eigenfunction of the Fourier transform and leads, after some calculations, to the other eigenfunctions (by way of the Hermite polynomials). We shall see below that the Gaussian plays a useful role in summability questions for the Fourier transform. It is a recurring feature of the subject.

Lemma A2.3.5

$$\int_{\mathbb{R}^N} e^{-|x|^2}\,dx = (\sqrt{\pi})^N.$$

Remark: Although the proof presented at the end of the chapter is the most common method for evaluating $\int e^{-|x|^2}\,dx$, several other approaches are provided in [HEI].

Corollary A2.3.6

$$\int_{\mathbb{R}^N} \pi^{-N/2} e^{-|x|^2}\,dx = 1.$$

Now let us calculate the Fourier transform of $e^{-|x|^2}$. It is slightly more convenient to calculate the Fourier transform of $f(x) = e^{-|x|^2/2}$, and this we do.

It suffices to treat the 1-dimensional case because

$$\left(e^{-|x|^2/2}\right)^{\widehat{\ }}(\xi) = \int_{\mathbb{R}^N} e^{-|x|^2/2} e^{ix\cdot\xi}\,dx$$

$$= \int_{\mathbb{R}} e^{-x_1^2/2} e^{ix_1\xi_1}\,dx_1 \cdots \int_{\mathbb{R}} e^{-x_N^2/2} e^{ix_N\xi_N}\,dx_N.$$

We thank J. Walker for providing the following argument (see also [FOL3, p. 242]):

By Proposition A2.1.3,

$$\frac{d\widehat{f}}{d\xi} = \int_{-\infty}^{\infty} ixe^{-x^2/2} e^{ix\xi}\,dx.$$

We now integrate by parts, with $dv = xe^{-x^2/2}\,dx$ and $u = ie^{ix\xi}$. The boundary terms vanish because of the presence of the rapidly decreasing exponential $e^{-x^2/2}$. The result is then

$$\frac{d\widehat{f}}{d\xi} = -\xi \int_{-\infty}^{\infty} e^{-x^2/2} e^{ix\xi}\,dx = -\xi\widehat{f}(\xi).$$

This is just a first-order linear ordinary differential equation for $\widehat{f}$. It is easily solved using the integrating factor $e^{\xi^2/2}$, and we find that

$$\widehat{f}(\xi) = \widehat{f}(0)\cdot e^{-\xi^2/2}.$$

But Lemma A2.3.5 (and a change of variable) tells us that $\widehat{f}(0) = \int f(x)\,dx = \sqrt{2\pi}$. In summary, on $\mathbb{R}^1$,

$$\widehat{e^{-x^2/2}}(\xi) = \sqrt{2\pi}\,e^{-\xi^2/2}; \tag{A2.3.7}$$

in $\mathbb{R}^N$ we therefore have

$$\widehat{(e^{-|x|^2/2})}(\xi) = (\sqrt{2\pi})^N e^{-|\xi|^2/2}.$$

We can, if we wish, scale this formula to obtain

$$(\widehat{e^{-|x|^2}})(\xi) = \pi^{N/2} e^{-|\xi|^2/4}.$$

The function $G(x) = (2\pi)^{-N/2} e^{-|x|^2/2}$ is called the *Gauss–Weierstrass kernel*, or sometimes just the *Gaussian*. It is a summability kernel (see [KAT]) for the Fourier transform. We shall flesh out this assertion in what follows. Observe that

$$\widehat{G}(\xi) = e^{-|\xi|^2/2} = (2\pi)^{N/2} G(\xi). \tag{A2.3.8}$$

On $\mathbb{R}^N$ we define

$$G_\epsilon(x) = \alpha^{\sqrt{\epsilon}}(G)(x) = \epsilon^{-N/2} (2\pi)^{-N/2} e^{-|x|^2/(2\epsilon)}.$$

Then

$$\widehat{G_\epsilon}(\xi) = \left(\alpha^{\sqrt{\epsilon}} G\right)^{\wedge}(\xi) = \alpha_{\sqrt{\epsilon}} \widehat{G}(\xi) = e^{-\epsilon|\xi|^2/2},$$

$$\widehat{\widehat{G_\epsilon}}(\xi) = \left(e^{-\epsilon|x|^2/2}\right)^{\wedge}(\xi)$$

$$= \left(\alpha_{\sqrt{\epsilon}} e^{-|x|^2/2}\right)^{\wedge}(\xi)$$

$$= \alpha^{\sqrt{\epsilon}} \left(e^{-|x|^2/2}\right)^{\wedge}(\xi)$$

$$= \epsilon^{-N/2} (2\pi)^{N/2} e^{-|\xi|^2/(2\epsilon)}$$

$$= (2\pi)^N G_\epsilon(\xi).$$

Observe that $\widehat{\widehat{G_\epsilon}}$ is, except for the factor of $(2\pi)^N$, the same as G_ϵ. This fact anticipates the Fourier inversion formula that we are gearing up to prove.

Now assume that $f \in C_c^{N+1}$. This implies in particular that $f, \widehat{f}$ are in L^1 and continuous (see the proof of the Riemann–Lebesgue lemma, Proposition A2.1.4). We apply Proposition A2.2.7 with $g = \widehat{G_\epsilon} \in L^1$ to obtain

$$\int f(x) \widehat{\widehat{G_\epsilon}}(x) \, dx = \int \widehat{f}(\xi) \widehat{G_\epsilon}(\xi) \, d\xi.$$

In other words,

$$\int f(x)(2\pi)^N G_\epsilon(x) \, dx = \int \widehat{f}(\xi) e^{-\epsilon|\xi|^2/2} \, d\xi. \tag{A2.3.9}$$

Now $e^{-\epsilon|\xi|^2/2} \to 1$ as $\epsilon \to 0^+$, uniformly on compact sets. Thus

$$\int \widehat{f}(\xi) e^{-\epsilon|\xi|^2/2} \, d\xi \to \int \widehat{f}(\xi) \, d\xi.$$

That concludes our analysis of the right-hand side of (A2.3.9).

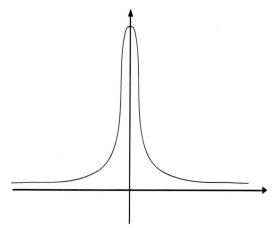

Figure A2.1. The Gaussian.

Next observe that for any $\epsilon > 0$,

$$\int G_\epsilon(x)\,dx = \int G_\epsilon(x)e^{ix\cdot 0}\,dx = \widehat{G_\epsilon}(0) = 1.$$

Thus the left side of (A2.3.7) equals

$$(2\pi)^N \int f(x)G_\epsilon(x)\,dx = (2\pi)^N \int f(0)G_\epsilon(x)\,dx$$

$$+ (2\pi)^N \int [f(x) - f(0)]G_\epsilon(x)\,dx$$

$$\equiv A_\epsilon + B_\epsilon$$

Now it is crucial to observe that the function G_ϵ has total mass 1, and that mass is more and more concentrated near the origin as $\epsilon \to 0^+$. Refer to Figure A2.1. As a result, $A_\epsilon \to (2\pi)^N \cdot f(0)$ and $B_\epsilon \to 0$. Altogether then,

$$\int f(x)(2\pi)^N G_\epsilon(x)\,dx \to (2\pi)^N f(0)$$

as $\epsilon \to 0^+$. Thus we have evaluated the limits of the left and right-hand sides of (A2.3.9). [Note that the argument just given is analogous to some of the proofs for the summation of Fourier series that were presented in Chapter 1.] We have proved the following:

Prelude: This next proposition (and the theorem following) is our first glimpse of the role of the Gaussian in summability results. It is simple and elegant and makes the theory proceed smoothly.

Proposition A2.3.10 (Gauss–Weierstrass summation) *Suppose that $f \in C_c^{N+1}$ $(\mathbb{R}^N)$ (this hypothesis is included to guarantee that $\widehat{f} \in L^1$). Then*

$$f(0) = \lim_{\epsilon \to 0^+} (2\pi)^{-N} \int \widehat{f}(\xi) e^{-\epsilon |\xi|^2/2} \, d\xi.$$

Limiting arguments may be used to show that the formula in Proposition A2.3.10 is valid for *any* $f \in L^1$, even when $\widehat{f}$ is not integrable. We shall omit the details of this assertion.

The method of Gauss–Weierstrass summation will prove to be crucial in some of our later calculations. However, in practice, it is convenient to have a result with a simpler and less technical statement. If f, $\widehat{f}$ are both known to be in L^1 (this is true, for example, if f has $(N + 1)$ derivatives in L^1), then a limiting argument gives the following standard result:

Theorem A2.3.11 *If f, $\widehat{f} \in L^1$ (and both are therefore continuous), then*

$$f(0) = (2\pi)^{-N} \int \widehat{f}(\xi) \, d\xi. \qquad\qquad \text{(A2.3.11.1)}$$

Of course there is nothing special about the point $0 \in \mathbb{R}^N$. We now exploit the compatibility of the Fourier transform with translations to obtain a more generally applicable formula. We apply formula (A2.3.11.1) in our theorem to $\tau_{-y} f$ for $f \in L^1$, $\widehat{f} \in L^1$, $y \in \mathbb{R}^N$. The result is

$$\left(\tau_{-y} f \right)(0) = (2\pi)^{-N} \int \left(\tau_{-y} f \right)^{\vee}(\xi) \, d\xi.$$

We summarize the key point in a theorem:

Theorem A2.3.12 (The Fourier inversion formula) *If f, $\widehat{f} \in L^1$ (and hence both f, $\widehat{f}$ are continuous), then for any $y \in \mathbb{R}^N$ we have*

$$f(y) = (2\pi)^{-N} \int \widehat{f}(\xi) e^{-iy \cdot \xi} \, d\xi.$$

Observe that this theorem tells us something that we already know implicitly: that if f, $\widehat{f}$ are both L^1, then f (being the inverse Fourier transform of an L^1 function) can be corrected on a set of measure zero to be continuous.

Prelude: We reap here the benefit of our development of the Gaussian in Fourier analysis. Now we see that the Fourier transform is univalent, and this sets the stage for the inverse Fourier transform, which we shall develop below.

Corollary A2.3.13 *The Fourier transform is one-to-one. That is, if $f, g \in L^1$ and $\widehat{f} \equiv \widehat{g}$, then $f = g$ almost everywhere.*

Even though we do not know the Fourier transform to be a surjective operator (in fact, see the next theorem) onto the continuous functions vanishing at infinity, it is convenient to be able to make explicit reference to the inverse operation. Thus we *define*

$$g^{\vee}(x) = (2\pi)^{-N} \int g(\xi) e^{-ix \cdot \xi} \, d\xi$$

whenever $g \in L^1(\mathbb{R}^N)$. We call the operation $^\vee$ the *inverse Fourier transform*. Notice that the operation expressed in the displayed equation makes sense for any $g \in L^1$, regardless of the operation's possible role as an inverse to the Fourier transform. It is convenient to give the operation the italicized name.

In particular, observe that if the hypotheses of the Fourier inversion formula are in force, then we have that the following result:

$$f(x) = (\widehat{f})^\vee(x).$$

Since the Fourier transform is one-to-one, it is natural to ask whether it is onto. We have the following result:

Proposition A2.3.14 *The operator*

$$^\vee : L^1 \to C_0$$

is not onto.

Exercise for the Reader: Imitate the proof of this last result to show that the mapping

$$\widehat{} : L^1(\mathbb{T}) \to c_0,$$

which assigns to each integrable function on the circle its sequence of Fourier coefficients (a bi-infinite sequence that vanishes at ∞), is not onto.

It is actually possible to write down explicitly a bi-infinite complex sequence that vanishes at ∞ and is not the sequence of Fourier coefficients of any L^1 function on the circle. See [KAT] for the details.

The Fourier inversion formula is troublesome. First, it represents the function f as the superposition of *uncountably many* basis elements $e^{-ix\cdot\xi}$, none of which is in any L^p class. In particular, the Fourier transform does not localize well. The theory of wavelets (see [MEY1], [MEY2], [MEY3], [KRA5]) is designed to address some of these shortcomings. We shall not treat wavelets in the present book.

Proofs of the Results in Appendix 2

Proof of Proposition A2.1.1: Observe that for any $\xi \in \mathbb{R}^N$,

$$|\widehat{f}(\xi)| \le \int |f(t)|dt. \qquad \square$$

Proof of Proposition 2.1.2: Integrate by parts: If $f \in C_c^\infty$ (the C^∞ functions with compact support), then

$$\left(\frac{\partial f}{\partial x_j}\right)^{\widehat{}}(\xi) = \int \frac{\partial f}{\partial t_j} e^{it\cdot\xi}\, dt$$

$$= \int \cdots \int \left[\int \frac{\partial f}{\partial t_j} e^{it \cdot \xi} \, dt_j \right] dt_1 \cdots dt_{j-1} dt_{j+1} \cdots dt_N$$

$$= -\int \cdots \int f(t) \left(\frac{\partial}{\partial t_j} e^{it \cdot \xi} \right) dt_j dt_1 \cdots dt_{j-1} dt_{j+1} \cdots dt_N$$

$$= -i\xi_j \int \cdots \int f(t) e^{it \cdot \xi} \, dt$$

$$= -i\xi_j \widehat{f}(\xi).$$

[Of course the "boundary terms" in the integration by parts vanish since $f \in C_c^\infty$.]
The general case follows from a limiting argument. $\qquad\square$

Proof of Proposition A2.1.3: Differentiate under the integral sign. $\qquad\square$

Proof of Proposition A2.1.4: First assume that $g \in C_c^2(\mathbb{R}^N)$. We know that

$$\|\widehat{g}\|_{L^\infty} \le \|g\|_{L^1} \le C$$

and, for each j,

$$\left\| \xi_j^2 \widehat{g} \right\|_{L^\infty} = \left\| \left[\left(\frac{\partial^2}{\partial x_j^2} \right) g \right]^{\widehat{}} \right\|_{L^\infty} \le \left\| \left(\frac{\partial^2}{\partial x_j^2} \right) g \right\|_{L^1} = C_j'.$$

Then $(1 + |\xi|^2)\widehat{g}$ is bounded. Therefore

$$|\widehat{g}(\xi)| \le \frac{C''}{1 + |\xi|^2} \xrightarrow{|\xi| \to +\infty} 0.$$

This proves the result for $g \in C_c^2$. [Notice that the same argument also shows that if
$g \in C_c^{N+1}(\mathbb{R}^N)$ then $\widehat{g} \in L^1$.]

Now let $f \in L^1$ be arbitrary. Then there is a function $\psi \in C_c^2(\mathbb{R}^N)$ such that
$\|f - \psi\|_{L^1} < \epsilon/2$. Indeed, it is a standard fact from measure theory (see [RUD2])
that there is a C_c function (a continuous function with compact support) that so
approximates. Then one can convolve that C_c function with a Friedrichs mollifier
(see [KRA5]) to obtain the desired C_c^2 function.

By the result already established, choose M so large that when $|\xi| > M$ then
$|\widehat{\psi}(\xi)| < \epsilon/2$. Then, for $|\xi| > M$, we have

$$|\widehat{f}(\xi)| = |(f - \psi)^{\widehat{}}(\xi) + \widehat{\psi}(\xi)|$$

$$\le |(f - \psi)^{\widehat{}}(\xi)| + |\widehat{\psi}(\xi)|$$

$$\le \|f - \psi\|_{L^1} + \frac{\epsilon}{2}$$

$$< \frac{\epsilon}{2} + \frac{\epsilon}{2} = \epsilon.$$

This proves the result. $\qquad\square$

Proof of Proposition A2.1.5: Note that $\widehat{f}$ is continuous by the Lebesgue dominated convergence theorem:

$$\lim_{\xi \to \xi_0} \widehat{f}(\xi) = \lim_{\xi \to \xi_0} \int f(x)e^{ix \cdot \xi} \, dx = \int \lim_{\xi \to \xi_0} f(x)e^{ix \cdot \xi} \, dx = \widehat{f}(\xi_0).$$

Since $\widehat{f}$ also vanishes at ∞ (see 3.1.4), the result is immediate. $\square$

Proof of Proposition A2.2.1: Remembering that ρ is orthogonal and has determinant 1, we calculate that

$$\widehat{\rho f}(\xi) = \int (\rho f)(t)e^{it \cdot \xi} \, dt = \int f(\rho(t))e^{it \cdot \xi} \, dt$$

$$\stackrel{(s=\rho(t))}{=} \int f(s)e^{i\rho^{-1}(s) \cdot \xi} \, ds$$

$$= \int f(s)e^{i[^t \rho(s)] \cdot \xi} \, ds$$

$$= \int f(s)e^{is \cdot \rho(\xi)} \, ds$$

$$= \widehat{f}(\rho\xi) = \rho \widehat{f}(\xi).$$

The proof is complete. $\square$

Proof of Proposition A2.2.3: We calculate that

$$\widehat{(\alpha_\delta f)}(\xi) = \int (\alpha_\delta f)(t)e^{it \cdot \xi} \, dt$$

$$= \int f(\delta t)e^{it \cdot \xi} \, dt$$

$$\stackrel{(s=\delta t)}{=} \int f(s)e^{i(s/\delta) \cdot \xi} \delta^{-N} \, ds$$

$$= \delta^{-N} \widehat{f}(\xi/\delta)$$

$$= \left(\alpha^\delta(\widehat{f})\right)(\xi).$$

That proves the first assertion. The proof of the second is similar. $\square$

Proof of Proposition A2.2.4: For the first equality, we calculate that

$$\widehat{\tau_a f}(\xi) = \int_{\mathbb{R}^N} e^{ix \cdot \xi}(\tau_a f)(x) \, dx$$

$$= \int_{\mathbb{R}^N} e^{ix \cdot \xi} f(x - a) \, dx$$

$$\underset{=}{(x-a)=t} \int_{\mathbb{R}^N} e^{i(t+a)\cdot\xi} f(t)dt$$

$$= e^{ia\cdot\xi} \int_{\mathbb{R}^N} e^{it\cdot\xi} f(t)dt$$

$$= e^{ia\cdot\xi} \widehat{f}(\xi).$$

The second identity is proved similarly. □

Proof of Proposition A2.2.5: We calculate that

$$\widehat{\widetilde{f}}(\xi) = \int \widetilde{f}(t)e^{it\cdot\xi}\,dt = \int f(-t)e^{it\cdot\xi}\,dt$$

$$= \int f(t)e^{-it\cdot\xi}\,dt = \widehat{f}(-\xi) = \widetilde{\widehat{f}}(\xi).$$ □

Proof of Proposition A2.2.6: We calculate that

$$\widehat{\overline{f}}(\xi) = \int \overline{f}(t)e^{it\cdot\xi}\,dt = \overline{\int f(t)e^{-it\cdot\xi}\,dt} = \overline{\widehat{f}(-\xi)} = \overline{\widetilde{\widehat{f}}}(\xi).$$ □

Proof of Proposition A2.2.7:   This is a straightforward change in the order of integration:

$$\int \widehat{f}(\xi)g(\xi)\,d\xi = \iint f(t)e^{it\cdot\xi}\,dt\,g(\xi)d\xi$$

$$= \iint g(\xi)e^{it\cdot\xi}\,d\xi\,f(t)dt$$

$$= \int \widehat{g}(t)f(t)dt.$$

Here we have applied the Fubini–Tonelli theorem. □

Proof of Proposition A2.2.8: Calculate $\alpha_\delta(\widehat{f})$ using (3.2.8) and a change of variables. □

Proof of Proposition A2.3.1: We calculate that

$$\widehat{f * g}(\xi) = \int (f * g)(t)e^{it\cdot\xi}\,dt = \iint f(t-s)g(s)\,ds\,e^{it\cdot\xi}\,dt$$

$$= \iint f(t-s)e^{i(t-s)\cdot\xi}\,dt\,g(s)e^{is\cdot\xi}\,ds$$

$$= \widehat{f}(\xi) \cdot \widehat{g}(\xi).$$ (□)

Proof of Lemma A2.3.5:   Breaking the integral into a product of 1-dimensional integrals, we see that it is enough to treat the case $N = 1$. Set $I = \int_{-\infty}^{\infty} e^{-t^2}\,dt$. Then

$$I \cdot I = \int_{-\infty}^{\infty} e^{-s^2} ds \int_{-\infty}^{\infty} e^{-t^2} dt = \iint_{\mathbb{R}^2} e^{-|(s,t)|^2} ds\, dt$$

$$= \int_{0}^{2\pi} \int_{0}^{\infty} e^{-r^2} r\, dr\, d\theta = \pi.$$

Thus $I = \sqrt{\pi}$, as desired. $\square$

Proof of Theorem A2.3.11: For the proof, consider $g \equiv f * G_\epsilon$, note that $\widehat{g} = \widehat{f} \cdot e^{-\epsilon|\xi|^2/2}$ and apply the previous theorem. Note that Figure A2.2 shows how $\widehat{G_\epsilon}$ flattens out, uniformly on compact sets, as $\epsilon \to 0^+$. We leave the details of the argument to the reader. $\square$

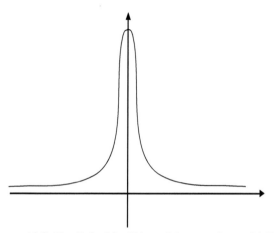

Figure A2.2. "Profile" of the Nth partial summation multiplier.

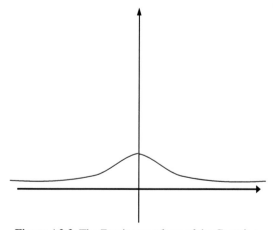

Figure A2.3. The Fourier transform of the Gaussian.

Proof of Corollary A2.3.13: Observe that $f - g \in L^1$ and $\widehat{f} - \widehat{g} \equiv 0 \in L^1$. The theorem allows us to conclude that $(f - g) \equiv 0$, and that is the result. □

Proof of Proposition A2.3.14: For simplicity, we restrict attention to $\mathbb{R}^1$. Seeking a contradiction, we suppose that the operator is in fact surjective. Then the open mapping principle guarantees that there is a constant $C > 0$ such that

$$\|f\|_{L^1} \leq C \cdot \|\widehat{f}\|_{\sup} \qquad \text{for all } f \in L^1. \tag{3.3.14.1}$$

On $\mathbb{R}^1$, let g be the characteristic function of the interval $[-1, 1]$. A calculation shows that the inverse Fourier transform of g is

$$\overset{\vee}{g}(t) = \frac{\sin t}{\pi t}.$$

We would like to say that $f = \overset{\vee}{g}$ violates (3.3.14.1). But this f is not in L^1, so such logic is flawed.

Instead we consider $h_\epsilon \equiv G_\epsilon * g$. Then h_ϵ, being the convolution of two L^1 functions, is certainly L^1. Moreover,

$$|h_\epsilon(x)| = \epsilon^{-1/2}(2\pi)^{-1/2} \int_{-1}^{1} e^{-|x-t|^2/2\epsilon} \, dt$$

$$= \epsilon^{-1/2}(2\pi)^{-1/2} e^{-x^2/2\epsilon} \int_{-1}^{1} e^{xt/\epsilon} e^{-t^2/2\epsilon} \, dt$$

$$\leq C_\epsilon e^{-x^2/4\epsilon}.$$

So $h_\epsilon \in L^1 \cap C_0$.

In addition,

$$h_\epsilon^{\vee}(t) = 2\pi \cdot \overset{\vee}{G_\epsilon}(t) \cdot \overset{\vee}{g} = e^{-\epsilon|t|^2/2} \cdot \overset{\vee}{g}(t) \to \overset{\vee}{g}(t)$$

pointwise. This convergence cannot be boundedly in L^1; otherwise, Fatou's lemma would show that $\overset{\vee}{g} \in L^1$; and we know that that is false.

A contradiction now arises because

$$\|2\pi \cdot \overset{\vee}{G_\epsilon} \cdot \overset{\vee}{g}\|_{L^1} = \|(G_\epsilon * g)^{\vee}\|_{L^1} \leq C \cdot \|G_\epsilon * g\|_{\sup} \leq C \cdot \|G_\epsilon\|_{L^1} \cdot \|g\|_{\sup} = C.$$

As noted, the left side blows up as $\epsilon \to 0^+$. □

Appendix 3: Pseudodifferential Operators

This appendix consists of material offered for cultural purposes. We have endeavored to present fractional integrals and singular integrals as stepping-stones to a comprehensive view of integral operators. What we were seeking in the first half of the twentieth century was a *calculus* of integral operators that can be used to construct parametrices for partial differential operators. Pseudodifferential operators are at least one answer to that quest. The brief introduction provided in this chapter will give the reader a glimpse of the culmination of this program.

Part of the interest of the present discussion is to acquaint the reader with the ideas of an "error term" and of the estimates that enable one to handle an error term. Certainly the Sobolev spaces, treated at the end of Chapter 2, will be of great utility in this treatment.

A3.1 Introduction to Pseudodifferential Operators

Consider the partial differential equation $\Delta u = f$. We wish to study the existence and regularity properties of solutions to this equation and equations like it. It turns out that in practice, existence follows with a little functional analysis from a suitable a priori regularity estimate (to be defined below). Therefore we shall concentrate for now on regularity.

The a priori regularity problem is as follows: If $u \in C_c^\infty(\mathbb{R}^N)$ and if

$$\Delta u = f, \tag{A3.1.1}$$

then how may we estimate u in terms of f? In particular, how can we estimate *smoothness* of u in terms of *smoothness* of f? Taking the Fourier transform of both sides of (A3.1.1) yields

$$\left(\Delta u\right)^\widehat{} = \widehat{f},$$

or

$$-\sum_{j} |\xi_j|^2 \widehat{u}(\xi) = \widehat{f}(\xi).$$

Arguing formally, we may solve this equation for $\widehat{u}$:

$$\widehat{u}(\xi) = -\frac{1}{|\xi|^2} \widehat{f}(\xi). \tag{A3.1.2}$$

Suppose for specificity that we are working in $\mathbb{R}^2$. Then $-1/|\xi|^2$ has an unpleasant singularity and we find that equation (A3.1.2) does not provide useful information.

The problem of studying existence and regularity for linear partial differential operators with constant coefficients was treated systematically in the 1950s by Ehrenpreis and Malgrange, among others. The approach of Ehrenpreis was to write

$$u(x) = c \cdot \int \widehat{u}(\xi) e^{-ix\cdot\xi}\, d\xi = c \cdot \int -\frac{1}{|\xi|^2} \widehat{f}(\xi) e^{-ix\cdot\xi}\, d\xi.$$

Using Cauchy theory, he was able to relate this last integral to

$$\int -\frac{1}{|\xi+i\eta|^2} \widehat{f}(\xi+i\eta) e^{-ix\cdot(\xi+i\eta)}\, d\xi$$

for $\eta > 0$. In this way he avoided the singularity at $\xi = 0$ of the right-hand side of (A3.1.2).

Malgrange's method, by contrast, was first to study (A3.1.1) for those f such that $\widehat{f}$ vanishes to some finite order at 0 and then to apply some functional analysis.

It is a basic fact that for the study of C^∞ regularity, the behavior of the Fourier transform on the finite part of space is of no interest. That is to say, the Paley–Wiener theorem (see [STG1] and our Section 10.3) tells us that the (inverse) Fourier transform of a compactly supported function (or distribution) is real analytic. Thus what is of greatest interest is the Fourier transform of that part of the function that lies outside every compact set.

Thus the philosophy of pseudodifferential operator theory is to replace the Fourier multiplier $1/|\xi|^2$ by the multiplier $(1 - \phi(\xi))/|\xi|^2$, where $\phi \in C_c^\infty(\mathbb{R}^N)$ is identically equal to 1 near the origin. Thus we define

$$(\widehat{Pg})(\xi) = -\frac{1 - \phi(\xi)}{|\xi|^2} \widehat{g}(\xi)$$

for any $g \in C_c^\infty$. Equivalently,

$$Pg = \left(-\frac{1 - \phi(\xi)}{|\xi|^2} \widehat{g}(\xi) \right)^{\vee}.$$

Now we look at $u - Pf$, where f is the function on the right of (A3.1.1) and u is the solution of that differential equation:

$$\left(u - Pf \right)\widehat{}(\xi) = \widehat{u} - \widehat{Pf}$$

$$= -\frac{1}{|\xi|^2}\hat{f} + \frac{1 - \phi(\xi)}{|\xi|^2}\hat{f}$$

$$= -\frac{\phi(\xi)}{|\xi|^2}\hat{f}.$$

Then $u - Pf$ is a distribution whose Fourier transform has compact support; that is, $u - Pf$ is C^∞. This means that for our purposes, $u - Pf$ is negligible: the function lies in *every* regularity class. So studying the regularity of u is equivalent to studying the regularity of Pf. This is precisely what we mean when we say that P is a *parametrix* for the partial differential operator Δ. And one of the main points is that P has symbol $-(1 - \phi)/|\xi|^2$, which is free of singularities.

Now we consider a very natural and more general situation. Let L be a partial differential operator with (smooth) variable coefficients:

$$L = \sum_\alpha a_\alpha(x)\left(\frac{\partial}{\partial x}\right)^\alpha.$$

The classical approach to studying such operators was to reduce to the constant-coefficient case by "freezing coefficients": Fix a point $z_0 \in \mathbb{R}^N$ and write

$$L = \sum_\alpha a_\alpha(x_0)\left(\frac{\partial}{\partial x}\right)^\alpha + \sum_\alpha (a_\alpha(x) - a_\alpha(x_0))\left(\frac{\partial}{\partial x}\right)^\alpha.$$

For a reasonable class of operators (elliptic) the second term turns out to be negligible because it has small coefficients. The principal term, the first, has constant coefficients.

The idea of freezing the coefficients is closely related to the idea of passing to the symbol of the operator L. We set

$$\ell(x, \xi) = \sum_\alpha a_\alpha(x)(-i\xi)^\alpha.$$

The motivation is that if $\phi \in D \equiv C_c^\infty$ and *if L has constant coefficients* then

$$L\phi(\xi) = c\int e^{-ix\cdot\xi}\ell(x, \xi)\widehat{\phi}(\xi)\,d\xi.$$

However, even in the variable coefficient case, we might hope that a parametrix for L is given by

$$\phi \longmapsto \left(\frac{1}{\ell(x, \xi)}\widehat{\phi}\right)^{\vee}.$$

Assume for simplicity that $\ell(x, \xi)$ vanishes only at $\xi = 0$ (in fact this is exactly what happens in the elliptic case). Let $\Phi \in C_c^\infty$ satisfy $\Phi(\xi) \equiv 1$ when $|\xi| \leq 1$ and $\Phi(\xi) \equiv 0$ when $|\xi| \geq 2$. Set

$$m(x, \xi) = \left(1 - \Phi(\xi)\right)\frac{1}{\ell(x, \xi)}.$$

We hope that m, acting as a Fourier multiplier by

$$T : f \longmapsto \left((m(x,\xi) \cdot \widehat{f}(x,\xi)) \right)^{\vee},$$

gives an approximate right inverse for L. More precisely, we hope that equations of the following form hold:

$$T \circ L = \mathrm{id} + \text{(negligible error term)},$$

$$L \circ T = \mathrm{id} + \text{(negligible error term)}.$$

In the constant coefficient case, composition of operators corresponds to multiplication of symbols, so that we would have

$$\left((L \circ T) f \right)^{\widehat{}} = \ell(\xi) \cdot \left(\frac{1 - \Phi(\xi)}{\ell(\xi)} \right) \widehat{f}(\xi)$$

$$= (1 - \Phi(\xi)) \widehat{f}(\xi)$$

$$= \widehat{f}(\xi) + [-\Phi(\xi) \widehat{f}(\xi)]$$

$$= If(\xi) + \mathcal{E} f(\xi).$$

Here, of course, I represents the identity operator. In the variable-coefficient case, we hope for an equation such as this but with a more elaborate error. However, the main point is that we can say something about the mapping properties of this more elaborate error term (it will typically be compact in a suitable Sobolev topology), so that it can still be handled roughly as in the constant coefficient case. This simple but subtle point is really at the heart of the calculus of pseudodifferential operators.

Of course a big part of what we must do, if we are going to turn these general remarks into a viable theory, is to decide what a "negligible error term" is. How do we measure negligibility? What does the error term mean? How do we handle it? It takes some significant ideas to address these questions and to make everything fit together.

A calculus of pseudodifferential operators is a collection of integral operators that contains all elliptic partial differential operators and their parametrices and is closed under composition and the taking of adjoints and inverses (and the adjoints and inverses are quickly and easily calculated). Once the calculus is in place, then, when one is given a partial or pseudodifferential operator, one can instantly write down a parametrix and obtain estimates. Pioneers in the development of pseudodifferential operators were Mikhlin [MIK1], [MIK2], and Calderón/Zygmund [CZ2]. Kohn–Nirenberg [KON] and Hörmander [HOR6] produced the first workable, modern theories.

One of the classical approaches to developing a calculus of operators finds it roots in the work of Hadamard [HAD] and Riesz [RIE] and Calderón–Zygmund [CZ1].

Here is a rather old attempt at a calculus of pseudodifferential operators:

Definition A3.1.3 A function $p(x, \xi)$ is said to be a *symbol* of order m if p is C^∞, has compact support in the z variable, and is homogeneous of degree m in ξ when ξ is large. That is, we assume that there is an $M > 0$ such that if $|\xi| > M$ and $\lambda > 1$ then

$$p(x, \lambda \xi) = \lambda^m p(x, \xi).$$

It is possible to show that symbols so defined, and the corresponding operators

$$T_p f(x) \equiv \int \widehat{f}(\xi) p(x, \xi) e^{-ix \cdot \xi} \, d\xi,$$

form an algebra in a suitable sense. These may be used to study elliptic operators effectively.

But the definition of symbol that we have just given is needlessly restrictive. For instance, the symbol of even a constant-coefficient partial differential operator is not generally homogeneous (because of lower-order terms) and we would have to deal with only the top-order terms. It was realized in the mid-1960s that homogeneity was superfluous to the intended applications. The correct point of view is to control the decay of derivatives of the symbol at infinity. In the next section we shall introduce the Kohn–Nirenberg approach to pseudodifferential operators.

Remark A3.1.4 It is worthwhile to make a few remarks about the Marcinkiewicz multiplier theorem, for which see [STE1]. Modern work in many different settings—ranging from partical differential equations to several complex variables to analysis on Lie groups—shows that this is just the right way to look at things. We state here a commonly used *consequence* of the Marcinkiewicz theorem. This is a strictly weaker result, but is quite useful in many contexts:

> **Theorem:** Consider a function m on $\mathbb{R}^N$ that is C^k in the complement of the origin with $k > N/2$. Assume that
>
> $$\left| \left(\frac{\partial^\alpha}{\partial x^\alpha} \right) m(x) \right| \leq C \cdot |x|^{-|\alpha|}$$
>
> for every multi-index α with $|\alpha| \leq k$. Then the Fourier integral operator
>
> $$T_m : f \mapsto \int_{\mathbb{R}^N} m(\xi) \widehat{f}(\xi) e^{ix \cdot \xi} \, d\xi$$
>
> is bounded on $L^p(\mathbb{R}^N)$, $1 < p < \infty$.

Note that the spirit of this result is that the Fourier multiplier m must decay at ∞ at a certain rate. The sharper version of Marcinkiewicz's theorem, which is more technical, may be found in [STE1, p. 108]. In any event, this circle of ideas is a motivation for the way that we end up defining the symbols in our calculus of pseudodifferential operators.

A3.2 A Formal Treatment of Pseudodifferential Operators

Now we give a careful treatment of an algebra of pseudodifferential operators. We begin with the definition of the symbol classes.

Definition A3.2.1 (Kohn–Nirenberg [KON1]) Let $m \in \mathbb{R}$. We say that a smooth function $\sigma(x, \xi)$ on $\mathbb{R}^N \times \mathbb{R}^N$ is a *symbol* of order m if there is a compact set $K \subseteq \mathbb{R}^N$ such that supp $\sigma \subseteq K \times \mathbb{R}^N$ and, for any pair of multi-indices α, β, there is a constant $C_{\alpha,\beta}$ such that

$$\left| D_\xi^\alpha D_x^\beta \sigma(x, \xi) \right| \leq C_{\alpha,\beta} \left(1 + |\xi| \right)^{m - |\alpha|}. \tag{A3.2.1}$$

We write $\sigma \in S^m$.

As a simple example, if $\Phi \in C_c^\infty(\mathbb{R}^N)$, $\Phi \equiv 1$ near the origin, define

$$\sigma(x, \xi) = \Phi(x)(1 - \Phi(\xi))(1 + |\xi|^2)^{m/2}.$$

Then σ is a symbol of order m. We leave it as an exercise for the reader to verify condition (A3.2.1).[5]

For our purposes, namely the interior regularity of elliptic partial differential operators, the Kohn–Nirenberg calculus will be sufficient. We shall study this calculus in detail. However, we should mention that there are several more-general calculi that have become important. Perhaps the most commonly used calculus is the Hörmander calculus [HOR2]. Its symbols are defined as follows:

Definition A3.2.2 Let $m \in \mathbb{R}$ and $0 \leq \rho, \delta \leq 1$. We say that a smooth function $\sigma(x, \xi)$ lies in the symbol class $S_{\rho,\delta}^m$ if

$$\left| D_\xi^\alpha D_x^\beta \sigma(x, \xi) \right| \leq C_{\alpha,\beta}(1 + |\xi|)^{m - \rho|\alpha| + \delta|\beta|}.$$

The Kohn–Nirenberg symbols are special cases of the Hörmander symbols with $\rho = 1$ and $\delta = 0$ and with the added convenience of restricting the x support to be compact. Hörmander's calculus is important for existence questions and for the study of the $\bar{\partial}$-Neumann problem (treated briefly in Section 10.5). In that context symbols of class $S_{1/2,1/2}^1$ arise naturally.

Even more general classes of operators that are spatially inhomogenous and nonisotropic in the phase variable ξ have been developed. Basic references are [BEF2], [BEA1], and [HOR5]. Pseudodifferential operators with "rough symbols" have been studied by Meyer [MEY4] and others.

The significance of the index m in the notation S^m is that it tells us how the corresponding pseudodifferential operator acts on certain function spaces. A pseudodifferential operator of order $m > 0$ "differentiates" to order m, while a pseudodifferential operator of order $m < 0$ "integrates" to order $-m$. While one may formulate

---

[5] In a more abstract treatment, one thinks of the x-variable as living in space and the ξ-variable as living in the cotangent space. This is because the ξ-variable *transforms* like a cotangent vector. This point of view is particularly useful in differential geometry. We shall be able to forgo such niceties.

results for C^k spaces, Lipschitz spaces, and other classes of functions, we find it most convenient at first to work with the Sobolev spaces. We have reviewed these spaces earlier in the text.

Theorem A3.2.3 *Let $p \in S^m$ (the Kohn–Nirenberg class) and define the associated pseudodifferential operator $P = Op(p) = T_p$ by*

$$P(\phi) = \int \widehat{\phi}(\xi) p(x, \xi) e^{-ix \cdot \xi} \, d\xi.$$

Then

$$P : H^s \to H^{s-m}$$

continuously.

Remark A3.2.4 Notice that if $m > 0$ then we lose smoothness under P. Likewise, if $m < 0$ then P is essentially a fractional integration operator and we gain smoothness. We say that the pseudodifferential operator T_p has *order m* precisely when its symbol is of order m.

Observe also that in the constant-coefficient case (which is misleadingly simple) we would have $p(x, \xi) = p(\xi)$ and the proof of the theorem would be as follows:

$$\|P(\phi)\|_{s-m}^2 = \int \left| \widehat{(P(\phi))}(\xi) \right|^2 (1 + |\xi|^2)^{s-m} \, d\xi$$

$$= \int |p(\xi) \widehat{\phi}(\xi)|^2 (1 + |\xi|^2)^{s-m} \, d\xi$$

$$\leq c \int |\widehat{\phi}(\xi)|^2 (1 + |\xi|^2)^s \, d\xi$$

$$= c\|\phi\|_s^2. \qquad \square$$

Remark A3.2.5 In the case that P is a partial differential operator

$$P = \sum_\alpha a_\alpha(x) \frac{\partial^\alpha}{\partial x^\alpha},$$

the symbol is

$$\sigma(P) = \sum_\alpha a_\alpha(x)(-i\xi)^\alpha.$$

Exercise for the Reader: Calculate the symbol of the linear operator (on the real line $\mathbb{R}$)

$$f \longmapsto \int_{-\infty}^x f(t) \, dt.$$

To prove the theorem in full generality is rather more difficult than the case of the constant-coefficient partial differential operator. We shall break the argument up into several lemmas.

Lemma A3.2.6 *For any complex numbers a, b we have*

$$\frac{1 + |a|}{1 + |b|} \leq 1 + |a - b|.$$

Proof: We have

$$1 + |a| \leq 1 + |a - b| + |b|$$
$$\leq 1 + |a - b| + |b| + |b|\,|a - b|$$
$$= (1 + |a - b|)(1 + |b|). \qquad \square$$

Lemma A3.2.7 *If $p \in S^m$, then for any multi-index α and integer $k > 0$, we have*

$$\left| \mathcal{F}_x\left(D_x^\alpha p(x, \xi) \right)(\eta) \right| \leq C_{k,\alpha} \frac{(1 + |\xi|)^m}{(1 + |\eta|)^k}.$$

Here $\mathcal{F}_x$ denotes the Fourier transform in the x-variable.

Proof: If α is any multi-index and γ is any multi-index such that $|\gamma| = k$ then

$$|\eta^\gamma| \left| \mathcal{F}_x\left(D_x^\alpha p(x, \xi) \right) \right| = \left| \mathcal{F}_x\left(D_x^\gamma D_x^\alpha p(x, \xi) \right)(\eta) \right|$$

$$\leq \| D_x^{\alpha + \gamma} p(x, \xi) \|_{L^1(x)} \leq C_{k,\alpha} \cdot \left(1 + |\xi| \right)^m.$$

As a result, since p is compactly supported in x,

$$\left(|\eta^\gamma| + 1 \right) \left| \mathcal{F}_x\left(D_x^\alpha p(x, \xi) \right) \right| \leq \left(C_{0,\alpha} + C_{k,\alpha} \right) \cdot \left(1 + |\xi| \right)^m.$$

This is what we wished to prove. $\qquad \square$

Lemma A3.2.8 *We have that*

$$\left(H^s \right)^* = H^{-s}.$$

Proof: Observe that

$$H^s = \{ g : \widehat{g} \in L^2\left((1 + |\xi|^2)^s \, d\xi \right).$$

But then H^s and H^{-s} are clearly dual to each other by way of the pairing

$$\langle f, g \rangle \equiv \int \widehat{f}(\xi) \, \widehat{g}(\xi) \, d\xi. \qquad \square$$

The upshot of the last lemma is that in order to estimate the H^s norm of a function (or Schwartz distribution) ϕ, it is enough (by the Hahn–Banach theorem) to prove an inequality of the form

$$\left| \int \phi(x)\overline{\psi}(x)\,dx \right| \le C\|\psi\|_{H^{-s}}$$

for every $\psi \in D$.

Proof of Theorem A3.2.3: Fix $\phi \in D$. Let $p \in S^m$ and let $P = \mathrm{Op}(p)$. Then

$$P\phi(x) = \int e^{-ix\cdot\xi} p(x,\xi)\widehat{\phi}(\xi)\,d\xi.$$

Define

$$\widehat{S}_x(\lambda,\xi) = \int e^{ix\cdot\lambda} p(x,\xi)\,dx.$$

This function is well defined since p is compactly supported in x. Then

$$\widehat{P\phi}(\eta) = \iint e^{-ix\cdot\xi} p(x,\xi)\widehat{\phi}(\xi)\,d\xi\, e^{i\eta\cdot x}\,dx$$

$$= \iint p(x,\xi)\widehat{\phi}(\xi)e^{ix\cdot(\eta-\xi)}\,dx\,d\xi$$

$$= \int \widehat{S}_x(\eta-\xi,\xi)\widehat{\phi}(\xi)\,d\xi.$$

We want to estimate $\|P\phi\|_{s-m}$. By the remarks following Lemma A3.2.8, it is enough to show that for $\psi \in D$,

$$\left| \int P\phi(x)\overline{\psi}(x)\,dx \right| \le C\|\phi\|_{H^s}\|\psi\|_{H^{m-s}}.$$

We have

$$\left| \int P\phi(x)\overline{\psi}(x)\,dx \right| = \left| \int \widehat{P\phi}(\xi)\widehat{\overline{\psi}}(\xi)\,d\xi \right|$$

$$= \left| \int \left(\int \widehat{S}_x(\xi-\eta,\eta)\widehat{\phi}(\eta)\,d\eta \right)\widehat{\overline{\psi}}(\xi)\,d\xi \right|$$

$$= \iint \widehat{S}_x(\xi-\eta,\eta)(1+|\eta|)^{-s}(1+|\xi|)^{s-m}$$

$$\times \widehat{\overline{\psi}}(\xi)(1+|\xi|)^{m-s}\widehat{\phi}(\eta)(1+|\eta|)^s\,d\eta\,d\xi.$$

Define

$$K(\xi,\eta) = \left| \widehat{S}_x(\xi-\eta,\eta)(1+|\eta|)^{-s}(1+|\xi|)^{s-m} \right|.$$

We claim that

$$\int |K(\xi,\eta)|\,d\xi \le C$$

and

$$\int |K(\xi, \eta)| \, d\eta \leq C$$

(these are the hypotheses of Schur's lemma).

Assume the claim for the moment. Then

$$\left| \int P\phi(x)\overline{\psi}(x) \, dx \right| \leq \iint K(\xi, \eta)|\widehat{\psi}(\xi)|(1 + |\xi|)^{m-s}|\widehat{\phi}(\eta)|(1 + |\eta|)^s \, d\eta \, d\xi$$

$$\leq C \left(\iint K(\xi, \eta)(1 + |\xi|^2)^{m-s}|\widehat{\psi}(\xi)|^2 \, d\xi \, d\eta \right)^{1/2}$$

$$\times \left(\iint K(\xi, \eta)(1 + |\eta|^2)^s |\widehat{\phi}(\eta)|^2 \, d\xi \, d\eta \right)^{1/2},$$

where we have used the obvious estimates $(1 + |\xi|)^2 \approx (1 + |\xi|^2)$ and $(1 + |\eta|)^2 \approx (1 + |\eta|^2)$. Now this last is

$$\leq C \left(\int |\widehat{\psi}(\xi)|^2 (1 + |\xi|^2)^{m-s} \, d\xi \right)^{1/2}$$

$$\times \left(\int |\widehat{\phi}(\eta)|^2 (1 + |\eta|^2)^s \, d\eta \right)^{1/2}$$

$$= C\|\psi\|_{H^{m-s}} \cdot \|\phi\|_{H^s},$$

which is the desired estimate. It remains to prove the claim.

By Lemma A3.2.7 we know that

$$|\widehat{S}_x(\zeta, \xi)| \leq C_k(1 + |\xi|)^m \cdot (1 + |\zeta|)^{-k}.$$

But now, by Lemma A3.2.6, we have

$$|K(\xi, \eta)| \equiv |\widehat{S}_x(\xi - \eta, \eta)(1 + |\eta|)^{-s}(1 + |\xi|)^{s-m}|$$

$$\leq C_k(1 + |\eta|)^m(1 + |\xi - \eta|)^{-k}(1 + |\eta|)^{-s}(1 + |\xi|)^{s-m}$$

$$= C_k \left(\frac{1 + |\eta|}{1 + |\xi|} \right)^{m-s} \cdot (1 + |\xi - \eta|)^{-k}$$

$$\leq C_k(1 + |\xi - \eta|)^{m-s}(1 + |\xi - \eta|)^{-k}.$$

We may specify k as we please, so we choose it so large that $m - s - k \leq -N - 1$. Then the claim is obvious and the theorem is proved. $\square$

A3.3 The Calculus of Pseudodifferential Operators

The three central facts about our pseudodifferential operators are these:

(1) If $p \in S^m$ then $T_p : H^s \to H^{s-m}$ for any $s \in \mathbb{R}$.

(2) If $p \in S^m$ then $(T_p)^*$ is "essentially" $T_{\overline{p}}$. In particular, the symbol of $(T_p)^*$ lies in S^m.

(3) If $p \in S^m, q \in S^n$, then $T_p \circ T_q$ is "essentially" T_{pq}. In particular, the symbol of $T_p \circ T_q$ lies in S^{m+n}.

We have already proved (1); in this section we shall give precise formulations to (2) and (3) and we shall prove them. Along the way, we shall give a precise explanation of what we mean by "essentially."

Remark: We begin by considering (2), and for motivation consider a simple example. Let $A = a(x)(\partial/\partial x_1)$. Observe that the symbol of A is $-a(x)i\xi_1$. Let us calculate A^*. If $\phi, \psi \in C_c^\infty$ then

$$\langle A^*\phi, \psi \rangle_{L^2} = \langle \phi, A\psi \rangle_{L^2}$$

$$= \int \phi(x)\overline{\left(a(x)\frac{\partial \psi}{\partial x_1}(x)\right)} dx$$

$$= -\int \frac{\partial}{\partial x_1}\left(\overline{a}(x)\phi(x)\right)\overline{\psi}(x) dx$$

$$= \int \left(-\overline{a}(x)\frac{\partial}{\partial x_1} - \frac{\partial \overline{a}}{\partial x_1}(x)\right)\phi(x) \cdot \overline{\psi}(x) dx.$$

Then

$$A^* = -\overline{a}(x)\frac{\partial}{\partial x_1} - \frac{\partial \overline{a}}{\partial x_1}(x)$$

$$= \mathrm{Op}\left(-\overline{a}(x)(-i\xi_1) - \frac{\partial \overline{a}}{\partial x_1}(x)\right)$$

$$= \mathrm{Op}\left(i\xi_1\overline{a}(x)\right) + \left(-\frac{\partial \overline{a}}{\partial x_1}(x)\right).$$

Thus we see in this example that the "principal part" of the adjoint operator (that is, the term with the highest-degree monomial in ξ of the symbol of A^*), and this is just the conjugate of the symbol of A. In particular, this lead term has order 1 as a pseudodifferential operator. The second term, which we think of as an "error term," is of zero order—it is simply the operator corresponding to multiplication by a function of x.

In general, it turns out that the symbol of A^* for a general pseudodifferential operator A is given by the asymptotic expansion

$$\sum_{\alpha} D_x^{\alpha} \left(\frac{\partial}{\partial \xi}\right)^{\alpha} \overline{\sigma(A)} \frac{1}{\alpha!}. \tag{A3.3.1}$$

Here $D_x^{\alpha} \equiv (i\partial/\partial x)^{\alpha}$. We learn more about asymptotic expansions later. The basic idea of an asymptotic expansion is that in a given application, the asymptotic expansion may be written in more precise form as

$$\sum_{|\alpha| \le k} D_x^{\alpha} \left(\frac{\partial}{\partial \xi}\right)^{\alpha} \overline{\sigma(A)} \frac{1}{\alpha!} + \mathcal{E}_k.$$

One selects k so large that the sum contains all the key information and the error term $\mathcal{E}_k$ is negligible.

If we apply this asymptotic expansion to the operator $a(x)\partial/\partial x_1$ that was just considered, it yields that

$$\sigma(A^*) = i\xi_1 \bar{a}(x) - \frac{\partial \bar{a}}{\partial x_1}(x),$$

which is just what we calculated by hand.

Now let us look at an example to motivate how compositions of pseudo differential operators will behave. Let the dimension N be 1 and let

$$A = a(x)\frac{d}{dx} \quad \text{and} \quad B = b(x)\frac{d}{dx}.$$

Then $\sigma(A) = a(x)(-i\xi)$ and $\sigma(B) = b(x)(-i\xi)$. Moreover, if $\phi \in D$, then

$$(A \circ B)(\phi) = \left(a(x)\frac{d}{dx}\right)\left(b(x)\frac{d\phi}{dx}\right)$$

$$= \left(a(x)\frac{db}{dx}(x)\frac{d}{dx} + a(x)b(x)\frac{d^2}{dx^2}\right)\phi.$$

Thus we see that

$$\sigma(A \circ B) = a(x)\frac{db}{dx}(x)(-i\xi) + a(x)b(x)(-i\xi)^2.$$

Notice that the principal part of the symbol of $A \circ B$ is

$$a(x)b(x)(-i\xi)^2 = \sigma(A) \cdot \sigma(B).$$

This term has order 2, as it should.

In general, the Kohn–Nirenberg formula says (in $\mathbb{R}^N$) that

$$\sigma(A \circ B) = \sum_{\alpha} \frac{1}{\alpha!} \left(\frac{\partial}{\partial \xi}\right)^{\alpha} (\sigma(A)) \cdot D_x^{\alpha}(\sigma(B)). \tag{A3.3.2}$$

Recall that the *commutator*, or Poisson bracket, of two operators is

$$[A, B] \equiv AB - BA.$$

Here juxtaposition of operators denotes composition. A corollary of the Kohn–Nirenberg formula is that

$$\sigma([A, B]) = \sum_{|\alpha|>0} \frac{(\partial/\partial\xi)^\alpha \sigma(A) D_x^\alpha \sigma(B) - (\partial/\partial\xi)^\alpha \sigma(B) D_x^\alpha \sigma(A)}{\alpha!}$$

(notice here that the $\alpha = 0$ term cancels out), so that $\sigma([A, B])$ has order strictly less than $(\text{order}(A) + \text{order}(B))$. This phenomenon is illustrated concretely in $\mathbb{R}^1$ by the operators $A = a(x)d/dx$, $B = b(x)d/dx$. One calculates that

$$AB - BA = \left(a(x)\frac{db}{dx}(x) - b(x)\frac{da}{dx}(x)\right)\frac{d}{dx},$$

which has order *one* instead of two.

Our final key result in the development of pseudodifferential operators is the asymptotic expansion for a symbol. We shall first have to digress a bit on the subject of asymptotic expansions.

Let f be a C^∞ function defined in a neighborhood of 0 in $\mathbb{R}$. Then

$$f(x) \sim \sum_0^\infty \frac{1}{n!}\frac{d^n f}{dx^n}(0)\,x^n. \tag{A3.3.3}$$

We are certainly not asserting that the Taylor expansion of an arbitrary C^∞ function converges back to the function, or even that it converges at all (generically just the opposite is true).

This formal expression (A3.3.3) means instead the following: Given an $N > 0$ there exists an $M > 0$ such that whenever $m > M$ and x is small then the partial sum S_m of the series (A3.3.3) satisfies

$$|f(x) - S_m| < C|x|^N.$$

Certainly the Taylor formula is historically one of the first examples of an asymptotic expansion.

Now we present a notion of asymptotic expansion that is related to this one, but is specially adapted to the theory of pseudodifferential operators:

Definition A3.3.4 Let $\{a_j\}_{j=1}^\infty$ be symbols in $\cup_m S^m$. We say that another symbol a satisfies

$$a \sim \sum_j a_j$$

if, for every $L \in \mathbb{R}^+$, there is an $M \in \mathbb{Z}^+$ such that

$$a - \sum_{j=1}^M a_j \in S^{-L}.$$

Definition A3.3.5 Let $K \subset\subset \mathbb{R}^N$ be a fixed compact set. Let Ψ_K be the set of symbols with z-support in K. If $p \in \Psi_K$ then we will think of the corresponding pseudodifferential operator $P = \mathrm{Op}(p)$ as

$$P : C_c^\infty(K) \to C_c^\infty(K).$$

This makes sense because

$$P\phi(x) = \int e^{-ix\cdot\xi} p(x,\xi)\widehat{\phi}(\xi)\, d\xi.$$

Now we have the tools assembled, and the motivation set in place, so that we can formulate and prove our principal results. Our first main result is the following Theorem:

Theorem A3.3.6 *Fix a compact set K and pick $p \in S^m \cap \Psi_K$. Let $P = \mathrm{Op}(p)$. Then P^* has symbol in $S^m \cap \Psi_K$ given by*

$$\sigma(P^*) \sim \sum_\alpha D_z^\alpha \left(\frac{\partial}{\partial\xi}\right)^\alpha \overline{p(x,\xi)} \cdot \frac{1}{\alpha!}.$$

Example A3.3.7 It is worthwhile to look at an example. But a caveat is in order. A general pseudodifferential operator is a rather abstract object. It is given by a Fourier multiplier and corresponding Fourier integral. If we want concrete, simple examples then we tend to look at (partial) *differential operators*. When we are endeavoring to illustrate algebraic ideas about pseudodifferential operators, this results in no loss of generality.

So let

$$P = \sum_\alpha a_\alpha(x)\frac{\partial^\alpha}{\partial x^\alpha}.$$

Here the sum is taken over multi-indices α. Observe that

$$\sigma(P) = \sum_\alpha a_\alpha(x)(-i\xi)^\alpha.$$

We calculate P^* by

$$\langle Pf, g\rangle = \left\langle \sum_\alpha \left(a_\alpha(x)\frac{\partial^\alpha}{\partial x^\alpha}\right) f(x), \overline{g(x)} \right\rangle$$

$$= \int \sum_\alpha \left(a_\alpha(x)\frac{\partial^\alpha}{\partial x^\alpha}\right) f(x), \overline{g(x)}\, dx$$

$$= \sum_\alpha \int \left(a_\alpha(x)\frac{\partial^\alpha}{\partial x^\alpha}\right) f(x), \overline{g(x)}\, dx$$

$$= \sum_\alpha \int \left(\frac{\partial^\alpha}{\partial x^\alpha}\right) f(x), [a_\alpha(x)\overline{g(x)}]\, dx$$

$$\overset{\text{(parts)}}{=} \sum_\alpha (-1)^{|\alpha|} \int f(x) \frac{\partial^\alpha}{\partial x^\alpha} [a_\alpha(x)\overline{g(x)}] \, dx$$

$$= \int f(x) \left[\sum_\alpha (-1)^{|\alpha|} \frac{\partial^\alpha}{\partial x^\alpha} [a_\alpha(x)\overline{g(x)}] \right] dx$$

$$\equiv \langle f, P^*g \rangle.$$

From this we conclude that

$$P^*g = \sum_\alpha (-1)^{|\alpha|} \frac{\partial^\alpha}{\partial x^\alpha} [a_\alpha(x)\overline{g(x)}].$$

Comparison with the result of the theorem shows that this answer is consistent with that more general result.

We shall prove this theorem in stages. There is a technical difficulty that arises almost immediately: Recall that if an operator T is given by integration against a kernel $K(x, y)$ then the roles of x and y are essentially symmetric. If we attempt to calculate the adjoint of T by formal reasoning, there is no difficulty in seeing that T^* is given by integration against the kernel $\overline{K(y, x)}$. However, at the symbol level matters are different. Namely, in our symbols $p(x, \xi)$, the role of x and ξ is *not* symmetric. In an abstract setting, x lives in space and ξ lives in the cotangent space. They transform differently. If we attempt to calculate the symbol of $\mathrm{Op}(p)$ by a formal calculation then this lack of symmetry serves as an obstruction.

It was Hörmander who determined a device for dealing with the problem just described. We shall now indicate his method. We introduce a new class of symbols $r(x, \xi, y)$. Such a smooth function on $\mathbb{R}^N \times \mathbb{R}^N \times \mathbb{R}^N$ is said to be in the symbol class T^m if there is a compact set K such that

$$\operatorname*{supp}_x r(x, \xi, y) \subseteq K$$

and

$$\operatorname*{supp}_y r(x, \xi, y) \subseteq K$$

and for any multi-indices α, β, γ there is a constant $C_{\alpha,\beta,\gamma}$ such that

$$\left| \left(\frac{\partial}{\partial \xi} \right)^\alpha \left(\frac{\partial}{\partial x} \right)^\beta \left(\frac{\partial}{\partial y} \right)^\gamma r(x, \xi, y) \right| \le C_{\alpha,\beta,\gamma} |\xi|^{m-|\alpha|}.$$

The corresponding operator R is defined by

$$R\phi(x) = \iint e^{i(y-x)\cdot\xi} r(x, \xi, y)\phi(y) \, dy \, d\xi \qquad \text{(A3.3.8)}$$

for ϕ a testing (that is, a $C_c^\infty = D$) function. Notice that the integral is not absolutely convergent and must therefore be interpreted as an iterated integral.

Proposition A3.3.9 *Let* $r \in T^m$ *have* x- *and* y-*supports contained in a compact set* K. *Then the operator* R *defined as in* (A3.3.7) *defines a pseudodifferential operator of Kohn–Nirenberg type with symbol* $p \in \Psi_K$ *having an asymptotic expansion*

$$p(x, \xi) \sim \sum_\alpha \frac{1}{\alpha!} \partial_\xi^\alpha D_y^\alpha r(x, \xi, y) \Big|_{y=x}.$$

Proof: Let ϕ be a testing function. We calculate that

$$\int e^{iy \cdot \xi} r(x, \xi, y) \phi(y) \, dy = \left(r(x, \xi, \cdot) \phi(\cdot) \right)\widehat{}(\xi)$$

$$= (\widehat{r_3}(x, \xi, \cdot) * \widehat{\phi}(\cdot))(\xi).$$

Here $\widehat{r_3}$ indicates that we have taken the Fourier transform of r in the third variable. By the definition of $R\phi$ we have

$$R\phi(x) = \iint e^{i(-x+y) \cdot \xi} r(x, \xi, y) \phi(y) \, dy \, d\xi$$

$$= \int e^{-ix \cdot \xi} \left[\widehat{r_3}(x, \xi, \cdot) * \widehat{\phi}(\cdot) \right](\xi) \, d\xi$$

$$= \iint \widehat{r_3}(x, \xi, \xi - \eta) \widehat{\phi}(\eta) \, d\eta \, e^{-ix \cdot \xi} \, d\xi$$

$$= \iint \widehat{r_3}(x, \xi, \xi - \eta) e^{-ix \cdot (\xi - \eta)} \, d\xi \, \widehat{\phi}(\eta) e^{-ix \cdot \eta} \, d\eta$$

$$\equiv \int p(x, \eta) \widehat{\phi}(\eta) e^{-ix \cdot \eta} \, d\eta.$$

We see that

$$p(x, \eta) \equiv \int \widehat{r_3}(x, \xi, \xi - \eta) e^{-ix(\xi - \eta)} \, d\xi$$

$$= \int e^{-ix \cdot \xi} \widehat{r_3}(x, \eta + \xi, \xi) \, d\xi.$$

Now if we expand the function $\widehat{r_3}(x, \eta + \cdot, \xi)$ in a Taylor expansion in powers of ξ then it is immediate that p has the claimed asymptotic expansion. In particular, one sees that $p \in S^m$. In detail, we have

$$\widehat{r_3}(x, \eta + \xi, \xi) = \sum_{|\alpha| < k} \partial_\eta^\alpha \widehat{r_3}(x, \eta, \xi) \frac{\xi^\alpha}{\alpha!} + \mathcal{R}.$$

Thus (dropping the ubiquitous c from the Fourier integrals)

$$p(x, \eta) = \sum_{|\alpha|<k} \int e^{-ix\cdot\xi} \partial_\eta^\alpha \widehat{r_3}(x, \eta, \xi) \frac{\xi^\alpha}{\alpha!} \, d\xi + \int \mathcal{R} \, d\xi$$

$$= \sum_{|\alpha|<k} \frac{1}{\alpha!} \partial_\eta^\alpha D_y^\alpha r(x, \eta, y)\Big|_{y=x} + \int \mathcal{R} \, d\xi.$$

The rest is formal checking.   □

With Hörmander's result in hand, we may now prove our first main result:

Proof of Theorem A3.3.6: Let $p \in \Psi_K \cap S^m$ and choose $\phi, \psi \in D$. Then, with P the pseudodifferential operator corresponding to the symbol p, we have

$$\langle \phi, P^*\psi \rangle \equiv \langle P\phi, \psi \rangle$$

$$= \int \left[\int e^{-ix\cdot\xi} p(x, \xi) \widehat{\phi}(\xi) \, d\xi \right] \overline{\psi}(x) \, dx$$

$$= \iiint e^{-i(x-y)\cdot\xi} \phi(y) \, dy \, p(x, \xi) \, d\xi \overline{\psi}(x) \, dx.$$

Let us suppose for the moment—just as a convenience—that p is compactly supported in ξ. With this extra hypothesis the integral is absolutely convergent and we may write

$$\langle \phi, P^*\psi \rangle = \int \phi(y) \left[\overline{\iint e^{i(x-y)\cdot\xi} \, \overline{p(x, \xi)} \psi(x) \, d\xi \, dx} \right] dy. \qquad \text{(A3.3.2.1)}$$

Thus we have

$$P^*\psi(y) = \iint e^{i(x-y)\cdot\xi} \, \overline{p(x, \xi)} \psi(x) \, d\xi \, dx.$$

Now let $\rho \in C_c^\infty$ be a real-valued function such that $\rho \equiv 1$ on K. Set

$$r(y, \xi, x) = \rho(y) \cdot p(x, \xi).$$

Then

$$P^*\psi(y) = \iint e^{i(x-y)\cdot\xi} \, \overline{p(x, \xi)} \rho(y) \psi(x) \, d\xi \, dx$$

$$= \int e^{i(x-y)\cdot\xi} r(y, \xi, x) \psi(x) \, d\xi \, dx$$

$$\equiv R\psi(y),$$

where we define R by means of the multiple symbol r. [Note that the roles of x and y here have unfortunately been reversed.]

By Proposition A3.3.8, P^* is then a classical pseudodifferential operator with symbol p^* whose asymptotic expansion is

$$p^*(x, \xi) \sim \sum_\alpha \partial_\xi^\alpha D_y^\alpha \left[\rho(x) \overline{p(y, \xi)} \right] \frac{1}{\alpha!} \bigg|_{y=x}$$

$$\sim \sum_\alpha \frac{1}{\alpha!} \partial_\xi^\alpha D_x^\alpha \overline{p(x, \xi)}.$$

We have used here the fact that $\rho \equiv 1$ on K. Thus the theorem is proved with the extra hypothesis of compact support of the symbol in ξ.

To remove the extra hypothesis, let $\phi \in C_c^\infty$ satisfy $\phi \equiv 1$ if $|\xi| \leq 1$ and $\phi \equiv 0$ if $|\xi| \geq 2$. Let

$$p_j(x, \xi) = \phi(\xi/j) \cdot p(x, \xi).$$

Observe that $p_j \to p$ in the C^k topology on compact sets for any k. Also, by the special case of the theorem already proved,

$$(\mathrm{Op}(p_j))^* \sim \sum_\alpha \partial_\xi^\alpha D_x^\alpha \overline{p_j(x, \xi)} \frac{1}{\alpha!}$$

$$\sim \sum_\alpha \partial_\xi^\alpha D_x^\alpha [\phi(\xi/j) p(x, \xi)] \frac{1}{\alpha!}.$$

The proof is completed now by letting $j \to \infty$.   □

Theorem A3.3.10 (Kohn–Nirenberg) *Let $p \in \Psi_K \cap S^m$, $q \in \Psi_K \cap S^n$. Let P, Q denote the pseudodifferential operators associated with p, q respectively. Then $P \circ Q = \mathrm{Op}(\sigma)$ where*

1. *$\sigma \in \Psi_K \cap S^{m+n}$;*
2. *$\sigma \sim \sum_\alpha \frac{1}{\alpha!} \partial_\xi^\alpha p(x, \xi) D_x^\alpha q(x, \xi)$.*

Proof: We may shorten the proof by using the following trick: write $Q = (Q^*)^*$ and recall that Q^* is defined by

$$Q^* \phi(y) = \iint e^{i(x-y) \cdot \xi} \phi(x) \overline{q(x, \xi)} \, dx \, d\xi$$

$$= \left(\int e^{ix \cdot \xi} \phi(x) \overline{q(x, \xi)} \, dx \right)^\vee (y).$$

Here we have used (A3.3.5).

Then

$$Q\phi(x) = \left(\int e^{iy \cdot \xi} \phi(y) \overline{q^*(y, \xi)} \, dy \right)^\vee (x), \tag{A3.3.10.1}$$

where q^* is the symbol of Q^* (note that q^* is *not* $\overline{q}$; however, we do know that $\overline{q}$ is the principal part of q^*). Then, using (A3.3.10.1), we may calculate that

$$(P \circ Q)(\phi)(x) = \int e^{-ix\cdot\xi} p(x,\xi)(\widehat{Q\phi})(\xi)\,d\xi$$

$$= \iint e^{-ix\cdot\xi} p(x,\xi)e^{iy\cdot\xi}\overline{q^*(y,\xi)}\phi(y)\,dy\,d\xi$$

$$= \iint e^{-i(x-y)\cdot\xi}[p(x,\xi)\overline{q^*(y,\xi)}]\phi(y)\,dy\,d\xi.$$

Set $\tilde{q} = \overline{q^*}$. Define

$$r(x,\xi,y) = p(x,\xi)\cdot\tilde{q}(y,\xi).$$

One verifies directly that $r \in T^{n+m}$. We leave this as an exercise. Thus R, the associated operator, equals $P \circ Q$. By Proposition A3.3.8, there is a classical symbol σ such that $R = \mathrm{Op}(\sigma)$ and

$$\sigma(x,\xi) \sim \sum_\alpha \frac{1}{\alpha!}\partial_\xi^\alpha D_y^\alpha r(x,\xi,y)\Big|_{y=x}.$$

Developing this last line, we obtain

$$\sigma(x,\xi) \sim \sum_\alpha \frac{1}{\alpha!}\partial_\xi^\alpha D_y^\alpha\left(p(x,\xi)\tilde{q}(y,\xi)\right)\Big|_{y=x}$$

$$\sim \sum_\alpha \frac{1}{\alpha!}\partial_\xi^\alpha[p(x,\xi)D_y^\alpha\tilde{q}(y,\xi)]\Big|_{y=x}$$

$$\sim \sum_\alpha \frac{1}{\alpha!}\partial_\xi^\alpha[p(x,\xi)D_x^\alpha\tilde{q}(x,\xi)]$$

$$\sim \sum_\alpha \frac{1}{\alpha!}\sum_{\alpha^1+\alpha^2=\alpha}\frac{\alpha!}{\alpha^1!\alpha^2!}[\partial_\xi^{\alpha^1}p(x,\xi)][\partial_\xi^{\alpha^2}D_x^\alpha\tilde{q}(x,\xi)]$$

$$\sim \sum_\alpha \sum_{\alpha^1+\alpha^2=\alpha}[\partial_\xi^{\alpha^1}p(x,\xi)][\partial_\xi^{\alpha^2}D_x^{\alpha^2}D_x^{\alpha^1}\tilde{q}(x,\xi)]\frac{1}{\alpha^1!\alpha^2!}$$

$$\sim \sum_{\alpha^1,\alpha^2}\frac{1}{\alpha^1!}[\partial_\xi^{\alpha^1}p(x,\xi)]\frac{1}{\alpha^2!}\left[\partial_\xi^{\alpha^2}D_x^{\alpha^2}D_x^{\alpha^1}\tilde{q}(x,\xi)\right]$$

$$\sim \left[\sum_{\alpha'}\frac{1}{\alpha^1!}\partial_\xi^{\alpha^1}p(x,\xi)\right]D_x^{\alpha^1}\left[\sum_{\alpha^2}\frac{1}{\alpha^2!}\partial_\xi^{\alpha^2}D_x^{\alpha^2}\tilde{q}(x,\xi)\right]$$

$$\sim \sum_{\alpha^1}\frac{1}{\alpha^1!}\partial_\xi^{\alpha^1}p(x,\xi)D_\xi^{\alpha^1}q(x,\xi).$$

Here we have used the fact that the expression inside the second set of brackets in the penultimate line is just the asymptotic expansion for the symbol of $(Q^*)^*$. That completes the proof. $\square$

Example A3.3.11 It is worth writing out the Kohn–Nirenberg result in the special case that P and Q are just partial differential operators. Thus take

$$P = \sum_\alpha a_\alpha(x) \frac{\partial^\alpha}{\partial x^\alpha}$$

and

$$Q = \sum_\beta b_\beta(x) \frac{\partial^\beta}{\partial x^\beta}.$$

Calculate $P \circ Q$, taking care to note that the coefficients b_β of Q will be differentiated by P. And compare this result with what is predicted by Theorem A3.3.10.

Next calculate $Q \circ P$.

Finally calculate $P \circ Q - Q \circ P$. If P is a partial differential operator of degree p and Q is a partial differential operator of degree q then of course $P \circ Q$ has degree $p + q$ and also $Q \circ P$ has degree $p + q$. But the "commutator" $P \circ Q - Q \circ P$ has degree $p + q - 1$. Explain in elementary terms why this is the case (i.e., the top-order terms vanish because their coefficients have not been differentiated).

The next proposition is a useful device for building pseudodifferential operators. Before we can state it we need a piece of terminology: we say that two pseudodifferential operators P and Q are equal *up to a smoothing operator* if $P - Q \in S^k$ for all $k < 0$. In this circumstance we write $P \sim Q$.

Proposition A3.3.12 Let p_j, $j = 0, 1, 2, \ldots$, be symbols of order m_j, with $m_j \searrow -\infty$. Then there is a symbol $p \in S^{m_0}$, unique modulo smoothing operators, such that

$$p \sim \sum_0^\infty p_j.$$

Proof: Let $\psi : \mathbb{R}^N \to [0, 1]$ be a C^∞ function such that $\psi \equiv 0$ when $|x| \leq 1$ and $\psi \equiv 1$ when $|x| \geq 2$. Let $1 < t_1 < t_2 < \cdots$ be a sequence of positive numbers that increases to infinity. We will specify these numbers later. Define

$$p(x, \xi) = \sum_{j=0}^\infty \psi(\xi/t_j) p_j(x, \xi).$$

Note that for every fixed x, ξ, the sum is finite; for $\psi(\xi/t_j) = 0$ as soon as $t_j > |\xi|$. Thus p is a well-defined C^∞ function.

Our goal is to choose the t_j's so that p has the correct asymptotic expansion. We claim that there exist $\{t_j\}$ such that

$$|D_x^\beta D_\xi^\alpha (\psi(\xi/t_j) p_j(x, \xi))| \leq 2^{-j}(1 + |\xi|)^{m_j - |\alpha|}.$$

Assume the claim for the moment. Then, for any multi-indices α, β, we have

$$|D_x^\beta D_\xi^\alpha p(x, \xi)| \leq \sum_{j=0}^{\infty} |D_x^\beta D_\xi^\alpha (\psi(\xi/t_j) p_j(x, \xi))|$$

$$\leq \sum_{j=0}^{\infty} 2^{-j}(1 + |\xi|)^{m_j - |\alpha|}$$

$$\leq C \cdot (1 + |\xi|)^{m_0 - |\alpha|}.$$

It follows that $p \in S^{m_0}$. Now we want to show that p has the right asymptotic expansion. Let $0 < k \in \mathbb{Z}$ be fixed. We will show that

$$p - \sum_{j=0}^{k-1} p_j$$

lives in S^{m_k}. We have

$$p - \sum_{j=0}^{k-1} p_j = \left[p(x, \xi) - \sum_{j=0}^{k-1} \psi(\xi/t_j) p_j(x, \xi) \right]$$

$$- \sum_{j=0}^{k-1} (1 - \psi(\xi/t_j)) p_j(x, \xi)$$

$$\equiv q(x, \xi) + s(x, \xi).$$

It follows directly from our construction that $q(x, \xi) \in S^{m_k}$. Since $[1 - \psi(\xi/t_j)]$ has compact support in $B(0, 2t_1)$ for every j, it follows that $s(x, \xi) \in S^{-\infty}$. Then

$$p - \sum_{j=0}^{k-1} p_j \in S^{m_k}$$

as we asserted.

We wish to see that p is unique modulo smoothing terms. Suppose that $q \in S^{m_0}$ and $q \sim \sum_{j=0}^{\infty} p_j$. Then

$$p - q = \left(p - \sum_{j<k} p_j \right) - \left(q - \sum_{j<k} p_j \right) \in S^{m_k}$$

for any k, which establishes the uniqueness.

It remains to prove the claim. First observe that for τ a multi-index with $|\tau| = j$,

$$|D_\xi^\tau \psi(\xi/t_j)| = \frac{1}{t_j^j} |(D_\xi^\tau \psi)(\xi/t_j)|$$

$$\leq \frac{1}{t_j^j} \left| \sup_{|\xi| \leq 2t_j} \left\{ (D^\tau \psi) \cdot (1 + |\xi|)^j \right\} \right| (1 + |\xi|)^{-j}$$

$$\leq C_j (1 + |\xi|)^{-j},$$

with C independent of j because $t_j \to +\infty$. Therefore

$$\left| D_\xi^\alpha \left(\psi(\xi/t_j) p_j(x, \xi) \right) \right| = \left| \sum_{\tau \leq \alpha} \binom{\alpha}{\tau} D_\xi^\tau \psi(\xi/t_j) D_\xi^{\alpha - \tau} p_j(x, \xi) \right|$$

$$\leq \sum_{\tau \leq \alpha} \binom{\alpha}{\tau} C_j (1 + |\xi|)^{-|\tau|} C_{j,\alpha} (1 + |\xi|)^{m_j - (|\alpha| - |\tau|)}$$

$$\leq C_{j,\alpha} (1 + |\xi|)^{m_j - |\alpha|}.$$

Consequently,

$$\left| D_x^\beta D_\xi^\alpha \left(\psi(\xi/t_j) p_j(x, \xi) \right) \right| \leq C_{j,\alpha,\beta} (1 + |\xi|)^{m_j - |\alpha|}$$

$$\leq C_j (1 + |\xi|)^{m_j - |\alpha|}$$

for every $j > |\alpha| + |\beta|$ (here we have set $C_j = \max\{C_{j,\alpha,\beta} : |\alpha| + |\beta| \leq j\}$).

Now recall that $\psi(\xi) = 0$ if $|\xi| \leq 1$. Then $\psi(\xi/t_j) \neq 0$ implies that $|\xi| \geq t_j$. Thus we choose t_j so large that $t_j > t_{j-1}$ and

$$|\xi| \geq t_j \quad \text{implies} \quad C_j (1 + |\xi|)^{m_j - m_{j-1}} \leq 2^{-j}$$

(remember that $t_j \to -\infty$). Then it follows that

$$\left| D_x^\beta D_\xi^\alpha \left(\psi(\xi/t_j) p_j(x, \xi) \right) \right| \leq 2^{-j} (1 + |\xi|)^{m_{j-1} - |\alpha|},$$

which establishes the claim and finishes the proof of the proposition. $\quad\square$

Remark A3.3.13 In the case that the p_j are just finitely many familiar partial differential operators, one obtains p by just adding up the p_j. In the case of *countably many* partial differential operators, the proposition already says something interesting. Basically we create p by adding up the tails of the symbols of the p_j. If we write

$$p(x, \xi) = \sum_{j=0}^{\infty} \psi(\xi/t_j) p_j(x, \xi)$$

in this case then the pseudodifferential operator corresponding to any partial sum $\sum_{j=0}^{K} \psi(\xi/t_j) p_j(x, \xi)$ obviously differs from the ordinary sum $\sum_{j=0}^{K} p_j$ by a smoothing operator.

References

[AHL] L. Ahlfors, *Complex Analysis*, 3rd ed., McGraw-Hill, 1979.

[ARO] N. Aronsjan, Theory of reproducing kernels, *Trans. Am. Math. Soc.* 68(1950), 337–404.

[ASH1] J. M. Ash, Multiple trigonometric series, in *Studies in Harmonic Analysis*, J. M. Ash, ed., Mathematical Association of America, Washington, D.C., 1976, 76–96.

[ASH2] J. M. Ash, A new proof of uniqueness for trigonometric series, *Proc. AMS* 107(1989), 409–410.

[ASH3] J. M. Ash, Uniqueness of representation by trigonometric series, *Am. Math. Monthly* 96(1989), 873–885.

[BAC] G. Bachmann, *Elements of Abstract Harmonic Analysis*, Academic Press, New York, 1964.

[BAS] F. Bagemihl and W. Seidel, Some boundary properties of analytic functions, *Math. Zeitschr.* 61(1954), 186–199.

[BAK] J. A. Baker, Integration over spheres and the divergence theorem for balls, *Am. Math. Monthly* 104(1997), 36–47.

[BAR] S. R. Barker, Two theorems on boundary values of analytic functions, *Proc. A. M. S.* 68(1978), 48–54.

[BASE] R. Basener, Peak points, barriers, and pseudoconvex boundary points, *Proc. Am. Math. Soc.* 65(1977), 89–92.

[BEA1] R. Beals, A general calculus of pseudo-differential operators, *Duke Math. J.* 42(1975), 1–42.

[BEF2] R. Beals and C. Fefferman, Spatially inhomogeneous pseudo-differential operators I, *Comm. Pure Appl. Math.* 27(1974), 1–24.

[BEC] W. Beckner, Inequalities in Fourier analysis, *Annals of Math.* 102(1975), 159–182.

[BED] E. Bedford, The Dirichlet problem for some overdetermined systems on the unit ball in $\mathbb{C}^n$, *Pac. Jour. Math.* 51(1974), 19–25.

[BEF] E. Bedford and P. Federbush, Pluriharmonic boundary values, *Tohoku Math. Jour.* 26(1974), 505–510.

[BEDF1] E. Bedford and J. E. Fornæss, A construction of peak functions on weakly pseudoconvex domains, *Ann. Math.* 107(1978), 555–568.

[BEDF2] E. Bedford and J. E. Fornæss, Biholomorphic maps of weakly pseudoconvex domains, *Duke Math. Jour.* 45(1978), 711–719.

[BET1] E. Bedford and B. A. Taylor, Variational properties of the complex Monge–Ampère equation. I. Dirichlet principle, *Duke Math. J.* 45(1978), 375–403.

[BET2] E. Bedford and B. A. Taylor, Variational properties of the complex Monge–Ampère equation. II. Intrinsic norms, *Amer. J. Math.* 101(1979), 1131–1166.

[BET3] E. Bedford an B. A. Taylor, A new capacity for plurisubharmonic functions, *Acta Math.* 149(1982), 1–40.

[BEL] S. R. Bell, Biholomorphic mappings and the $\bar{\partial}$ problem, *Ann. Math.*, 114(1981), 103–113.

[BEB] S. R. Bell and H. Boas, Regularity of the Bergman projection in weakly pseudo-convex domains, *Math. Annalen* 257(1981), 23–30.

[BEK] S. R. Bell and S. G. Krantz, Smoothness to the boundary of conformal mappings, *Rocky Mountain J. Math.* 17(1987), 23–40.

[BELL] S. R. Bell and E. Ligocka, A simplification and extension of Fefferman's theorem on biholomorphic mappings, *Invent. Math.* **57** (1980), 283–289.

[BERL] J. Bergh and J. Löfström, *Interpolation Spaces: An Introduction*, Springer-Verlag, Berlin, 1976.

[BERG] S. Bergman, *The Kernel Function and Conformal Mapping*, Am. Math. Soc., Providence, RI, 1970.

[BERS] S. Bergman and M. Schiffer, *Kernel Functions and Elliptic Differential Equations in Mathematical Physics*, Academic Press, New York, 1953.

[BER] L. Bers, *Introduction to Several Complex Variables*, New York Univ. Press, New York, 1964.

[BES] A. Besicovitch, On Kakeya's problem and a similar one, *Math. Z.* 27(1928), 312–320.

[BLO] T. Bloom, C^∞ peak functions for pseudoconvex domains of strict type, *Duke Math. J.* 45(1978), 133–147.

[BLG] T. Bloom and I. Graham, A geometric characterization of points of type m on real submanifolds of $\mathbb{C}^n$, *J. Diff. Geom.* 12(1977), 171–182.

[BOC1] S. Bochner, Orthogonal systems of analytic functions, *Math. Z.* 14(1922), 180–207.

[BOC2] S. Bochner, *Lectures by S. Bochner on Fourier Analysis*, Edwards Brothers, Ann Arbor, Michigan, 1937.

[BOM] S. Bochner and W. T. Martin, *Several Complex Variables*, Princeton Univ. Press, Princeton, 1948.

[BOF] T. Bonneson and W. Fenchel, *Theorie der konvexen Körper*, Springer-Verlag, Berlin, 1934.

[BOU] J. Bourgain, Spherical summation and uniqueness of multiple trigonometric series, *Internat. Math. Res. Notices* 3(1996), 93–107.

[BMS] L. Boutet de Monvel and J. Sjöstrand, Sur la singularité des noyaux de Bergman et Szegő, *Soc. Mat. de France Asterisque* 34-35(1976), 123–164.

[BGS] D. Burkholder, R. Gundy, M. Silverstein, A maximal function characterization of the class H^p, *Trans. Am. Math. Soc.* 157(1971), 137–153.

[CAL] A. P. Calderón, Intermediate spaces and interpolation, the complex method, *Studia Math.* 24(1964), 113–190.

[CZ1] A. P. Calderón and A. Zygmund, On the existence of certain singular integrals, *Acta Math.* 88 (1952), 85–139.

[CZ2] A. P. Calderón and A. Zygmund, Singular integral operators and differential equations, *Am. Jour. Math.* 79(1957), 901–921.

[CAR] L. Carleson, On convergence and growth of partial sums of Fourier series, *Acta Math.* 116(1966), 135–157.

[CCP] G. Carrier, M. Crook, and C. Pearson, *Functions of a Complex Variable*, McGraw-Hill, New York, 1966.

[CAT1]   D. Catlin, Global regularity for the $\bar{\partial}$-Neumann problem, *Proc. Symp. Pure Math.* v. 41 (Y. T. Siu ed.), Am. Math. Soc., Providence, 1984.

[CAT2]   D. Catlin, Necessary conditions for subellipticity of the $\bar{\partial}$-Neumann problem, *Ann. Math.* 117(1983), 147–172.

[CAT3]   D. Catlin, Subelliptic estimates for the $\bar{\partial}$-Neumann problem, *Ann. Math.* 126(1987), 131–192.

[CEG]   U. Cegrell, *Capacities in Complex Analysis*, Vieweg, Braunschweig, 1988.

[CHKS1] D. C. Chang, S. G. Krantz, and E. M. Stein, Hardy spaces and elliptic boundary value problems, Proceedings of a Conference in Honor of Walter Rudin, Am. Math. Society, 1992.

[CHKS2] D. C. Chang, S. G. Krantz, and E. M. Stein, H^p theory on a smooth domain in $\mathbb{R}^N$ and applications to partial differential equations, *Jour. Funct. Anal.* 114(1993), 286–347.

[CNS]   D. C. Chang, A. Nagel, and E. M. Stein, Estimates for the $\bar{\partial}$-Neumann problem in pseudoconvex domains of finite type in $\mathbb{C}^2$, *Acta Math.* 169(1992), 153–228.

[CHE]   S.-C. Chen, A counterexample to the differentiability of the Bergman kernel, *Proc. AMS* 124(1996), 1807–1810.

[CHS]   S.-C. Chen and M.-C. Shaw, *Partial Differential Equations in Several Complex Variables*, American Mathematical Society, Providence, RI, 2001.

[CHM]   S. S. Chern and J. Moser, Real hypersurfaces in complex manifolds, *Acta Math.* 133(1974), 219–271.

[CHR1]   M. Christ, Regularity properties of the $\bar{\partial}_b$-equation on weakly pseudoconvex CR manifolds of dimension 3, *Jour. AMS* 1(1988), 587–646.

[CHR2]   M. Christ, On the $\bar{\partial}_b$ equaton and Szegő projection on CR manifolds, *Harmonic Analysis and PDE's* (El Escorial, 1987), Springer Lecture Notes vol. 1384, Springer-Verlag, Berlin, 1989.

[CHR3]   M. Christ, Estimates for fundamental solutions of second-order subelliptic differential operators, *Proc. AMS* 105(1989), 166–172.

[CHO]   G. Choquet, *Lectures on Analysis*, in three volumes, edited by J. Marsden, T. Lance, and S. Gelbart, Benjamin, New York, 1969.

[CHR]   F. M. Christ, *Singular Integrals*, CBMS, Am. Math. Society, Providence, 1990.

[COI]   R. R. Coifman, A real variable characterization of H^p, *Studia Math.* 51(1974), 269–274.

[COIW1] R. R. Coifman and G. Weiss, *Analyse Harmonique Noncommutative sur Certains Espaces Homogenes*, Springer Lecture Notes vol. 242, Springer-Verlag, Berlin, 1971.

[COIW2] R. R. Coifman and G. Weiss, Extensions of Hardy spaces and their use in analysis, *Bull. AMS* 83(1977), 569–645.

[CON]   B. Connes, Sur les coefficients des séries trigonometriques convergents sphériquement, *Comptes Rendus Acad. Sci. Paris* Ser. A 283(1976), 159–161.

[COO]   R. Cooke, A Cantor–Lebesgue theorem in two dimensions, *Proc. Am. Math. Soc.* 30(1971), 547–550.

[COR1]   A. Cordoba, The Kakeya maximal function and the spherical summation multipliers, *Am. Jour. Math.* 99(1977), 1–22.

[COR2]   A. Cordoba, The multiplier problem for the polygon, *Annals of Math.* 105(1977), 581–588.

[CORF1] A. Cordoba and R. Fefferman, On the equivalence between the boundedness of certain classes of maximal and multiplier operators in Fourier analysis, *Proc. Nat. Acad. Sci. USA* 74(1977), 423–425.

[CORF2]   A. Cordoba and R. Fefferman, On differentiation of integrals, *Proc. Nat. Acad. Sci. USA* 74(1977), 2221–2213.

[COT]   M. Cotlar, A combinatorial inequality and its applications to L^2-spaces, *Revista Math. Cuyana* 1(1955), 41–55.

[COP]   M. Cotlar and R. Panxone, On almost orthogonal operators in L^p-spaces, *Acta Sci. Math. Szeged* 19(1958), 165–171.

[COH]   R. Courant and D. Hilbert, R. Courant and D. Hilbert, *Methods of Mathematical Physics*, 2nd ed., Interscience, New York, 1966.

[CUN]   F. Cunningham, The Kakeya problem for simply connected and for star-shaped sets, *Am. Math. Monthly* 78(1971), 114–129.

[DAN1]   J. P. D'Angelo, Real hypersurfaces, orders of contact, and applications, *Annals of Math.* 115(1982), 615–637.

[DAN2]   J. P. D'Angelo, Intersection theory and the $\bar{\partial}$-Neumann problem, *Proc. Symp. Pure Math.* 41(1984), 51–58.

[DAN3]   J. P. D'Angelo, Finite type conditions for real hypersurfaces in $\mathbb{C}^n$, in *Complex Analysis Seminar*, Springer Lecture Notes vol. 1268, Springer-Verlag, 1987, 83–102.

[DAN4]   J. P. D'Angelo, Iterated commutators and derivatives of the Levi form, in *Complex Analysis Seminar*, Springer Lecture Notes vol. 1268, Springer-Verlag, 1987, 103–110.

[DAN5]   J. P. D'Angelo, *Several Complex Variables and the Geometry of Real Hypersurfaces*, CRC Press, Boca Raton, 1992.

[DAU]   I. Daubechies, *Ten Lectures on Wavelets*, Society for Industrial and Applied Mathematics, Philadelphia, 1992.

[DAVJ]   G. David and J. L. Journé, A boundedness criterion for generalized Calderón–Zygmund operators, *Ann. Math* 120(1984), 371–397.

[DAY]   C. M. Davis and Yang-Chun Chang, *Lectures on Bochner–Riesz Means*, London Mathematical Society Lecture Note Series 114, Cambridge University Press, Cambridge–New York, 1987.

[DIB]   F. Di Biase, *Fatou Type Theorems. Maximal Functions and Approach Regions*, Birkhäuser Publishing, Boston, 1998.

[DIK]   F. Di Biase and S. G. Krantz, *The Boundary Behavior of Holomorphic Functions*, Birkhäuser Publishing, Boston, 2010, to appear.

[DIF]   K. Diederich and J. E. Fornæss, Pseudoconvex domains: An example with nontrivial Nebenhülle, *Math. Ann.* 225(1977), 275–292.

[DUS]   N. Dunford and J. Schwartz, *Linear Operators*, Wiley-Interscience, New York, 1958, 1963, 1971.

[DRS]   P. Duren, Romberg, and A. Shields, Linear functionals on H^p spaces with $0 < p < 1$, *J. Reine Angew. Math.* 238(1969), 32–60.

[EVG]   L. C. Evans and R. Gariepy, *Measure Theory and Fine Properties of Functions*, CRC Press, Boca Raton, FL, 1992.

[FAT]   P. Fatou, Séries trigonométriques et séries de Taylor, *Acta Math.* 30(1906), 335–400.

[FED]   H. Federer, *Geometric Measure Theory*, Springer-Verlag, Berlin and New York, 1969.

[FEF1]   C. Fefferman, Inequalities for strongly singular convolution operators, *Acta Math.* 124(1970), 9–36.

[FEF2]   C. Fefferman, The multiplier problem for the ball, *Annals of Math.* 94(1971), 330–336.

[FEF3]   C. Fefferman, The uncertainty principle, *Bull. AMS*(2) 9(1983), 129–206.

[FEF4]   C. Fefferman, Pointwise convergence of Fourier series, *Annals of Math.* 98(1973), 551–571.

[FEF5]   C. Fefferman, On the convergence of Fourier series, *Bull. AMS* 77(1971), 744–745.

[FEF6]   C. Fefferman, On the divergence of Fourier series, *Bull. AMS* 77(1971), 191–195.

[FEF7]   C. Fefferman, The Bergman kernel and biholomorphic mappings of pseudoconvex domains, *Invent. Math.* 26(1974), 1–65.

[FEF8]   C. Fefferman, Parabolic invariant theory in complex analysis, *Adv. Math.* 31(1979), 131–262.

[FEK1]   C. Fefferman, Hölder estimates on domains of complex dimension two and on three-dimensional CR-manifolds, *Adv. in Math.* 69(1988), 223–303.

[FEK2]   C. Fefferman, Estimates of kernels on three-dimensional CR manifolds, *Rev. Mat. Iboamericana*, 4(1988), 355–405.

[FES]   C. Fefferman and E. M. Stein, H^p spaces of several-variables, *Acta Math.* 129(1972), 137–193.

[FEN]   W. Fenchel, Convexity through the ages, in *Convexity and Its Applications*, Birkhäuser, Basel, 1983, 120–130.

[FOL1]   G. B. Folland, The tangential Cauchy–Riemann complex on spheres, *Trans. Am. Math. Soc.* 171(1972), 83–133.

[FOL2]   G. B. Folland, A fundamental solution for a subelliptic operator, *Bull. Am. Math. Soc.* 79(1973), 373–376.

[FOL3]   G. B. Folland, *Real Analysis: Modern Techniques and Their Applications*, John Wiley and Sons, New York, 1984.

[FOL2]   G. B. Folland, *A Course in Abstract Harmonic Analysis*, CRC Press, Boca Raton, 1995.

[FOK]   G. B. Folland and J.J. Kohn, *The Neumann Problem for the Cauchy–Riemann Complex*, Princeton University Press, Princeton, 1972.

[FOLS]   G. B. Folland and A. Sitram, The uncertainty principle: a mathematical survey, *J. Fourier Anal. and Appl.* 3(1997), 207–238.

[FOST1]   G. B. Folland and E. M. Stein, Estimates for the $\overline{\partial}_b$ complex and analysis on the Heisenberg group, *Comm. Pure Appl. Math.* 27(1974), 429–522.

[FOST2]   G. B. Folland and E. M. Stein, *Hardy Spaces on Homogeneous Groups*, Princeton University Press, Princeton, NJ, 1982.

[FOR]   F. Forelli, Pluriharmonicity in terms of harmonic slices, *Math. Scand.* 41(1977), 358–364.

[FOR1]   J. E. Fornæss, Strictly pseudoconvex domains in convex domains, *Am. J. Math.* 98(1976), 529–569.

[FOR2]   J. E. Fornæss, Peak points on weakly pseudoconvex domains, *Math. Ann.* 227(1977), 173–175.

[FOR3]   J. E. Fornæss, Sup norm estimates for $\overline{\partial}$ in $\mathbb{C}^2$, *Annals of Math.* 123(1986), 335–346.

[FOM]   J. E. Fornæss and J. McNeal, A construction of peak functions on some finite type domains, *Am. J. Math.* 116(1994), 737–755.

[FOSI]   J. E. Fornæss and N. Sibony, Construction of plurisubharmonic functions on weakly pseudoconvex domains, *Duke Math. J.* 58(1989), 633–655.

[FORS]   F. Forstneric, On the boundary regularity of proper mappings, *Ann. Scuola Norm. Sup. Pisa Cl Sci.* 13(1986), 109–128.

[FOU]   J. Fourier, *The Analytical Theory of Heat*, Dover, New York, 1955.

[GAM]   T. Gamelin, *Uniform Algebras*, Prentice-Hall, Englewood Cliffs, 1969.

[GAC]   J. Garcia-Cuerva, Weighted H^p spaces, MA thesis, Washington University in St. Louis, 1975.

[GAR]   J. Garnett, *Bounded Analytic Functions*, Academic Press, New York, 1981.

[GAJ]   J. Garnett and P. W. Jones, The distance in *BMO* to L^∞, *Annals of Math.* 108(1978), 373–393.

[GAL]   J. Garnett and R. Latter, The atomic decomposition for Hardy spaces in several complex variables, *Duke Math. J.* 45(1978), 815–845.

[GEL1]   D. Geller, Fourier analysis on the Heisenberg group, *Proc. Nat. Acad. Sci. USA* 74(1977), 1328–1331.

[GEL2]   D. Geller, Fourier analysis on the Heisenberg group, *Lie Theories and Their Applications*, Queen's University, Kingston, Ontario, 1978, 434–438.

[GEL3]   D. Geller, Fourier analysis on the Heisenberg group. I. Schwartz space, *J. Functional Anal.* 36(1980), 205–254.

[GELS]   D. Geller and E. M. Stein, Convolution operators on the Heisenberg group, *Bull. Am. Math. Soc.* 6(1982), 99–103.

[GIL]   S. Gilbarg and N. Trudinger, *Elliptic Partial Differential Equations of Second-Order*, Springer, Berlin, 1977.

[GRA1]   C. R. Graham, The Dirichlet problem for the Bergman Laplacian. I., *Comm. Partial Differential Equations* 8(1983), no. 5, 433–476.

[GRA2]   C. R. Graham, The Dirichlet problem for the Bergman Laplacian. II., *Comm. Partial Differential Equations* 8(1983), no. 6, 563–641.

[GRK1]   R. E. Greene and S. G. Krantz, Stability properties of the Bergman kernel and curvature properties of bounded domains, *Recent Progress in Several Complex Variables*, Princeton University Press, Princeton, 1982, 179–198.

[GRK2]   R. E. Greene and S. G. Krantz, Deformation of complex structures, estimates for the $\bar{\partial}$ equation, and stability of the Bergman kernel, *Adv. Math.* 43(1982), 1–86.

[GRK3]   R. E. Greene and S. G. Krantz, The automorphism groups of strongly pseudoconvex domains, *Math. Annalen* 261(1982), 425–446.

[GRK4]   R. E. Greene and S. G. Krantz, The stability of the Bergman kernel and the geometry of the Bergman metric, *Bull. Am. Math. Soc.* 4(1981), 111–115.

[GRK5]   R. E. Greene and S. G. Krantz, Stability of the Carathéodory and Kobayashi metrics and applications to biholomorphic mappings, *Proc. Symp. in Pure Math.*, Vol. 41 (1984), 77–93.

[GRK6]   R. E. Greene and S. G. Krantz, Normal families and the semicontinuity of isometry and automorphism groups, *Math. Zeitschrift* 190(1985), 455–467.

[GRK7]   R. E. Greene and S. G. Krantz, characterizations of certain weakly pseudo-convex domains with noncompact automorphism groups, in *Complex Analysis Seminar*, Springer Lecture Notes 1268(1987), 121–157.

[GRK8]   R. E. Greene and S. G. Krantz, characterization of complex manifolds by the isotropy subgroups of their automorphism groups, *Indiana Univ. Math. J.* 34(1985), 865–879.

[GRK9]   R. E. Greene and S. G. Krantz, Biholomorphic self-maps of domains, *Complex Analysis II* (C. Berenstein, ed.), Springer Lecture Notes, vol. 1276, 1987, 136–207.

[GRK10] R. E. Greene and S. G. Krantz, Techniques for Studying the Automorphism Groups of Weakly Pseudoconvex Domains, Proceedings of the Special Year at the Mittag-Leffler Institute (J. E. Fornæss and C. O. Kiselman, eds.) *Annals of Math. Studies,* Princeton Univ. Press, Princeton, 1992.

[GRK11] R. E. Greene and S. G. Krantz, Invariants of Bergman geometry and results concerning the automorphism groups of domains in $\mathbb{C}^n$, Proceedings of the 1989 Conference in Cetraro (D. Struppa, ed.), 1991.

[GRK12] R. E. Greene and S. G. Krantz, *Function Theory of One Complex Variable*, John Wiley and Sons, New York, 1997.

[GRE] P. Greiner, Subelliptic estimates for the $\bar{\partial}$-Neumann problem in $\mathbb{C}^2$, *J. Diff. Geom.* 9(1974), 239–250.

[GRL] L. Gruman and P. Lelong, *Entire Functions of Several Complex Variables*, Springer-Verlag, 1986.

[GUA] P. Guan, Hölder regularity of subelliptic pseudodifferential operators, *Duke Math. Jour.* 60(1990), 563–598.

[GUZ] M. de Guzman, *Differentiation of Integrals in $\mathbb{R}^n$*, Springer Lecture Notes, Springer-Verlag, Berlin and New York, 1975.

[HAD] J. Hadamard, Le problème de Cauchy et les équations aux dérivées partielles linéaires hyperboliques, Paris, 1932.

[HAS] M. Hakim and N. Sibony, Quelques conditions pour l'existence de fonctions pics dans les domains pseudoconvexes, *Duke Math. J.* 44(1977), 399–406.

[HAN] Y. S. Han, Triebel-Lizorkin spaces on spaces of homogeneous type, *Studia Math.* 108(1994), 247–273.

[HLP] G. H. Hardy, J. E. Littlewood, and G. Polya, *Inequalities*, 2nd ed., Cambridge University Press, Cambridge, 1988.

[HED] E. R. Hedrick, Functions that are nearly analytic, *Bull. AMS* 23(1917), 213.

[HEI] M. Heins, *Complex Function Theory*, Academic Press, New York, 1968.

[HEL] S. Helgason, *Differential Geometry and Symmetric Spaces*, Academic Press, New York, 1962.

[HERG] E. Hernandex and G. Weiss, *A First Course on Wavelets*, CRC Press, Boca Raton, 1996.

[HER] M. Hervé, *Analytic and Plurisubharmonic Functions in Finite and Infinite Dimensional Spaces*, Springer Lecture Notes, vol. 198, Springer-Verlag, Berlin, 1971.

[HERS] E. Hewitt and K. Ross, *Abstract Harmonic Analysis*, vol. 1, Springer-Verlag, Berlin, 1963.

[HIR] M. Hirsch, *Differential Topology*, Springer-Verlag, Berlin and New York, 1976.

[HOF] K. Hoffman, *Banach Spaces of Analytic Functions*, Prentice-Hall, Englewood Cliffs, 1962.

[HOR1] L. Hörmander, L^p estimates for (pluri-)subharmonic functions, *Math. Scand.* 20(1967), 65–78.

[HOR2] L. Hörmander, *The Analysis of Linear Partial Differential Operators*, Springer-Verlag, Berlin, 1985.

[HOR3] L. Hörmander, Estimates for translation invariant operators in L^p spaces, *Acta Math.* 104(1960), 93–140.

[HOR4] L. Hörmander, L^p estimates for (pluri-) subharmonic functions, *Math. Scand* 20(1967), 65–78.

[HOR5] L. Hörmander, The Weyl calculus of pseudo-differential operators, *Comm. Pure Appl. Math.* 32(1979), 359–443.

[HOR6] L. Hörmander, Pseudo-differential operators, *Comm. Pure and Appl. Math.* 18(1965), 501–517.

[HUA] L.-K. Hua, *Harmonic Analysis of Functions of Several Complex Variables in the Classical Domains*, American Mathematical Society, Providence, 1963.

[HUN] R. Hunt, On the convergence of Fourier series, *1968 Orthogonal Expansions and Their Continuous Analogues* (Proc. Conf., Edwardsville, Illinois, 1967), pp. 235–255. Southern Illinois University Press, Carbondale, Illinois.

[HUT] R. Hunt and M. Taibleson, Almost everywhere convergence of Fourier series on the ring of integers of a local field, *SIAM J. Math. Anal.* 2(1971), 607–625.

[HUW]   W. Hurewicz and H. Wallman, *Dimension Theory*, Princeton University Press, Princeton, NJ, 1948.

[JON]   F. John and L. Nirenberg, On functions of bounded mean oscillation, *Comm. Pure and Appl. Math.* 14(1961), 415–426.

[JOS]   B. Josefson, On the equivalence between locally polar and globally polar sets for plurisubharmonic functions on $\mathbb{C}^n$, *Ark. Mat.* 16(1978), 109–115.

[JOUR] J. L. Journé, Calderón–Zygmund operators, pseudodifferential operators and the Cauchy integral of Calderón, *Springer Lecture Notes* 994, Springer–Verlag, Berlin, 1983.

[KAT]   Y. Katxnelson, *Introduction to Harmonic Analysis*, Wiley, New York, 1968.

[KEL]   O. Kellogg, *Foundations of Potential Theory*, Dover, New York, 1953.

[KER]   N. Kerzman, The Bergman kernel function. Differentiability at the boundary, *Math. Ann.* 195(1972), 149–158.

[KST1] N. Kerzman and E. M. Stein, The Cauchy kernel, the Szegő kernel, and the Riemann mapping function, *Math. Ann.* 236(1978), 85–93.

[KST2] N. Kerzman and E. M. Stein, The Szegő kernel in terms of Cauchy–Fantappiè kernels, *Duke Math. J.* 45(1978), 197–224.

[KIS]   C. Kiselman, A study of the Bergman projection in certain Hartogs domains, *Proc. Symposia in Pure Math.*, American Math. Society, Providence, RI, 52, III(1991), 219–231.

[KLE]   P. Klembeck, Kähler metrics of negative curvature, the Bergman metric near the boundary and the Kobayashi metric on smooth bounded strictly pseudoconvex sets, *Indiana Univ. Math. J.* 27(1978), 275–282.

[KLI]   M. Klimek, *Pluripotential Theory*, Oxford University Press, Oxford, 1991.

[KNS1] A. Knapp and E. M. Stein, Singular integrals and the principal series, *Proc. Nat. Acad. Sci. USA* 63(1969), 281–284.

[KNS2] A. Knapp and E. M. Stein, Intertwining operators for semisimple Lie groups, *Annals of Math.* 93(1971), 489–578.

[KOB]   S. Kobayashi, *Hyperbolic Manifolds and Holomorphic Mappings*, Marcel Dekker, New York, 1970.

[KOH1] J.J. Kohn, Boundary behavior of $\bar{\partial}$ on weakly pseudoconvex manifolds of dimension two, *J. Diff. Geom.* 6(1972), 523–542.

[KOH2] J.J. Kohn, Global regularity for $\bar{\partial}$ on weakly pseudoconvex manifolds, *Trans. AMS* 181(1973), 273–292.

[KON1] J. J. Kohn and L. Nirenberg, On the algebra of pseudodifferential operators, *Comm. Pure and Appl. Math.* 18(1965), 269–305.

[KON2] J.J. Kohn and L. Nirenberg, A pseudo-convex domain not admitting a holomorphic support function, *Math. Ann.* 201(1973), 265–268.

[KONS] S. V. Konyagin, On divergence of trigonometric Fourier series over cubes, *Acta Sci. Math.* (Szeged) 61(1995), 305–329.

[KOO]   P. Koosis, *Lectures on H^p Spaces*, Cambridge University Press, Cambridge, UK, 1980.

[KOR1] A. Koranyi, Harmonic functions on Hermitian hyperbolic space, *Trans. AMS* 135(1969), 507–516.

[KOR2] A. Koranyi, Boundary behavior of Poisson integrals on symmetric spaces, *Trans. AMS* 140(1969), 393–409.

[KOR3] A. Koranyi, The Poisson integral for generalized half-planes and bounded symmetric domains, *Annals of Math.* 82(1965), 332–350.

[KOR]   T. J. Körner, *Fourier Analysis*, 2nd ed., Cambridge University Press, Cambridge, 1989.

[KRA1]   S. G. Krantz, *Real Analysis and Foundations*, CRC Press, Boca Raton, 1992.

[KRA2]   S. G. Krantz, Fractional integration on Hardy spaces, *Studia Math.* 73(1982), 87–94.

[KRA3]   S. G. Krantz, *Partial Differential Equations and Complex Analysis*, CRC Press, Boca Raton, 1992.

[KRA4]   S. G. Krantz, *Function Theory of Several Complex Variables*, 2nd ed., American Mathematical Society, Providence, RI, 2001.

[KRA5]   S. G. Krantz, *A Panorama of Harmonic Analysis*, Mathematical Association of America, Washington, D.C., 1999.

[KRA6]   S. G. Krantz, Lipschitz spaces, smoothness of functions, and approximation theory, *Expositiones Math.* 3(1983), 193–260.

[KRA7]   S. G. Krantz, Invariant metrics and the boundary behavior of holomorphic functions on domains in $\mathbb{C}^n$, *Jour. Geometric. Anal.* 1(1991), 71–98.

[KRA8]   S. G. Krantz, Convexity in complex analysis, *Proc. Symp. in Pure Math.*, vol. 52 (E. Bedford, J. D'Angelo, R. Greene, and S. Krantz eds.), American Mathematical Society, Providence, 1991.

[KRA9]   S. G. Krantz, The boundary behavior of holomorphic functions: global and local results, *Asian Journal of Mathematics*, 11(2007), 179–199.

[KRA10] S. G. Krantz, Calculation and estimation of the Poisson kernel, *Journal of Math. Analysis and Applications* 302(2005), 143–148.

[KRA11] S. G. Krantz, Lipschitz spaces on stratified groups, *Trans. Am. Math. Soc.* 269(1982), 39–66.

[KRA12] S. G. Krantz, Characterization of various domains of holomorphy via $\bar{\partial}$ estimates and applications to a problem of Kohn, *Ill. J. Math.* 23(1979), 267–286.

[KRA13] S. G. Krantz, Optimal Lipschitz and L^p regularity for the equation $\bar{\partial}u = f$ on strongly pseudo-convex domains, *Math. Annalen* 219(1976), 233–260.

[KRA14] S. G. Krantz, pseudoconvexity, analytic disks, and invariant metrics, *The Bulletin of the Allahabad Mathematical Society*, to appear.

[KRL1]   S. G. Krantz and S.-Y. Li, A Note on Hardy spaces and functions of bounded mean oscillation on domains in $\mathbb{C}^n$, *Michigan Jour. Math.* 41 (1994), 51–72.

[KRL2]   S. G. Krantz and S.-Y. Li, Duality theorems for Hardy and Bergman spaces on convex domains of finite type in $\mathbb{C}^n$, *Ann. Inst. Fourier Grenoble* 45(1995), 1305–1327.

[KRL3]   S. G. Krantz and S.-Y. Li, On decomposition theorems for Hardy spaces on domains in $\mathbb{C}^n$ and applications, *Jour. Four. Anal. and Applics.* 2(1995), 65–107.

[KRL4]   S. G. Krantz and S.-Y. Li, Boundedness and compactness of integral operators on spaces of homogeneous type and applications, I, *Jour. Math. Anal. and Applic.*, 258(2001), 629–641.

[KRL5]   S. G. Krantz and S.-Y. Li, Boundedness and compactness of integral operators on spaces of homogeneous type and applications, II, *Jour. Math. Anal. and Applic.*, 258(2001), 642–657.

[KRP1]   S. G. Krantz and H. R. Parks, *The Geometry of Domains in Space*, Birkhäuser, Boston, MA, 1999.

[KRP2]   S. G. Krantz and H. R. Parks, *The Implicit Function Theorem*, Birkhäuser, Boston, MA, 2002.

[KRP3]   S. G. Krantz and H. R. Parks, *A Primer of Real Analytic Functions*, 2nd ed., Birkhäuser Publishing, Boston, MA, 2002.

[KRPE]   S. G. Krantz and M. M. Peloso, Analysis and geometry on worm domains, *J. Geom. Anal.* 18(2008), 478–510.

[LAT]   ] M. Lacey and C. Thiele, A proof of boundedness of the Carleson operator, *Math. Res. Letters* 7(2000), 361–370.

[LANG]   S. Lang, *Algebra*, 2nd ed., Addison-Wesley, Reading, 1984.

[LAN]   R. E. Langer, *Fourier Series: The Genesis and Evolution of a Theory*, Herbert Ellsworth Slaught Memorial Paper I, *Am. Math. Monthly* 54(1947).

[LATT]   R. H. Latter, A characterization of $H^p(\mathbb{R}^N)$ in terms of atoms, *Studia Math.* 62(1978), 93–101.

[LAY]   S. R. Lay, *Convex Sets and Their Applications*, John Wiley and Sons, New York, 1982.

[LEF]   L. Lee and J. E. Fornæss, Asymptotic behavior of the Kobayashi metric in the worm direction, *Math. Z.*, to appear.

[LEM]   L. Lempert, Boundary behavior of meromorphic functions of several complex variables, *Acta Math.* 144(1980), 1–26.

[LIG]   E. Ligocka, Remarks on the Bergman kernel function of a worm domain, *Studia Math.* 130(1998), 109–113.

[LIP]   J. L. Lions and J. Peetre, Sur une classe d'espaces d'interpolation, *Inst. Hautes Études Sci. Publ. Math.* 19(1964), 5–68.

[LOS]   L. Loomis and S. Sternberg, *Advanced Calculus*, Addison-Wesley, Reading, 1968.

[LUZ]   N. Luzin, The evolution of "Function," Part I, Abe Shenitzer, ed., *Am. Math. Monthly* 105(1998), 59–67.

[MAS]   R. A. Macias and C. Segovia, Lipschitz functions on spaces of homogeneous type, *Adv. Math.* 33(1979), 257–270.

[MCN1]   J. McNeal, Boundary behavior of the Bergman kernel function in $\mathbb{C}^2$, *Duke Math. J.* 58(1989), 499–512.

[MCN2]   J. McNeal, Local geometry of decoupled pseudoconvex domains, *Proceedings of a Conference for Hans Grauert*, Vieweg, 1990.

[MEY1]   Y. Meyer, *Wavelets and Operators*, Translated from the 1990 French original by D. H. Salinger, Cambridge Studies in Advanced Mathematics 37, Cambridge University Press, Cambridge, 1992.

[MEY2]   Y. Meyer, *Wavelets. Algorithms and Applications*, translated from the original French and with a forward by Robert D. Ryan, SIAM, Philadelphia, 1993.

[MEY3]   Y. Meyer, Wavelets and operators, in *Analysis at Urbana 1*, E. R. Berkson, N. T. Peck, and J. Uhl eds., London Math. Society Lecture Note Series, Vol. 137, Cambridge University Press, Cambridge, 1989.

[MEY4]   Y. Meyer, Estimations L^2 pour les opérateurs pseudo-differentiels, *Séminaire d'Analyse Harmonique (1977/1978)*, 47-53, *Publ. Math. Orsay* 78, 12, Univ. Paris XI, Orsay, 1978.

[MIK1]   S. G. Mikhlin, On the multipliers of Fourier integrals, *Dokl. Akad. Nauk SSSR* 109(1956), 701–703.

[MIK2]   S. G. Mikhlin, *Multidimensional Singular Integral Equations*, Pergammon Press, New York, 1965.

[MOS1]   J. Moser, A new proof of de Giorgi's theorem concerning the regularity problem for elliptic differential equations, *Comm. Pure and Appl. Math.* 13(1960), 457–468.

[MOS2]   J. Moser, On Harnack's theorem for elliptic differential equations, *Comm. Pure and Appl. Math.* 14(1961), 577–591.

[NRSW]   A. Nagel, J.-P. Rosay, E. M. Stein, and S. Wainger, Estimates for the Bergman and Szegő kernels in $\mathbb{C}^2$, *Ann. Math.* 129(1989), 113–149.

[NAS]   A. Nagel and E. M. Stein, *Lectures on Pseudodifferential Operators: Regularity Theorems and Applications to Nonelliptic Problems*, Mathematical Notes, 24.

Princeton University Press, Princeton, N.J.; University of Tokyo Press, Tokyo, 1979.

[NSW]   A. Nagel, E. M. Stein, and S. Wainger, Balls and metrics defined by vector fields. I. Basic properties, *Acta Math.* 155(1985), 103–147.

[ONE]   B. O'Neill, *Elementary Differential Geometry,* Academic Press, New York, 1966.

[PAW]   R. E. A. C. Paley and N. Wiener, *Fourier Transforms in the Complex Domain,* Reprint of the 1934 original, American Mathematical Society Colloquium Publications, 19, American Mathematical Society, Providence, RI, 1987.

[RIE]   M. Riesz, L'intégrale de Riemann–Liouville et le problème de Cauchy, *Acta Math.* 81(1949), 1–223.

[RUD1] W. Rudin, *Principles of Mathematical Analysis*, 3rd ed., McGraw-Hill, New York, 1976.

[RUD2] W. Rudin, *Real and Complex Analysis*, McGraw-Hill, New York, 1966.

[RUD3] W. Rudin, *Functional Analysis*, McGraw-Hill, New York, 1973.

[RUD4] W. Rudin, *Function Theory in the Unit Ball of* $\mathbb{C}^n$, Springer-Verlag, New York, 1980.

[SAD]   C. Sadosky, *Interpolation of Operators and Singular Integrals: An Introduction to Harmonic Analysis*, Marcel Dekker, New York, 1979.

[SAR]   D. Sarason, Functions of vanishing mean oscillation, *Trans. AMS* 207(1975), 391–405.

[SCH]   L. Schwartz, *Théorie des distributions*, Hermann, Paris, 1957.

[SER]   J.-P. Serre, *Lie Algebras and Lie Groups*, Benjamin, New York, 1965.

[GOS1] G. Sheng, *Singular Integrals in Several Complex Variables*, Oxford University Press, 1991.

[GOS2] G. Sheng, *Harmonic Analysis on Classical Groups*, Springer-Verlag, 1991.

[SIB]   N. Sibony, Sur le plongement des domaines faiblement pseudoconvexes dans des domaines convexes, *Math. Ann.* 273(1986), 209–214.

[SIU]   Y. T. Siu, The Levi problem, *Proc. Symposia on Pure Math.*, vol. XXX, Americal Mathematical Society, Providence, 1977.

[SJO1] P. Sjölin, An inequality of Paley and convergence a.e. of Walsh-Fourier series, *Ark. Math.* 7(1969), 551–570.

[SJO2] P. Sjölin, On the convergence almost everywhere of certain singular integrals and multiple Fourier series, *Ark. Math.* 9(1971), 65–90.

[SMI]   K. T. Smith, A generalization of an inequality of Hardy and Littlewood, *Can. Jour. Math.* 8(1956), 157–170.

[SPI]   M. Spivak, *Calculus on Manifolds*, Benjamin, New York, 1965.

[STE1] E. M. Stein, *Singular Integrals and Differentiability Properties of Functions*, Princeton University Press, Princeton, 1970.

[STE2] E. M. Stein, *Harmonic Analysis: Real-Variable Methods, Orthogonality and Oscillatory Integrals*, Princeton University Press, Princeton, 1993.

[STE3] E. M. Stein, Harmonic analysis on $\mathbb{R}^N$, in *Studies in Harmonic Analysis*, J. M. Ash, ed., Math. Assn. of America, Washington, D.C., 1976, pp. 97–135.

[STE4] E. M. Stein, *Boundary Behavior of Holomorphic Functions of Several Complex Variables*, Princeton University Press, Princeton, NJ, 1972.

[STE5] E. M. Stein, Note on the class $L \log L$, *Studia Math.* 32(1969), 305–310.

[STE6] E. M. Stein, On limits of sequences of operators, *Annals of Math.* 74(1961), 140–170.

[STG1] E. M. Stein and G. Weiss, *Introduction to Fourier Analysis on Euclidean Spaces*, Princeton University Press, Princeton, 1971.

[STG2]   E. M. Stein and G. Weiss, On the theory of harmonic functions of several-variables, I. The theory of H^p spaces, *Acta Math.* 103(1960), 25–62.

[STR1]   R. Strichartz, *A Guide to Distribution Theory and Fourier Transforms*, CRC Press, Boca Raton, 1993.

[STR2]   R. Strichartz, Para-differential operators—another step forward for the method of Fourier, *Notices of the AMS* 29(1982), 402–406.

[STR3]   R. Strichartz, How to make wavelets, *Am. Math. Monthly* 100(1993), 539–556.

[SxE]   G. Szegő, Über orthogonalsysteme von Polynomen, *Math. Z.* 4(1919), 139–151.

[TAW]   M. Taibleson and G. Weiss, The molecular characterization of certain Hardy spaces, *Representation Theorems for Hardy Spaces*, pp. 67–149, *Astérisque* 77, Soc. Math. France, Paris, 1980.

[TEV]   N. Tevxadxe, On the convergence of double Fourier series of quadratic summable functions, *Soobšč. Akad. Nauk. Gruzin. SSR.* 5(1970), 277–279.

[TRE]   F. Treves, *Introduction to Pseudodifferential and Fourier Integral Operators*, 2 vols., Plenum, New York, 1980.

[TSU]   M. Tsuji, *Potential Theory in Modern Function Theory*, Maruzen, Tokyo, 1959.

[UCH]   A. Uchiyama, A maximal function characterization of H^p on the space of homogeneous type, *Trans. AMS* 262(1980), 579–592.

[VAL]   F. Valentine, *Convex Sets*, McGraw-Hill, New York, 1964.

[VLA]   V. Vladimirov, *Methods of the Theory of Functions of Several Complex Variables*, MIT Press, Cambridge, 1966.

[WAL]   J. S. Walker, Fourier analysis and wavelet analysis, *Notices* of the AMS 44(1997), 658–670.

[WAT]   G. N. Watson, *A Treatise on the Theory of Bessel Functions*, Cambridge University Press, Cambridge, 1922.

[WEL]   R. O. Wells, *Differential Analysis on Complex Manifolds*, 2nd ed., Springer-Verlag, 1979.

[WHI1]   H. Whitney, Differentiable functions defined in closed sets. I, *Trans. AMS* 36(1934), 369–387.

[WHI2]   H. Whitney, Differentiable functions defined in arbitrary subsets of Euclidean space, *Trans. AMS* 40(1936), 309–317.

[WHI3]   H. Whitney, *Geometric Integration Theory*, Princeton University Press, Princeton, 1957.

[WHW]   E. Whittaker and G. Watson, *A Course of Modern Analysis*, 4th ed., Cambridge Univ. Press, London, 1935.

[WIC]   M. V. Wickerhauser, *Adapted Wavelet Analysis from Theory to Software*, A. K. Peters, Wellesley, 1994.

[WID]   K. O. Widman, On the boundary behavior of solutions to a class of elliptic partial differential equations, *Ark. Mat.* 6(1966), 485–533.

[WIE]   J. Wiegerinck, Separately subharmonic functions need not be subharmonic, *Proc. AMS* 104(1988), 770–771.

[WIE]   N. Wiener, *The Fourier Integral and Certain of Its Applications*, Dover, New York, 1958.

[YU]   J. Yu, MA, Thesis, Washington University, 1992.

[ZYG]   A. Zygmund, *Trigonometric Series*, 2nd ed., Cambridge Univ. Press, Cambridge, 1968.

Index

a priori estimate, 315
Abel summation, 26, 273
abelian subgroup, 180, 184
action
    of a pseudodifferential operator of order
      m, 320
    of a pseudodifferential operator on a
      Sobolev space, 321
    of Heisenberg group is biholomorphic,
      186
admissible approach region
    in an arbitrary domain, 161
    in the ball, 158
admissible
    boundary limits for holomorphic functions
      on arbitrary domains, 165
    limits for holomorphic functions, 159
    maximal function of a holomorphic
      function, 161
algebraic invariant for finite type, 263
The Analytical Theory of Heat, 4
analytic convexity
    and geometric convexity, 72
    strong, 70
    weak, 70
analytic disk, 86
    at a strongly pseudoconvex point, order of
      contact of, 87
    boundary of, 86
    in the boundary of a strongly pseudocon-
      vex domain, 87
    in the boundary of the ball, 87

analytic structure as an obstruction to $\bar{\partial}$
    regularity, 259
analytic type, 255, 257
analytically convex domain, 70
approximate right inverse, 318
assembling pseudodifferential operators into
    a single operator, 334
asymptotic expansion, 326–328
    for a symbol, 327
    for the symbol of the adjoint operator, 325
    for pseudodifferential operators, 336
asymptotics, 47
atom, 173
    definition of, 173
atomic
    decomposition of a Hardy space function,
      174
    theory of Hardy spaces, 172
atoms and functions of bounded mean
    oscillation, pairing of, 177
automorphism group, 183
    action, 181
    of the ball, 183
    of the upper half-plane, 180
averaging the partial summation operators,
    272
axioms of K.T. Smith, 161

balls, 62
    in the boundary of a complex domain, 157
    in the Heisenberg group, 192
Banach–Alaoglu theorem, 136
Behnke–Stein theorem, 107
Bell's theorem, 253

Applied and Numerical Harmonic Analysis

J.M. Cooper: *Introduction to Partial Differential Equations with MATLAB*
(ISBN 978-0-8176-3967-9)

C.E. D'Attellis and E.M. Fernández-Berdaguer: *Wavelet Theory and Harmonic Analysis in Applied Sciences* (ISBN 978-0-8176-3953-2)

H.G. Feichtinger and T. Strohmer: *Gabor Analysis and Algorithms*
(ISBN 978-0-8176-3959-4)

T.M. Peters, J.H.T. Bates, G.B. Pike, P. Munger, and J.C. Williams: *The Fourier Transform in Biomedical Engineering* (ISBN 978-0-8176-3941-9)

A.I. Saichev and W.A. Woyczyński: *Distributions in the Physical and Engineering Sciences* (ISBN 978-0-8176-3924-2)

R. Tolimieri and M. An: *Time-Frequency Representations* (ISBN 978-0-8176-3918-1)

G.T. Herman: *Geometry of Digital Spaces* (ISBN 978-0-8176-3897-9)

A. Procházka, J. Uhlíř, P.J.W. Rayner, and N.G. Kingsbury: *Signal Analysis and Prediction* (ISBN 978-0-8176-4042-2)

J. Ramanathan: *Methods of Applied Fourier Analysis* (ISBN 978-0-8176-3963-1)

A. Teolis: *Computational Signal Processing with Wavelets*
(ISBN 978-0-8176-3909-9)

W.O. Bray and Č.V. Stanojević: *Analysis of Divergence* (ISBN 978-0-8176-4058-3)

G.T Herman and A. Kuba: *Discrete Tomography* (ISBN 978-0-8176-4101-6)

J.J. Benedetto and P.J.S.G. Ferreira: *Modern Sampling Theory*
(ISBN 978-0-8176-4023-1)

A. Abbate, C.M. DeCusatis, and P.K. Das: *Wavelets and Subbands*
(ISBN 978-0-8176-4136-8)

L. Debnath: *Wavelet Transforms and Time-Frequency Signal Analysis*
(ISBN 978-0-8176-4104-7)

K. Gröchenig: *Foundations of Time-Frequency Analysis* (ISBN 978-0-8176-4022-4)

D.F. Walnut: *An Introduction to Wavelet Analysis* (ISBN 978-0-8176-3962-4)

O. Bratteli and P. Jorgensen: *Wavelets through a Looking Glass*
(ISBN 978-0-8176-4280-8)

H.G. Feichtinger and T. Strohmer: *Advances in Gabor Analysis*
(ISBN 978-0-8176-4239-6)

O. Christensen: *An Introduction to Frames and Riesz Bases*
(ISBN 978-0-8176-4295-2)

L. Debnath: *Wavelets and Signal Processing* (ISBN 978-0-8176-4235-8)

J. Davis: *Methods of Applied Mathematics with a MATLAB Overview*
(ISBN 978-0-8176-4331-7)

G. Bi and Y. Zeng: *Transforms and Fast Algorithms for Signal Analysis and Representations* (ISBN 978-0-8176-4279-2)

J.J. Benedetto and A. Zayed: *Sampling, Wavelets, and Tomography*
(ISBN 978-0-8176-4304-1)

E. Prestini: *The Evolution of Applied Harmonic Analysis* (ISBN 978-0-8176-4125-2)

O. Christensen and K.L. Christensen: *Approximation Theory*
(ISBN 978-0-8176-3600-5)

L. Brandolini, L. Colzani, A. Iosevich, and G. Travaglini: *Fourier Analysis and Convexity* (ISBN 978-0-8176-3263-2)

W. Freeden and V. Michel: *Multiscale Potential Theory* (ISBN 978-0-8176-4105-4)

O. Calin and D.-C. Chang: *Geometric Mechanics on Riemannian Manifolds*
(ISBN 978-0-8176-4354-6)

J.A. Hogan and J.D. Lakey: *Time-Frequency and Time-Scale Methods*
(ISBN 978-0-8176-4276-1)

C. Heil: *Harmonic Analysis and Applications* (ISBN 978-0-8176-3778-1)

K. Borre, D.M. Akos, N. Bertelsen, P. Rinder, and S.H. Jensen: *A Software-Defined GPS and Galileo Receiver* (ISBN 978-0-8176-4390-4)

T. Qian, V. Mang I, and Y. Xu: *Wavelet Analysis and Applications*
(ISBN 978-3-7643-7777-9)

G.T. Herman and A. Kuba: *Advances in Discrete Tomography and Its Applications*
(ISBN 978-0-8176-3614-2)

M.C. Fu, R.A. Jarrow, J.-Y. J. Yen, and R.J. Elliott: *Advances in Mathematical Finance* (ISBN 978-0-8176-4544-1)

O. Christensen: *Frames and Bases* (ISBN 978-0-8176-4677-6)

P.E.T. Jorgensen, K.D. Merrill, and J.A. Packer: *Representations, Wavelets, and Frames* (ISBN 978-0-8176-4682-0)

M. An, A.K. Brodzik, and R. Tolimieri: *Ideal Sequence Design in Time-Frequency Space* (ISBN 978-0-8176-4737-7)

S.G. Krantz: *Explorations in Harmonic Analysis* (ISBN 978-0-8176-4668-4)

Breinigsville, PA USA
24 November 2009
227837BV00002BE/2/P

9 780817 646684